Advances in Experimental Medicine and Biology

Volume 1193

More information about this series at http://www.springer.com/series/5584

Jun Ren • Yingmei Zhang • Junbo Ge
Editors

Aldehyde Dehydrogenases

From Alcohol Metabolism to Human Health and Precision Medicine

Springer

Editors
Jun Ren
Department of Cardiology,
Shanghai Institute of Cardiovascular Diseases
Zhongshan Hospital, Fudan University
Shanghai, China

Center for Cardiovascular Research and Alternative Medicine
University of Wyoming College of Health Sciences
Laramie, WY, USA

Yingmei Zhang
Department of Cardiology,
Shanghai Institute of Cardiovascular Diseases
Zhongshan Hospital, Fudan University
Shanghai, China

Center for Cardiovascular Research and Alternative Medicine
University of Wyoming College of Health Sciences
Laramie, WY, USA

Junbo Ge
Department of Cardiology,
Shanghai Institute of Cardiovascular Diseases
Zhongshan Hospital, Fudan University
Shanghai, China

Institute of Biomedical Science
Fudan University
Shanghai, China

ISSN 0065-2598 ISSN 2214-8019 (electronic)
Advances in Experimental Medicine and Biology
ISBN 978-981-13-6262-0 ISBN 978-981-13-6260-6 (eBook)
https://doi.org/10.1007/978-981-13-6260-6

This Springer imprint is published by the registered company Springer Nature Singapore Pte Ltd.
The registered company address is: 152 Beach Road, #21-01/04 Gateway East, Singapore 189721, Singapore

Foreword

It is with great pleasure that I introduce this impressive collection of ALDH2 in cardiovascular and other chronic diseases edited by Professors Jun Ren, Junbo Ge, and Yingmei Zhang. This book is timely given the ever-rising concept of "precision medicine" in clinical diagnosis and therapeutics and the abundant human populations with ALDH2 genetic mutations worldwide. With the increasing use of new techniques for gene-targeted clinical diagnosis and drug therapy, the link between a given genetic polymorphism and disease prevalence is more and more appreciated. In this context, the pathophysiology of cardiovascular and other chronic diseases in association with ALDH2 genetic polymorphism needs to be examined in depth to better guide the therapeutic approaches.

This book addresses a number of mechanistic advances in our understanding of the role for ALDH2 in various cardiovascular diseases including diabetes, obesity, hypertension, sepsis, ischemic injury, myocardial infarction, and heart failure. In addition, the clinically established association between ALDH2 gene polymorphism and carcinogenesis and chemotherapy efficacy is also discussed in depth. A number of regulatory mechanisms are reviewed for ALDH2 including autophagy, oxidative stress, mitochondria, and crosstalk with cell survival machineries (such as apoptosis). Finally, this volume discusses potential drugs targeting ALDH2 in cardiovascular therapeutics (such as Alda-1) which represents an exciting emerging but rather challenging field. To this end, there is a fundamental need to understand the processes inherent in ALDH2 dehydrogenase as well as reductase enzymatic profiles, so that ALDH2 can be regulated with pharmacological or natural agents.

Our knowledge on ALDH2 polymorphism on important disease topics is summarized in this book, with contributions from experts from around the world. Both researchers and health-care providers (as well as students) in the medical field should find this volume to be particularly valuable.

Helmut K. Seitz MD, AGAF

Distinguished Professor of Medicine,
Gastroenterology and Alcohol Research
University of Heidelberg,
Heidelberg, Germany

Preface

Much progress has been made for the management of various cardiovascular and chronic diseases over the past decades, although the therapeutic efficacy still displays a geographical and ethnic disparity. Several scenarios have been indicated for the apparent therapeutic disparity in cardiovascular and other chronic diseases such as socioeconomic, genetic, and epigenetic factors as well as policy- and decision-making in health care. The surface of the term "precision medicine" recently has called upon the necessity for a thorough understanding of individual's genetic profile in clinical therapeutic strategy. To this end, it is quite pertinent to improve our knowledge for genetic polymorphism of a given gene in the clinical assessment and therapeutics of cardiovascular and other chronic diseases such as cancer and metabolic disorders.

Aldehyde dehydrogenases (ALDHs) belong to a superfamily of 19 isozymes and play important physiological roles including detoxification of endogenous and exogenous aldehyde substrates. However, recent findings have depicted a unique role for ALDH family, in particular mitochondrial ALDH (ALDH2), in a variety of human pathologies independent of their alcohol metabolism property. Moreover, mutation of ALDH2 (termed ALDH2*2) represents the single most abundant human genetic mutation among 4000+ human genetic diseases. Many disease processes may be impacted by ALDH2 genetic mutation including ischemia heart diseases, diabetes mellitus, hypertension, Fanconi anemia, pain, sepsis, cancer, osteoporosis, stroke, and aging. The heterogeneity and genetic polymorphism of ALDH2 enzyme may affect the clinical outcome of pharmacological agents such as nitroglycerin. Despite the available studies and clinical trials in diseases, the underlying mechanisms behind ALDH2 polymorphism in disease etiology, progression, and treatment still remain somewhat elusive. To optimize individualized therapeutic regimes, it is imperative to better understand the role for ALDH2 in the regulation of both physiological and pathophysiological processes. While scientists and clinicians have a good understanding and grasps of human disease processes, they may be less conversant for the role of genetic variation of ALDH2 in these comorbidities. To optimize therapeutic regimes against cardiovascular and other chronic diseases, it is imperative to understand the role for ALDH2 polymorphism in the pathogenesis and therapeutics of various human diseases. Fifteen chapters are dedicated in this book. In chapter "Role of Alcohol Oxidative Metabolism in Its Cardiovascular and Autonomic Effects", El-Mas and Abdel-Rahman briefly summarized alcohol metabolism in cardiovascular and autonomic systems and discussed ALDH2 polymorphism in alcohol-related disease such as hypertension. Chapter "Environmental Aldehyde Sources and the Health Implications of Exposure" by Sinharoy and coworkers offered a nice overview of how ALDH2 detoxifies the cytotoxic, mutagenic, and carcinogenic aldehydes and how ALDH2 variants may be predisposed to various environmental health issues. Mochly-Rosen and associates then discussed the role of ALDH2 in cardiovascular diseases in chapter "ALDH2 and Cardiovascular Disease" while Matsumoto laid out a systematic review of ALDH2 polymorphism in general human health and precision medicine in chapter "Aldehyde Dehydrogenases: From Alcohol Metabolism to Human Health and Precision Medicine". In chapter "Aldehyde Dehydrogenase 2 and Heart Failure", Li and colleagues presented the link between ALDH2 and heart failure, while Ding et al. discussed the role of ALDH2 in myocardial ischemic and ischemia-reperfusion injury in chapter "Mitochondrial Aldehyde Dehydrogenase in Myocardial Ischemic and Ischemia-Reperfusion Injury". Yasue and associates discussed ALDH2 mutation in various forms of coronary artery spasm and acute myocardial infarction more from the clinical perspective in chapter "East Asian Variant Aldehyde Dehydrogenase 2 Genotype (ALDH2*2*) Contributes to Coronary Artery Spasm". Given the global pandemics of obesity in the twenty-first century, Hu reviewed ALDH2 mutation and obesity prevalence by revisiting a number of relevant clinical and epidemiological reports in chapter "Aldehyde Dehydrogenases Genetic Polymorphism and Obesity: From Genomics to Behavior and Health". Likewise, Palaniyandi and coworkers focused on the correlation between ALDH2 and diabetic heart diseases in chapter "Aldehyde Dehydrogenase (ALDH) 2 in Diabetic Heart Diseases". Chapter "The Role of ALDH2 in Sepsis and the To-Be-Discovered Mechanisms" by Pang and colleagues reviewed the role of ALDH2 and its genetic mutation in sepsis and septic organ injury. The next chapter by Xu et al. examined the possible role of ALDH2 in stroke prevalence, while Ma and Cao revisited ALDH2 polymorphism in hypertension in chapter "Targeting ALDH2 in Atherosclerosis: Molecular Mechanisms and Therapeutic Opportunities". Wang and Wu reviewed the role for ALDH2 polymorphism in cancer and discussed its potential effect on cancer therapy, in particular in chemotherapeutics with anthracycline (chapter "ALDH2 and Cancer Therapy"). To support this link in carcinogenesis from the prospective of genetic-environmental interaction, Yang and coworkers discussed how ALDH2 polymorphism and ethanol consumption may affect the dogma of cancer prevention and treatment in chapter "ALDH2 Polymorphism and Ethanol Consumption: A Genetic-Environmental Interaction in Carcinogenesis". In the final chapter, Wu and Ren offered both pros and cons for ALDH2 in aging and aging-related human chronic diseases. In general, this book may shed some lights toward a better understanding of ALDH2 genetic polymorphism in various human diseases; it is noteworthy that the topics covered here cannot cover many other comorbidities (such as pulmonary and renal diseases) with a close tie with ALDH2 polymorphism. More importantly, our current contributed volume probably raised much more questions than answers. First, ALDH2 genetic profile isn't necessarily the benchmark for its enzymatic activity given the possible existence of epigenetic and posttranslational modifications. Therefore, using ALDH2 genotype as the sole indicator for the denominator of disease prevalence may be misleading. Second, similar to all animal models, experimental data (in vivo or in vitro) suffer from limitation to mimic pathological changes in actual clinical settings. Special caution is warranted when applying knowledge from bench-side to the bed-side practice. Last but not least, given the complexity in the etiology of human chronic diseases, it is rather difficult to pinpoint ALDH2 genetic mutation as the main drive for disease prevalence or therapeutic efficacy, letting alone the possible gene-gene or gene-environment interaction for ALDH2 that is not included in our book. We certainly hope that this contributed volume will assist physicians and scientists to apply the ALDH2 genetic polymorphism concept in the therapeutics of cardiovascular and other chronic diseases.

Shanghai, China

Jun Ren
Junbo Ge
Yingmei Zhang

Acknowledgments

We wish to express our appreciation to the editorial staff members at Springer Nature for their professional and courteous guidance. Becky Zhao, senior editor, piloted the initial launch of this book, and Uma Maheswari Srinivasan, editorial project manager, offered assistance throughout the entire process of this project. We wish to give credits to junior research scientists who trained our labs working on ALDH2. Of course, our greatest dedication should be given to our colleagues over the past 2 years covering these 15 chapters, and without their hard work and enthusiasm, this contributed volume would have not been possible.

Contents

Contributors

Abdel A. Abdel-Rahman Department of Pharmacology and Toxicology, the Brody School of Medicine, East Carolina University, Greenville, NC, USA

Feng Cao Department of Cardiology & National Clinical Research Center of Geriatrics Disease, Chinese PLA General Hospital, Beijing, China

Che-Hong Chen Department of Chemical and Systems Biology, Stanford University, School of Medicine, Stanford, CA, USA

Yuguo Chen Department of Emergency Medicine and Chest Pain Center, Qilu Hospital of Shandong University, Jinan, China

Clinical Research Center for Emergency and Critical Care Medicine of Shandong Province, Institute of Emergency and Critical Care Medicine of Shandong University, Qilu Hospital of Shandong University, Jinan, China

Key Laboratory of Emergency and Critical Care Medicine of Shandong Province, Key Laboratory of Cardiopulmonary-Cerebral Resuscitation Research of Shandong Province, Qilu Hospital of Shandong University, Jinan, China

The Key Laboratory of Cardiovascular Remodeling and Function Research, Chinese Ministry of Education, Chinese Ministry of Health and Chinese Academy of Medical Sciences, The State and Shandong Province Joint Key Laboratory of Translational Cardiovascular Medicine, Qilu Hospital of Shandong University, Jinan, China

Jie Ding State Key Laboratory of Cardiovascular Disease, Anesthesia Department, Fuwai Hospital, National Center for Cardiovascular Diseases, Chinese Academy of Medical Science and Peking Union Medical College, Beijing, China

Mahmoud M. El-Mas Department of Pharmacology and Toxicology, Faculty of Pharmacy, Alexandria University, Alexandria, Egypt

Julio C. B. Ferreira Department of Anatomy, Institute of Biomedical Sciences University of Sao Paulo, Sao Paulo, Brazil

Junbo Ge Department of Cardiology, Shanghai Institute of Cardiovascular Diseases, Zhongshan Hospital Fudan University, Shanghai, China

Institute of Biomedical Science, Fudan University, Shanghai, China

Eric R. Gross Department of Anesthesiology, Perioperative and Pain Medicine, Stanford University School of Medicine, Stanford, CA, USA

Qi Han Department of Emergency Medicine and Chest Pain Center, Qilu Hospital of Shandong University, Jinan, China

Clinical Research Center for Emergency and Critical Care Medicine of Shandong Province, Institute of Emergency and Critical Care Medicine of Shandong University, Qilu Hospital of Shandong University, Jinan, China

Key Laboratory of Emergency and Critical Care Medicine of Shandong Province, Key Laboratory of Cardiopulmonary-Cerebral Resuscitation Research of Shandong Province, Qilu Hospital of Shandong University, Jinan, China

The Key Laboratory of Cardiovascular Remodeling and Function Research, Chinese Ministry of Education, Chinese Ministry of Health and Chinese Academy of Medical Sciences, The State and Shandong Province Joint Key Laboratory of Translational Cardiovascular Medicine, Qilu Hospital of Shandong University, Jinan, China

Eisaku Harada Division of Cardiovascular Medicine, Kumamoto Kinoh Hospital, Kumamoto Aging Research Institute, Kumamoto City, Japan

Cheng Hu Shanghai Diabetes Institute, Shanghai Key Laboratory of Diabetes Mellitus, Shanghai Clinical Center for Diabetes, Shanghai Jiao Tong University Affiliated Sixth People's Hospital, Shanghai, China

Institute for Metabolic Diseases, Shanghai Jiao Tong University Affiliated Sixth People's Hospital South Campus, Shanghai, China

Wenjia Li Shanghai Institute of Cardiovascular Diseases, Zhongshan Hospital, Fudan University, Shanghai, China

Heng Ma Department of Physiology and Pathophysiology, Fourth Military Medical University, Xi'an, China

Sai Ma Department of Cardiology, & National Clinical Research Center of Geriatrics Disease, Chinese PLA General Hospital, Beijing, China

Jinling Hospital Department of Cardiology, Nanjing University, School of Medicine, Nanjing, China

Akiko Matsumoto Department of Social Medicine, Saga University School of Medicine, Saga, Japan

Stacy L. McAllister Department of Anesthesiology, Perioperative and Pain Medicine, Stanford University School of Medicine, Stanford, CA, USA

Yuji Mizuno Division of Cardiovascular Medicine, Kumamoto Kinoh Hospital, Kumamoto Aging Research Institute, Kumamoto City, Japan

Daria Mochly-Rosen Department of Chemical and Systems Biology, Stanford University, School of Medicine, Stanford, CA, USA

Srikar Munukutla Division of Hypertension and Vascular Research, Department of Internal Medicine, Henry Ford Health System, Detroit, MI, USA

Suresh S. Palaniyandi Division of Hypertension and Vascular Research, Department of Internal Medicine, Henry Ford Health System, Detroit, MI, USA

Department of Physiology, Wayne State University, Detroit, MI, USA

Jiaojiao Pang Department of Emergency Medicine and Chest Pain Center, Qilu Hospital of Shandong University, Jinan, China

Clinical Research Center for Emergency and Critical Care Medicine of Shandong Province, Institute of Emergency and Critical Care Medicine of Shandong University, Qilu Hospital of Shandong University, Jinan, China

Key Laboratory of Emergency and Critical Care Medicine of Shandong Province, Key Laboratory of Cardiopulmonary-Cerebral Resuscitation Research of Shandong Province, Qilu Hospital of Shandong University, Jinan, China

The Key Laboratory of Cardiovascular Remodeling and Function Research, Chinese Ministry of Education, Chinese Ministry of Health and Chinese Academy of Medical Sciences, The State and Shandong Province Joint Key Laboratory of Translational Cardiovascular Medicine, Qilu Hospital of Shandong University, Jinan, China

Guodong Pan Division of Hypertension and Vascular Research, Department of Internal Medicine, Henry Ford Health System, Detroit, MI, USA

Jun Ren Department of Cardiology, Shanghai Institute of Cardiovascular Diseases, Zhongshan Hospital, Fudan University, Shanghai, China

Center for Cardiovascular Research and Alternative Medicine, University of Wyoming College of Health Sciences, Laramie, WY, USA

Cheng Shen Department of Cardiology, The Affiliated Hospital of Jining Medical University, Jining, Shandong, China

Pritam Sinharoy Department of Anesthesiology, Perioperative and Pain Medicine, Stanford University School of Medicine, Stanford, CA, USA

Aijun Sun Shanghai Institute of Cardiovascular Diseases, Zhongshan Hospital, Fudan University, Shanghai, China

Institute of Biomedical Science, Fudan University, Shanghai, China

Ruru Sun Department of Emergency Medicine and Chest Pain Center, Qilu Hospital of Shandong University, Jinan, China

Clinical Research Center for Emergency and Critical Care Medicine of Shandong Province, Institute of Emergency and Critical Care Medicine of Shandong University, Qilu Hospital of Shandong University, Jinan, China

Key Laboratory of Emergency and Critical Care Medicine of Shandong Province, Key Laboratory of Cardiopulmonary-Cerebral Resuscitation Research of Shandong Province, Qilu Hospital of Shandong University, Jinan, China

The Key Laboratory of Cardiovascular Remodeling and Function Research, Chinese Ministry of Education, Chinese Ministry of Health and Chinese Academy of Medical Sciences, The State and Shandong Province Joint Key Laboratory of Translational Cardiovascular Medicine, Qilu Hospital of Shandong University, Jinan, China

Megana Vasu Department of Anesthesiology, Perioperative and Pain Medicine, Stanford University School of Medicine, Stanford, CA, USA

Jiali Wang Department of Emergency Medicine and Chest Pain Center, Qilu Hospital of Shandong University, Jinan, China

Clinical Research Center for Emergency and Critical Care Medicine of Shandong Province, Institute of Emergency and Critical Care Medicine of Shandong University, Qilu Hospital of Shandong University, Jinan, China

Key Laboratory of Emergency and Critical Care Medicine of Shandong Province, Key Laboratory of Cardiopulmonary-Cerebral Resuscitation Research of Shandong Province, Qilu Hospital of Shandong University, Jinan, China

The Key Laboratory of Cardiovascular Remodeling and Function Research, Chinese Ministry of Education, Chinese Ministry of Health and Chinese Academy of Medical Sciences, The State and Shandong Province Joint Key Laboratory of Translational Cardiovascular Medicine, Qilu Hospital of Shandong University, Jinan, China

Li-Shun Wang Institute of Fudan-Minhang Academic Health System, Minhang Branch, Zhongshan Hospital, Fudan University, Shanghai, China

Ne Natalie Wu Department of Cardiology, Shanghai Institute of Cardiovascular Diseases, Zhongshan Hospital, Fudan University, Shanghai, China

Zhao-Xia Wu Institute of Fudan-Minhang Academic Health System, Minhang Branch, Zhongshan Hospital, Fudan University, Shanghai, China

Feng Xu Department of Emergency Medicine and Chest Pain Center, Qilu Hospital of Shandong University, Jinan, China

Clinical Research Center for Emergency and Critical Care Medicine of Shandong Province, Institute of Emergency and Critical Care Medicine of Shandong University, Qilu Hospital of Shandong University, Jinan, China

Key Laboratory of Emergency and Critical Care Medicine of Shandong Province, Key Laboratory of Cardiopulmonary-Cerebral Resuscitation Research of Shandong Province, Qilu Hospital of Shandong University, Jinan, China

The Key Laboratory of Cardiovascular Remodeling and Function Research, Chinese Ministry of Education, Chinese Ministry of Health and Chinese Academy of Medical Sciences, The State and Shandong Province Joint Key Laboratory of Translational Cardiovascular Medicine, Qilu Hospital of Shandong University, Jinan, China

Haixia Xu Department of Cardiology and Shanghai Institute of Cardiovascular Diseases, Zhongshan Hospital, Fudan University, Shanghai, People's Republic of China

Mingjie Yang Department of Cardiology and Shanghai Institute of Cardiovascular Diseases, Zhongshan Hospital, Fudan University, Shanghai, People's Republic of China

Zheng Yang Department of Physiology and Pathophysiology, Fourth Military Medical University, Xi'an, China

Hirofumi Yasue Division of Cardiovascular Medicine, Kumamoto Kinoh Hospital, Kumamoto Aging Research Institute, Kumamoto City, Japan

Lei Yin Department of Urology, Shanghai Tenth People's Hospital, Tongji University, Shanghai, China

Hao Zhang Heart Center and Shanghai Institution of Pediatric Congenital Heart Disease, Shanghai Children's Medical Center, National Children's Medical Center, Shanghai Jiaotong University School of Medline, Shanghai, China

Ying Zhang Department of Emergency Medicine and Chest Pain Center, Qilu Hospital of Shandong University, Jinan, China

Clinical Research Center for Emergency and Critical Care Medicine of Shandong Province, Institute of Emergency and Critical Care Medicine of Shandong University, Qilu Hospital of Shandong University, Jinan, China

Key Laboratory of Emergency and Critical Care Medicine of Shandong Province, Key Laboratory of Cardiopulmonary-Cerebral Resuscitation Research of Shandong Province, Qilu Hospital of Shandong University, Jinan, China

The Key Laboratory of Cardiovascular Remodeling and Function Research, Chinese Ministry of Education, Chinese Ministry of Health and Chinese Academy of Medical Sciences, The State and Shandong Province Joint Key Laboratory of Translational Cardiovascular Medicine, Qilu Hospital of Shandong University, Jinan, China

Yingmei Zhang Department of Cardiology, Shanghai Institute of Cardiovascular Diseases, Zhongshan Hospital, Fudan University, Shanghai, China

Center for Cardiovascular Research and Alternative Medicine, University of Wyoming College of Health Sciences, Laramie, WY, USA

Yue Zheng Department of Emergency Medicine and Chest Pain Center, Qilu Hospital of Shandong University, Jinan, China

Clinical Research Center for Emergency and Critical Care Medicine of Shandong Province, Institute of Emergency and Critical Care Medicine of Shandong University, Qilu Hospital of Shandong University, Jinan, China

Key Laboratory of Emergency and Critical Care Medicine of Shandong Province, Key Laboratory of Cardiopulmonary-Cerebral Resuscitation Research of Shandong Province, Qilu Hospital of Shandong University, Jinan, China

The Key Laboratory of Cardiovascular Remodeling and Function Research, Chinese Ministry of Education, Chinese Ministry of Health and Chinese Academy of Medical Sciences, The State and Shandong Province Joint Key Laboratory of Translational Cardiovascular Medicine, Qilu Hospital of Shandong University, Jinan, China

Role of Alcohol Oxidative Metabolism in Its Cardiovascular and Autonomic Effects

Mahmoud M. El-Mas and Abdel A. Abdel-Rahman

Abstract Several review articles have been published on the neurobehavioral actions of acetaldehyde and other ethanol metabolites as well as in major alcohol-related disorders such as cancer and liver and lung disease. However, very few reviews dealt with the role of alcohol metabolism in the adverse cardiac and autonomic effects of alcohol and their potential underlying mechanisms, particularly in vulnerable populations. In this chapter, following a brief overview of the dose-related favorable and adverse cardiovascular effects of alcohol, we discuss the role of ethanol metabolism in its adverse effects in the brainstem and heart. Notably, current knowledge dismisses a major role for acetaldehyde in the adverse autonomic and cardiac effects of alcohol because of its low tissue level in vivo. Contrary to these findings in men and male rodents, women and hypertensive individuals are more sensitive to the adverse cardiac effects of similar amounts of alcohol. To understand this discrepancy, we discuss the autonomic and cardiac effects of alcohol and its metabolite acetaldehyde in a model of hypertension, the spontaneously hypertensive rat (SHR) and female rats. We present evidence that enhanced catalase activity, which contributes to cardioprotection in hypertension (compensatory) and in the presence of estrogen (inherent), becomes detrimental due to catalase catalysis of alcohol metabolism to acetaldehyde. Noteworthy, studies in SHRs and in estrogen deprived or replete normotensive rats implicate acetaldehyde in triggering oxidative stress in autonomic nuclei and the heart via (i) the Akt/extracellular signal-regulated kinases (ERK)/nitric oxide synthase (NOS) cascade and (ii) estrogen receptor-alpha (ERα) mediation of the higher catalase activity, which generates

M. M. El-Mas
Department of Pharmacology and Toxicology, Faculty of Pharmacy, Alexandria University, Alexandria, Egypt
e-mail: mahmoud.elmas@alexu.edu.eg

A. A. Abdel-Rahman (✉)
Department of Pharmacology and Toxicology, The Brody School of Medicine, East Carolina University, Greenville, NC, USA
e-mail: abdelrahmana@ecu.edu

J. Ren et al. (eds.), *Aldehyde Dehydrogenases*, Advances in Experimental Medicine and Biology 1193, https://doi.org/10.1007/978-981-13-6260-6_1

higher ethanol-derived acetaldehyde in female heart. The latter is supported by the ability of ERα blockade or catalase inhibition to attenuate alcohol-evoked myocardial oxidative stress and dysfunction. More mechanistic studies are needed to further understand the mechanisms of this public health problem.

1 Introduction

Alcohol is the most commonly used mood-altering drug worldwide. The proportions of pleasant and unpleasant effects of alcohol experienced by any individual depend largely on the amount of alcohol consumed and the pattern of alcohol use [6]. Studies have shown that alcohol exerts both favorable and detrimental cardiovascular effects (Fig. 1), possibly contributing to the J- or U-shaped association between alcohol consumption and overall disease and cardiovascular risks. Specifically, mild-to-moderate alcohol consumption is associated with protection against coronary artery disease and reduced incidence of heart failure, atherosclerosis, and peripheral vascular disease. The beneficial cardiovascular effects of mild/moderate alcohol consumption have been attributed to increased high-density lipoprotein cholesterol, improved endothelial function, anti-inflammatory effect, and decreased platelet aggregation [116, 155, 200].

On the other hand, excessive or binge drinking is one of the leading causes of higher cardiovascular morbidity and mortality worldwide [80, 150, 186]. Alcohol-induced cardiomyopathy is characterized by cardiac hypertrophy, interstitial fibro-

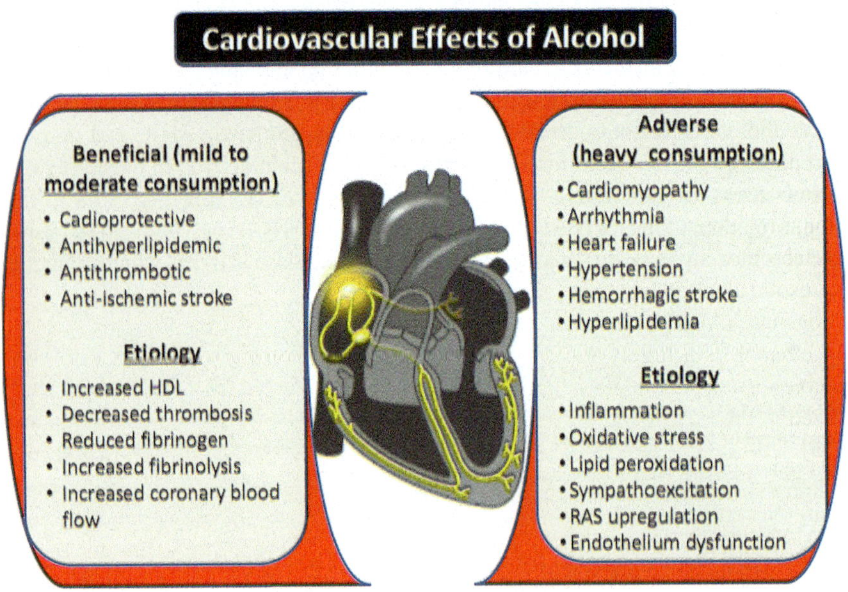

Fig. 1 Beneficial and harmful cardiovascular effects of alcohol and contributing etiologies

sis, and compromised myocardial contractile capacity [167, 187, 202]. The cardiodepressant effect of alcohol often results from chronic and excessive exposure to alcohol, e.g., a daily alcohol dose that exceeds 40 g [184]. It is characterized by the loss of contractile capacity, cardiomegaly, derangement of myofibrillary architecture, and increased incidence of heart failure, stroke, and hypertension [167, 179, 187, 202]. Cardiac arrhythmia is another consequence of excessive alcohol consumption, which may not be related to other common alcohol-related comorbidities such as liver disease, hypertension, diabetes, renal failure, and peripheral vascular disease [157, 179, 212].

In addition to cardiomyopathy, a tight association exists between alcohol consumption and elevated arterial pressure. Most of habitual alcohol users exhibit higher systolic blood pressures and hypertension compared with nondrinkers or light to moderate drinkers [103, 106, 150, 219]. In such cases, alcohol withdrawal or at least reduced intake is one of the clinical recommendations for hypertension management [14, 150]. The relationship between alcohol intake and blood pressure appears to be independent of any other confounding pathological variables such as diabetes mellitus, coronary heart disease, age, cigarette smoking, and dyslipidemia. Reported experimental findings confirmed epidemiological evidence that alcohol use positively correlates with the incidence of hypertension. Several underlying mechanisms have been implicated in the hypertensive action of alcohol such as impairment of arterial baroreceptor activity, vascular endothelial dysfunction, oxidative stress, cardiovascular inflammation, sympathetic and renin-angiotensin-aldosterone system hyperactivities, and increases in cortisol levels and vascular reactivity [43–45, 95, 150].

2 Peripheral and Central Metabolism of Alcohol

Accumulating evidence suggests that at least 95% of alcohol is eliminated via metabolism, with the remaining fraction being excreted unchanged through exhalation, sweating, or urinary excretion [156]. As depicted in Fig. 2, alcohol is mainly metabolized by hepatic oxidative degradation into acetaldehyde and acetate by cytosolic alcohol dehydrogenase (ADH) enzyme and mitochondrial aldehyde dehydrogenase (ALDH) enzymes, respectively. Acetaldehyde, the first oxidative product of ethanol, is a highly toxic molecule and is probably 10 times more toxic than alcohol itself [17]. In the liver, the conversion of ethanol into acetaldehyde is catalyzed by ADH, cytochrome P450 2E1 (CYP2E1), and, to a lesser extent, by catalase [34]. The two latter enzymes (CYP2E1 and catalase) constitute what is known as the non-ADH pathway, which contributes minimally to the hepatic alcohol metabolism. Nevertheless, the metabolic role of this pathway is magnified under circumstances of excessively high blood alcohol levels or chronic alcohol exposure [91] and under settings of increased catalase activity, which might explain the higher estrogen-related acetaldehyde level in women [73].

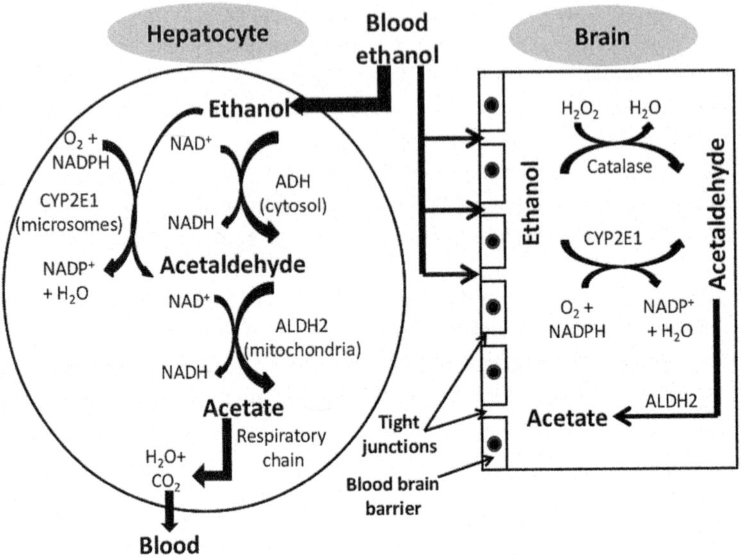

Fig. 2 Scheme of alcohol oxidative metabolism in the liver and brain

Ethanol, but not acetaldehyde, can easily reach the brain after crossing the blood-brain barrier. The cellular components of the blood-brain barrier such as endothelial cells and oligodendrocytes contain ALDH [220, 224], which breaks down circulating acetaldehyde of peripheral origin to acetate and limits its passage into brain tissues [93]. Nonetheless, acetaldehyde can be formed locally in the brain from alcohol that has crossed the blood-brain barrier. That said, the enzymatic pattern of central ethanol metabolism is different from the peripheral one. As mentioned above, the hepatic oxidative metabolism of ethanol into acetaldehyde is catalyzed mainly by ADH [34, 130, 156]. This enzymatic profile of ethanol metabolism is not visualized in the brain tissues where catalase and, to a lesser extent, CYP2E1, serve as the principal metabolizing enzymes [223] [221] (Fig. 2). Protein expression studies showed that catalase is expressed in all neural cells of the brain [93]. Moreover, immunohistochemical catalase-positive staining is particularly prominent in brain areas containing aminergic neuronal bodies [222].

Conflicting data are reported on the role of acetaldehyde in the neurobiological and epigenetic changes caused by ethanol. Numerous studies have implicated acetaldehyde in behavioral, reinforcing, and neurotoxic effects of ethanol. For example, microinjection of the catalase knockdown (shRNA) in ventral tegmental area of rats virtually abolishes the voluntary consumption of alcohol, suggesting that central metabolism of ethanol into acetaldehyde is necessary for generating reward and reinforcement [107]. Ethanol and acetaldehyde cause similar disruption of cellular differentiation and growth, with subsequent abnormalities in fetal development [123]. Locomotor stimulation induced by alcohol administration into the hypothalamic arcuate nucleus is abolished after pharmacologic catalase inhibition [153]. By contrast, evidence obtained from other studies failed to establish such causal relationship between ethanol effects and its oxidative product acetaldehyde or at least reached a conclusion that the two materials produce similar actions, but the underlying cellular mechanisms are probably different [162, 164]. Although acetaldehyde is considered the prime perpetrator for ethanol-induced organ damage, other products of ethanol metabolism, e.g., fatty acid ethyl esters, may also contribute to the onset and progression of alcoholic organ injury. The transport of these fatty acids from intracellular sites to mitochondrial membranes causes mitochondrial injury and loss of its energy generation capacity [118, 160, 217].

Several review articles on the role of acetaldehyde and other oxidative products in the neurobehavioral actions of ethanol have been published over the last few years [164] [154]. Similarly, reviews on the involvement of oxidative products and enzymes of ethanol in major alcohol-related disorders such as cancer [72, 147, 166, 176], liver disease [185, 189], lung disease [178], and alcoholism and addiction [139, 163] are also available. Yet, little or no reviews have been recently published that summarize reported findings on the role of acetaldehyde and synthesizing and degrading enzymes in cardiovascular complications induced by alcoholism and potential cellular and molecular mechanisms of these interactions.

3 Role of Central Metabolizing Enzymes in Alcohol-Induced Hypertension

3.1 Alcohol-Induced Hypertension

Epidemiological evidence supports a strong association between alcohol use and hypertension [11, 95, 150, 159]. Experimentally, ethanol elicits hypertension after its acute, parenteral, or oral administration [43, 51]. Gender, route of administration, rat strain, and arousal state are considerable factors that modify the blood pressure response elicited by alcohol [46, 48, 50, 51, 63, 122]. However, the effect of ethanol on blood pressure is not dependent on obesity, cigarette smoking, or physical activity [11, 15] and is observed in both normotensive and in hypertensive patients [159, 161]. The ethanol-induced pressor effect is reversible because it disappears when ethanol intake is stopped [159, 161].

Specifically, the role of sympathoexcitation in the ethanol-evoked hypertension is evidenced by the rises in plasma norepinephrine [98] and efferent sympathetic neural activity [175]. In a previous report from our laboratory [126], we employed microinjection and electrochemical protocols to determine neuronal substrates in the brainstem that underlie the hypertensive action of ethanol. The rostral ventrolateral medulla (RVLM) is the brainstem pressor region from which bulbospinal sympathetic neurons descend to the intermediolateral cell column of the spinal cord [26, 216] and is a major site for the sympathoexcitatory action of ethanol [136]. The C1 neurons of the RVLM contain essential NE-synthesizing enzymes such as tyrosine hydroxylase, dopamine hydroxylase, and phenylethanolamine *N*-methyltransferase [88]. Indeed, the quantity of the norepinephrine (NE)-containing neurons in the RVLM positively correlates with sympathetic neural activity [16]. Measurements of NE and its metabolites in the RVLM neurons by microdialysis or electrochemistry denote NE neuronal activity [16, 109, 136, 194].

The unilateral administration of ethanol (1–10 μg) into the RVLM causes dose-dependent increases in norepinephrine electrochemical signal and blood pressure in SHRs in contrast to much smaller pressor effects in Wistar Kyoto rats (WKYs) [126]. Similarly greater increases in blood pressure and RVLM norepinephrine are observed in SHRs after systemic ethanol administration [136]. Considering that the RVLM contains noradrenergic nerve terminals that originate from other brain areas [26, 36, 88] and that the activity of these neurons is proportionally related to the sympathetic neural activity [26, 136], these findings suggested a primary role for norepinephrine released from RVLM noradrenergic neurons in the sympathoexcitatory and pressor actions of ethanol in SHRs [126].

The impairment of arterial baroreceptor activity is a plausible mechanism for the sympathoexcitatory and the subsequent hypertensive actions of ethanol [43, 44, 47, 98, 199]. The arterial baroreflex function is one of the most rapidly acting homeostatic mechanisms for the regulation of blood pressure. Baroreceptor function is impaired in human hypertensives [85] and precedes the development of hypertension in some experimental models including the ethanol-induced hypertension [1, 82, 83, 175]. The cardiovascular nuclei of the brainstem such as the nucleus tractus solitarius are important neuroanatomical targets for the baroreflex depressant effect of ethanol [45, 56]. The ethanol enhancement of γ-aminobutyric acid–mediated (GABAergic) [198] and attenuation of glutamatergic [135] neurotransmission in the brainstem have been implicated in baroreflex dysfunction induced by ethanol.

3.2 *Acetaldehyde Mediates the Pressor Effect of Intra-RVLM Alcohol*

As indicated above, the local synthesis of acetaldehyde from alcohol in the brain is catalyzed mainly by catalase and CYP2E1. Catalase is present in all neural cells, and its expression is particularly evident in aminergic neuronal bodies [93, 222]. Further, central CYP2E1 expression is region-specific and is found in peroxisomes and mitochondria [181]. Although the role of acetaldehyde in the biological effects of ethanol has been extensively investigated, the specific involvement of the oxidative products of alcohol in ethanol-induced hypertension has only been recognized in recent reports. Further, considering a key role of brain catalase in the central ethanol disposition [210] [221] is challenging because reported findings on catalase activity in experimental models of genetic hypertension are limited and inconsistent. For example, a higher striatal [5], but not renal [158], catalase activity was reported in SHRs compared with WKYs.

Although the oxidative product acetaldehyde has been implicated in the behavioral effects of ethanol [113, 154, 164, 177], scarce data exist on acetaldehyde contribution to the ethanol-evoked pressor response. Integrative and electrochemical studies revealed heightened pressor and sympathoexcitatory effects of intra-RVLM ethanol in conscious SHRs compared with WKYs [126, 136]. In a series of recent and novel reports from our laboratory, we established compelling evidence that implicated acetaldehyde in this hypertension-specific pressor effect of ethanol and identified the molecular underpinnings of this effect [58, 61, 66]. We first demonstrated that local catalase-mediated oxidation, into acetaldehyde, accounted for the hypertensive response elicited by intra-RVLM ethanol in SHRs. It is not surprising, therefore, that an imbalance between ethanol-derived acetaldehyde and its further oxidation to acetate (via ALDH) results in acetaldehyde accumulation and a subsequent RVLM neuronal oxidative stress serves as underlying mechanisms for the heightened increases in central sympathetic tone and blood pressure in SHRs [9, 22].

This premise is further supported by: (1) compared to control WKYs, the SHR RVLM exhibits higher catalase activity, and (2) the pressor effect of ethanol was dramatically reduced in SHRs pretreated systemically with 3-amino-1,2,4-triazole (3-AT, catalase inhibitor) [58]. Further, the finding that similar increases in blood pressure are caused by acetaldehyde (2 μg) or ethanol (10 μg) microinjection into the RVLM supports the premise that acetaldehyde is the principal mediator of ethanol-evoked hypertension in SHRs [58]. In this circumstance, the blood pressure rises caused by acetaldehyde were similar, in duration and magnitude, to that caused by intra-RVLM ethanol. Considering the importance of discrete areas of the brainstem such as the RVLM in blood pressure control [26, 79], these findings suggested a centrally mediated effect of acetaldehyde on blood pressure. Because systemic 3-AT caused significant drop in blood pressure, it is possible that 3-AT counterbalanced the pressor effect of intra-RVLM ethanol perhaps via physiological antagonism. This possibility seems unlikely because under the same conditions, systemic 3-AT failed to alter the pressor response elicited by intra-RVLM acetaldehyde [58], and produced variable effects on blood pressure that include decreases [23], increases [215], or no changes [121].

Another evidence for the acetaldehyde hypothesis are the findings in normotensive (WKY) rats pretreated with an ALDH inhibitor. Whereas intra-RVLM acetaldehyde failed to alter blood pressure in WKY rats, substantial increases in blood pressure were seen when acetaldehyde was microinjected in the same rat strain pretreated with the ALDH inhibitor cyanamide [58]. It is plausible, therefore, that ALDH rapidly oxidizes acetaldehyde into acetate, thus preventing the accumulation of this neurotoxic aldehyde in the RVLM of the WKY. Conceivably, ALDH inhibition appears to have created RVLM environment conducive to acetaldehyde accumulation and the unraveling of its pressor effect in WKY rats. While these findings argue against a significant role for acetate, the next step in ethanol metabolism [165], in the pressor effect of ethanol or acetaldehyde, it is imperative to note that acetate contributes to other neurobiological/behavioral effects of ethanol [33, 35].

Recent neurochemical data reveal substantially higher catalase but similar ALDH activity in the RVLM of SHRs and WKY rats [58]. Such enzymatic profile is anticipated to provoke greater formation and accumulation of the ethanol-derived acetaldehyde in the RVLM of the SHRs. Studies on catalase activity in other brain areas of hypertensive rats demonstrate inconsistent effects. For example, the SHRs exhibit higher and lower catalase activity in the striatum and the whole brain homogenate, respectively, compared with normotensive controls [5, 158]. As discussed earlier, intra-RVLM acetaldehyde elevates blood pressure in SHRs, but not in WKY rats, despite the similar RVLM ALDH activity [58]. It is likely, therefore, that this ethanol-metabolizing enzymes profile or other intrinsic nonenzymatic mechanisms [164] contribute to such strain-dependent blood pressure effect of ethanol or its first metabolite, acetaldehyde.

It is noteworthy that the pressor response elicited by intra-RVLM ethanol or acetaldehyde is associated with significant decreases in the heart rate. While the pressor effect of ethanol relates to the enhancement of central sympathetic tone [126, 136], the mechanism of the associated bradycardia has not been elucidated. Speculatively, the bradycardic action of ethanol might erupt as a reflex response due to the activation of arterial baroreceptors in the aortic arch and carotid sinus [67, 131, 146]. However, this possibility is contradicted by the findings that intra-RVLM acetaldehyde elicited pressor responses in cyanamide-pretreated WKY rats or in 3-AT pretreated SHRs without affecting HR. Therefore, more studies are needed to characterize the underlying mechanism of the bradycardic effect of intra-RVLM ethanol [58].

It is important to comment on the clinical relevance of the microinjected alcohol dose. Given the similarity of the cardiovascular effects produced by systemic (1 g/kg) [44, 45, 198, 199] and intra-RVLM (10 μg) [126, 136] alcohol in SHRs, it is plausible that these two alcohol regimens would produce comparable levels of the drug in the RVLM. This premise is supported by the study by Robinson et al. [173], which showed similar blood and brain levels of ~25 mM following alcohol (1 g/kg, i.v). It is conceivable, therefore, that the 10 μg intra-RVLM dose of alcohol used in our studies might lead to tissue alcohol concentration of approximately 25 mM, which is consistent with blood levels achieved following social alcohol consumption [3, 98]. Moreover, the use of the 2 μg dose of acetaldehyde for intra-RVLM studies was based on reported relative potencies of acetaldehyde and ethanol in cardiovascular [58] and behavioral studies [164]. A higher intra-RVLM dose of acetaldehyde (4 μg) increased blood pressure to levels that were not different from those produced by the 2 μg dose [66].

Apart from neurohumoral pathways, a potential role for osmolality changes in the pressor effect of ethanol is likely because clinical and experimental studies showed that hyperosmolality increases sympathetic neural activity and blood pressure through the stimulation of central osmoreceptors [190]. These osmotically mediated effects have been attributed to the activation of vasopressinergic and glutamatergic projections to the RVLM sympathetic neurons [10, 18, 74]. Given that ethanol is osmotically active and that osmolarity modulation accounts for the ethanol-evoked changes in secretory [117] and disease states [133], future studies are warranted to investigate whether osmotic changes in RVLM neurons contribute to the sympathetic and blood pressure responses elicited by ethanol.

Notably, the reported pressor and sympathoexcitatory effects of acetaldehyde appear to be at odds with its direct effects on vascular reactivity. In vitro studies have shown that acute or chronic acetaldehyde exposure relaxes vascular smooth muscle and reduce responsiveness to vasoconstrictor stimuli [7, 19, 144]. Ren et al. [170] reported that aortic relaxations induced by acetaldehyde are dependent upon the endothelium and blood pressure states. Compared with normotensive counterparts, acetaldehyde-induced vasorelaxations are diminished and augmented in SHR aortas with intact and denuded endothelium, respectively [170]. Electrophysiological evidence suggests a vital role for the inhibition of voltage-dependent Ca^{2+} currents in the attenuating action of acetaldehyde on contractions induced by potassium depolarization in vascular smooth muscle cells [144]. High blood acetaldehyde concentrations seen in some Asians with a genetically low ALDH activity [71, 90] have been implicated in hypotension, flushing, and palpitation associated with alcohol use. It is likely, therefore, that the blood pressure response elicited by ethanol signifies the net of its vascular and neurohumoral effects at central and peripheral sites.

3.3 *ALDH2 Polymorphism Modulates Alcohol-Evoked Hypertension*

As discussed above, acetaldehyde is detoxified into acetate by ALDH2, which subsequently enters into the Krebs cycle to generate CO_2 and water as the end products of ethanol oxidation [78, 134]. Clinical data demonstrate elevated circulating acetaldehyde levels following acute or chronic alcohol use [149, 151]. Impairment or genetic mutations (polymorphism) of ALDH result in elevated circulating levels of acetaldehyde in alcoholics [149]. Compared to blood acetaldehyde concentrations of ~5 μM in Asians with intact ALDH2 activity, individuals with one mutant ALDH2 allele exhibit acetaldehyde levels of 30–125 μM and develop severe organ injury following alcohol intake [31, 148, 205].

The generation of free radicals and oxidative stress are the main mechanisms responsible for cytotoxicity induced by alcohol and acetaldehyde. Further, the oxidation of acetaldehyde into acetate may produce oxidative free radicals such as superoxide, acetyl, hydroxyl, and methyl radicals [145] through a number of subcellular enzymes and structures including aldehyde oxidase, xanthine oxidase, mitochondria, and microsomes [171]. ALDH2 mutation is linked to increased prevalence of hypertensive states including that induced by alcohol [8, 99, 193]. It should be remembered, however, that the influence of ALDH2 genotype on blood pressure is rather complex and depends on the amount of alcohol consumed, the timing of blood pressure measurement, and environmental and genetic factors. Collectively, while Iwai and colleagues identified a tie between the ALDH2*1/*1 genotype and the prevalence of hypertension [100], the precise role of ALDH2 polymorphism in the regulation of blood pressure, particularly in alcoholics, remains largely unclear.

3.4 *Central MAPK Signaling Contributes to Ethanol-/Acetaldehyde-Evoked Hypertension*

Mitogen-activated protein kinases (MAPKs) are a group of serine/threonine kinases that transfer extracellular stimuli into a wide range of cellular responses [24, 70]. Conventional MAPKs comprise ERK1/2, c-Jun amino (N)-terminal kinases (JNK1/2/3), and p38 [24]. The pathogenic intermediary roles of MAPKs in serious adverse effects of alcohol such as hepatotoxicity, pancreatitis, neurotoxicity, and increased cancer risk have been recognized [12, 180, 183]. MAPKs phosphorylation might be facilitated or inhibited by ethanol depending on factors such as the duration and dose of alcohol exposure, cell type, and particular MAPK isoform under consideration [12]. In vitro studies showed that vascular contractions and elevations in intracellular calcium caused by alcohol in cerebral smooth muscle cells are mediated via activation of ERK1/2 and p38 [207]. The alcohol-induced contractions of isolated aortas are suppressed after MEK1/2 inhibition, thereby implicating ERK1/2 in the alcohol effect [208]. In whole animal studies, pharmacologic inhibition of RVLM ERK1/2 attenuates the pressor response caused by the microinjection of angiotensin II into the same neuroanatomical area [182].

Evidence implicates MAPKs within the RVLM neurons in the pressor effect of alcohol or its metabolite acetaldehyde in SHRs [66] because (1) the RVLM area, which controls blood pressure and central sympathetic activity, is a neuroanatomical site for mediating the pressor effect of alcohol or its metabolite, acetaldehyde [58, 126, 136]; (2) RVLM MAPKs signaling is crucial for blood pressure control in normotensive and hypertensive states [182]; and (3) activation of RVLM ERK1/2 and p38 accounts for the higher sympathetic activity in stroke-prone SHRs [114] and in models of heart failure [77]. A causal role for enhanced ERK2 signaling in the pressor effect of intra-RVLM alcohol was supported by the observations that alcohol microinjection caused significant increases in local ERK2 phosphorylation and that prior RVLM ERK1/2 inhibition by PD98059 virtually abolished the pressor action of alcohol [66].

Unlike MAPK-ERK1/2, conflicting data are obtained regarding the roles of p38 and JNK signaling in alcohol-induced pressor effect. Whereas RVLM p38 phosphorylation is upregulated by alcohol, the associated pressor response is preserved after pharmacological inhibition of p38 (SB203580) [66], denoting no role for the increase in RVLM p38 activity in the pressor effect of alcohol. On the other hand, despite the lack of change in neuronal p-JNK expression in alcohol-treated SHRs, JNK inhibition (SP600125) compromises the pressor effect of alcohol. While these findings preclude a direct interaction of alcohol with JNK, the possibility that alcohol interacts with downstream effectors of JNK signaling cannot be overlooked. For instance, the immediate early gene c-jun and/or the transcription factor activator protein-1 [32, 213] might contribute to the alcohol-evoked pressor response. Interestingly, a causal link exists between increases in c-Jun or its message, *c-jun*, in the RVLM and sympathetic activity [52, 203].

Like alcohol, acetaldehyde has no effect on phosphatase activity but increases RVLM p-ERK2 level and caused p-ERK2, but not p38, dependent elevation in blood pressure. These data further support a critical role for acetaldehyde in the molecular and cardiovascular effects of alcohol. Notably, the effects of alcohol and acetaldehyde are not usually identical. One important difference is the dependence of pressor effect of alcohol, but not acetaldehyde, on enhanced RVLM JNK2/3 phosphorylation [66]. Thus, it is possible that while enhancement of JNK2/3 and ERK2 phosphorylation underlie alcohol-evoked pressor response, ERK2 phosphorylation plays the major role in the pressor effect of acetaldehyde. Our molecular findings are consistent with the observations that (1) enhanced RVLM ERK1/2 signaling mediates increases in sympathetic activity and blood pressure caused by the activation of RVLM angiotensin AT1 receptors [114] and (2) ERK1/2 inhibition in the REVLM causes hypotension and reduces the pressor response caused by intra-RVLM angiotensin II [182].

Based on the individual molecular profiles of alcohol and acetaldehyde, it is possible that JNK2/3 and ERK2 are phosphorylated consecutively in the signaling pathway leading to the acetaldehyde-dependent pressor effect of alcohol. Evidence suggests a facilitatory role for alcohol dehydrogenase in the alcohol-evoked JNK phosphorylation [152] in addition to its established role in alcohol metabolism [164]. Notably, the oxidative metabolism of alcohol into acetaldehyde by alcohol dehydrogenase/catalase mediates the pressor action of intra-RVLM alcohol in SHRs [58]. Unlike alcohol, the pressor effect of acetaldehyde involved the direct activation of ERK, thereby bypassing JNK and the alcohol metabolic pathway. These cellular mechanisms along with the proposed cascade of neuronal substrates involved in the MAPKs-related pressor effect of alcohol/acetaldehyde are illustrated in Fig. 3.

Protein phosphatases are enzymes that dephosphorylate amino acid residues of protein substrates including MAPKs and are involved in multiple regulatory processes such as DNA replication, metabolism, transcription, and development [24, 174]. Studies have shown that the inhibition of phosphatase activity in cell culture contributes to alcohol-evoked increase in p-JNK level [140] and that enhanced MAPKs signaling in the RVLM increases sympathetic activity [77, 114]. With this in mind, it would be expected that the inhibition of RVLM phosphatases might have contributed to the ethanol-evoked hypertension and enhancement of ERK1/2 phosphorylation was evaluated. Such argument seems unlikely because the measurement of neuronal phosphatase activity in RVLM tissues of rats treated with alcohol or its metabolite acetaldehyde revealed no differences from control tissues. Similar to alcohol, the inhibition of neuronal ser/thr phosphatase activity by intra-RVLM okadaic acid increased blood pressure and p-ERK2 expression [66], suggesting that RVLM phosphatases tonically limit the buildup of neuronal phosphorylated MAPKs, which enhance sympathetic activity [77, 114]. Collectively, the increases in blood pressure and RVLM p-ERK2 caused by okadaic acid lend credence to the concept that higher levels of phosphorylated kinases in the RVLM mediate, at least partly, the pressor effect of intra-RVLM alcohol in SHRs.

One potential limitation of microinjection studies of alcohol relates to the possibility that the microinjected alcohol dose into brainstem neurons might cause neuronal damage and nonspecific effects. The microinjected dose of alcohol (10 μg) has been used in brainstem microinjection studies in our laboratory [45, 136] and by

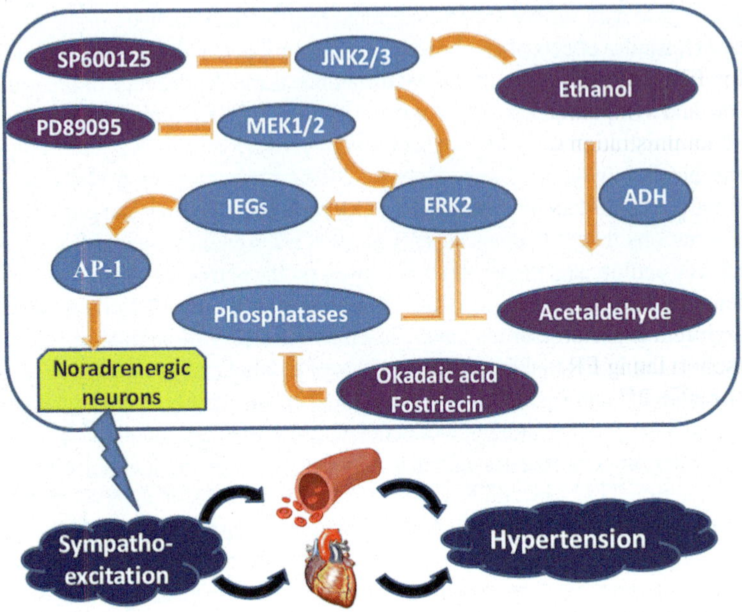

Fig. 3 Schematic paradigm of neuronal events in the RVLM that lead to the increase in central sympathetic outflow and elevation in blood pressure caused by intra-RVLM administration of ethanol or acetaldehyde in spontaneously hypertensive rats (SHRs). *ADH*, alcohol dehydrogenase; *IEGs*, immediate early genes; *AP-1*, activator protein-1; *MEK1/2*, mitogen-activated protein kinase kinase; *p-ERK2*, phosphorylated extracellular signal-regulated kinases

others [198, 199] to discern the role of these brain areas in the cardiovascular effects of alcohol. Alcohol, microinjected into brainstem nuclei, impairs arterial baroreflexes via interaction with specific subsets of central GABA and glutamate receptors, and these effects disappeared within 1–2 h [45, 198]. The findings that RVLM ERK and JNK, but not p38, contributed to pressor effect of alcohol [66] reflect its selectivity and, along with the reversibility in the observed effects, made it unlikely that the alcohol-evoked cardiovascular and molecular responses could be attributed to nonspecific neuronal damage.

3.5 *Brainstem Phosphatases Dampen Alcohol-Induced Hypertension*

As discussed above, the microinjection of ethanol or acetaldehyde into the RVLM only modestly increased local ERK phosphorylation and blood pressure in conscious normotensive, WKY, rats [58, 126]. To understand the underlying mechanisms that dampen these effects, studies were undertaken in conscious normotensive rats to test the hypothesis that RVLM phosphatases act tonically to dampen ethanol- or acetaldehyde-evoked ERK phosphorylation and subsequent increases in blood pressure. As is the case in SHRs [66], the ERK1/2 inhibitor PD98059 abolished the modest pressor effect caused by intra-RVLM ethanol in WKY rats, suggesting a key role for ERK1/2 phosphorylation in alcohol-evoked increases in blood pressure response and sympathetic activity [126]. More importantly, the simultaneous intra-RVLM administration of ethanol and okadaic acid (nonselective inhibitor of all ser/thr phosphatase isoforms, PP1 through PP6) or fostriecin (selective inhibitor of PP1 and PP2A) [192] caused exaggerated inhibition of phosphatase activity along with greater and more sustained elevations in blood pressure.

Similar exacerbation of the pressor response and local phosphatases inhibition are seen when acetaldehyde is combined with okadaic acid or fostriecin [61]. The data highlight a preferential role for phosphatases of the PP1 and PP2A types in dephosphorylating ERK1/2 and the dampening of acetaldehyde-dependent pressor effect of intra-RVLM alcohol in normotensive rats. In addition to its selective phosphatase inhibitory effect, fostriecin also offers a better experimental tool for phosphatase inhibition because of the reported cytotoxicity associated with the use of okadaic acid [92, 214]. Collectively, our observations reinforce a restraining influence of local RVLM phosphatases on alcohol-evoked pressor response [66].

4 Alcohol Metabolism Contributes to Its Sex-Dependent Cardiovascular Effects

4.1 Estrogen Provokes Cardiodepressant and Hypotensive Effects of Alcohol

Clinical and experimental findings demonstrate that the net effect of alcohol on blood pressure follows a J-shaped relationship due probably to the complex effects of alcohol on cardiovascular functions under different settings. Acute alcohol increases [43], decreases [110], or has no effect [4, 198] on blood pressure, while chronic alcohol increases [175] or decreases [54, 55, 168] blood pressure. The mechanisms by which alcohol elevates blood pressure include increase in sympathetic activity as indicated by the rise in plasma norepinephrine levels [27, 28, 98]. The observation that sympathetic neural activity is elevated in alcohol-fed, compared with pair-fed control, rats provides more direct evidence for the involvement of the sympathetic nervous system in alcohol-induced hypertension [175]. The attenuation by alcohol of the gain in arterial baroreceptor activity may also contribute to the increases in sympathetic activity and the subsequent pressor effect [43, 47].

On the other hand, ethanol elicits other cardiovascular actions that may counterbalance its sympathoexcitatory effects and result in a net drop in blood pressure such as direct myocardial depression [111, 217], reduction in cardiac output [49], vasodilation [195], and α-adrenoceptor blockade [3]. Alcohol use results in an increased incidence of cardiac morbidity and mortality. As recently reviewed [202], the alcohol-evoked cardiomyopathy manifests as ventricular dilation, reduced ventricular wall thickness, myofibrillary disarray, interstitial fibrosis, hypertrophy, and contractile dysfunction. The underlying mechanisms include ethanol/acetaldehyde toxicity [124], mitochondrial production of reactive oxygen species [138], oxidative injury, apoptosis [188], impaired myofilament Ca^{2+} sensitivity [104], and abnormalities in fatty acid deposition [206].

Experimental reports suggest fundamental roles for ovarian hormones in the sex-specific cardiovascular derangements caused by alcohol. Alcohol reduces cardiac output, stroke volume, and blood pressure in female rats during proestrus, which exhibits the highest level of circulating estrogen [49, 51]. Indices of myocardial contractility such as left ventricular pressure over time (dP/dt_{max}) and left ventricular developed pressure are also significantly reduced by alcohol. These hemodynamic effects of alcohol are minimal or absent in male rats and in ovarian hormone-deprived (ovariectomized) rats but are restored following estrogen replacement in both preparations [60, 96, 209]. Because reductions in myocardial contractility indices and blood pressure are tightly related, it is concluded that myocardial depression is largely responsible for the developed hypotension in these reported studies [60, 96, 209]. These findings have clinical implications because alcohol lowers blood pressure in young, but not in older women [115]. The measurements of left ventricular developed pressure (LVDP), dP/dt_{max} [40, 41, 201, 218] and sympathovagal control of the heart [42, 53] provide more direct assessment of cardiac contractility and autonomic control. Our reported findings implicated cardiac vagal dominance in the estrogen-dependent chronic hypotensive effect of alcohol in female rats [57, 59]. At the molecular level, facilitation of the myocardial PI3K/Akt/nNOS signaling cascade contributes, at least partly, to these effects [64].

Studies in proestrus rats provided evidence for a causal role for autonomic dysregulation in the sex-specific myocardial oxidative stress and dysfunction. Specifically, ethanol-evoked reductions in cardiac output [49, 57], blood pressure and indices of myocardial contractility, LVDP, and dP/dt_{max} were associated with prolongation of the left ventricular isovolumic relaxation constant Tau [96], which is a measure of cardiac diastolic function [69, 97]. Further, the power spectral analysis of heart rate variability revealed a shift toward vagal dominance in the presence of ethanol in estrogen-replete rats. Most compelling are the findings that pharmacological interruption of cardiac vagal innervation attenuated the estrogen-dependent myocardial oxidative stress/dysfunction. These findings suggest a pivotal role for enhanced vagal dominance in the cardiodepressant effect of alcohol in female rats [59, 64].

Pharmacological (iNOS inhibition) and molecular studies implicate vascular iNOS upregulation in the sex/estrogen-dependent hypotensive effect of alcohol [68]. The latter is associated with an adaptive increase in the gene expression of the immediate early gene c-jun in the RVLM [55], which reflects an increase in sympathetic neural activity to counterbalance the primary decreases in blood pressure. Notably, the estrogen dependence of the hypotensive effect of alcohol may be explained in view of the similarity of the vascular effects of the alcohol and estrogen because both inhibit calcium influx [25, 195], promote endothelial nitric oxide activity [84, 94], and reduce α-adrenoceptor responsiveness [2, 191]. It is possible, therefore, that ethanol may interact synergistically with estrogen to elicit vascular changes that trigger vasodilatation and subsequent falls in blood pressure.

The role of estrogen receptors (ER) in the estrogen-dependent hypotensive and myocardial depressant actions of alcohol has been investigated. The three ERs, ERα, ERβ, and G protein-coupled ER (GPER), are distributed throughout the cardiovascular system. They act as important regulators of myocardial function by genomic and non-genomic signaling mechanisms [141, 142]. The observations that non-genomic effects are involved in the cardiovascular effects of alcohol in female [60] and male [62] rats might link one or more of ER subtypes to the acute estrogen-dependent myocardial depressant effect of alcohol. It was also imperative to identify the ER subtype implicated in estrogen enhancement of the activity of two myocardial redox enzymes, catalase and ALDH2, which confer cardioprotection [20, 127] and also catalyze alcohol oxidation to acetaldehyde and acetate, respectively [34, 91]. Pharmacological loss-of-function studies using highly selective antagonists showed that compared with ERβ (PHTPP) or GPER (G15) blockade, ERα blockade (MPP) greatly attenuated the reductions in myocardial contractility indices (dP/dt_{max} and left ventricular developed pressure) and blood pressure produced by ethanol [209]. However, because PHTPP did reduce the cardiac, but not hypotensive, effect of alcohol during the last 30 min of the study, it is plausible that functional ERβ and GPER might be required for the demonstration of the delayed effects of alcohol on the heart. These findings are consistent with the interplay of ER subtypes in regulating cardiac functions [116, 155, 186]. The preserved hypotensive effect of ethanol in ERβ-blocked rats might be related to ERα/GPER-mediated enhancement of nitric oxide-dependent vasodilation [112, 172]. Discrepancies in the modulation of eNOS and nNOS by ER subtypes in the myocardium and vasculature [75, 125, 137] might account for the preservation of ethanol-evoked hypotension in ERβ-blocked rats. Mechanistically, our findings implicated the Akt/ERK1/2/p38 pathway to the estrogen-dependent myocardial depressant effect of ethanol [64, 209], which is also causally related to oxidative stress [76, 101, 196]. ERα and GPER, but not ERβ, contribute to these molecular events. These pharmacological and biochemical findings support a pro-oxidant role for estrogen, in the presence of ethanol, mediated via MAPK phosphorylation [96]. These findings were extended by pharmacological gain-of-function studies using highly selective ER subtype agonists in the absence or presence of alcohol in ovariectomized rats [211]. The latter study showed that compared with ERα or ERβ, GPER activation exhibited a lesser ability in uncovering the adverse hemodynamic effects of alcohol. Accordingly, replacement with selective GPER agonists might present a safer therapeutic alternative for estrogen in women with habitual alcohol consumption.

4.2 Enhanced Alcohol Metabolism Underlies Its Heightened Cardiodepressant Effect in Females

In peripheral tissues, alcohol is metabolized via alcohol dehydrogenase (ADH) and catalase to acetaldehyde, which is subsequently eliminated by aldehyde dehydrogenase (ALDH2) [17, 156, 164]. Acetaldehyde is largely responsible for myocardial dysfunction induced by alcohol and underlying cellular mechanisms [39, 169]. Because estrogen enhances alcohol metabolism to acetaldehyde [102], which is known to upregulate several molecular entities along the oxidative stress signaling [202], recent studies interrogated the role of acetaldehyde accumulation in the female myocardium in the exacerbated estrogen-dependent myocardial depressant effect of alcohol [96]. The elegant studies by Ren et al. [39] showed higher sensitivity of the isolated female cardiomyocytes to the myocyte contractile depressant effect of ethanol-derived acetaldehyde. These in vitro [39] and the aforementioned in vivo [49, 51] findings might explain the higher sensitivity of women to alcohol-induced cardiomyopathy. While a major role for locally generated acetaldehyde in the adverse effects of alcohol in peripheral tissues is debated, there are many findings that support this premise in the heart, particularly in the presence of estrogen. Unfortunately, direct measurements of acetaldehyde levels in the heart and linking such levels to cardiac function are rather challenging. Nonetheless, many alternative approaches support this argument starting with the findings that acetaldehyde being far more toxic and reactive than ethanol and is largely blamed for alcohol-induced cardiac damage [127]. The heart obtained from alcohol-treated rats exhibited higher ADH, but not ALDH2, activity reflecting higher myocardial acetaldehyde levels after alcohol administration [96]. These biochemical findings coincided with the (1) reductions in left ventricular pressure, dP/dt_{max}, and blood pressure, (2) declining concentrations of blood alcohol [64], and (3) increased cardiac accumulation of cardiotoxic aldehydes such as malondialdehyde and 4-hydroxy-2-nonenal adducts [64, 96]. The latter products are cardiotoxic and potent inducers of cardiac pathology and oxidative damage [17, 37]. Under this circumstance, the preservation of ALDH2 activity would be expected to facilitate the detoxification of acetaldehyde as well as other cardiotoxic aldehydes. Li et al. [127] suggested a therapeutic potential of ALDH2 in alcoholic complications because ALDH2 overexpression effectively alleviates acetaldehyde-induced injury in cardiomyocytes through an ERK1/2 and SPAK/JNK-dependent mechanism. The study by Budas et al. [20] has reached a similar conclusion and proposed that ALDH2 activation might be exploited to preserve the cardioprotective effect of ethanol while minimizing the side effects associated with alcohol consumption. Together, the worsened cardiac profile seen in alcohol-treated rats despite the preservation of ALDH2 activity may be attributed to the enhanced ADH activity and subsequent elevations in the cardiac levels of toxic acetaldehydes, which exceed the detoxification capacity of ALDH2.

The presumption that acetaldehyde mediates myocardial depression triggered by alcohol in proestrus rats receives support from earlier reports, which demonstrated that acetaldehyde is responsible for the higher female propensity to alcohol-induced myocardial dysfunction [39, 169, 197]. Also, similar amounts of alcohol produced higher acetaldehyde levels in premenopausal women and in women on oral contraceptives, compared to men [73]. Additionally, the overproduction of acetaldehyde that follows ADH overexpression accelerated cardiac dysfunction in female, more than in male, cardiomyocytes [39, 129]. Notably, contrary to our recent report [96], none of these reported studies linked the molecular events in the myocardium to the alcohol-evoked myocardial dysfunction at the integrative level. Our data have also shown that myocardial acetaldehyde accumulation (albeit indirectly assessed) triggers molecular events conducive to the generation of oxidative state. Alcohol caused remarkable increases in myocardial Akt/ERK1/2 and nicotinamide adenine dinucleotide phosphate oxidase (NADPHox) activation and reactive oxygen species production (Fig. 4). Tempol, a superoxide dismutase mimetic with antioxidant activity, abrogated myocardial oxidative stress induced by alcohol and preserved myocardial function [96].

These biochemical findings lend credence to earlier findings that alcohol elicits acetaldehyde-dependent activation of NADPHox [39, 89] and ERK1/2 [127] in other tissues, and PI3K/Akt signaling mediates NADPHox activation and reactive oxygen species (ROS) production in macrophages [204]. Moreover, these findings support and extend our earlier findings that implicated PI3K/Akt activation in the alcohol-evoked reductions in cardiac output and blood pressure in proestrus female rats [57, 65]. As illustrated in Fig. 4, p-Akt may also enhance ROS generation via endothelial nitric oxide synthase (eNOS) uncoupling [29, 108], which could be exacerbated by the accumulation of the cardiotoxic aldehyde adducts.

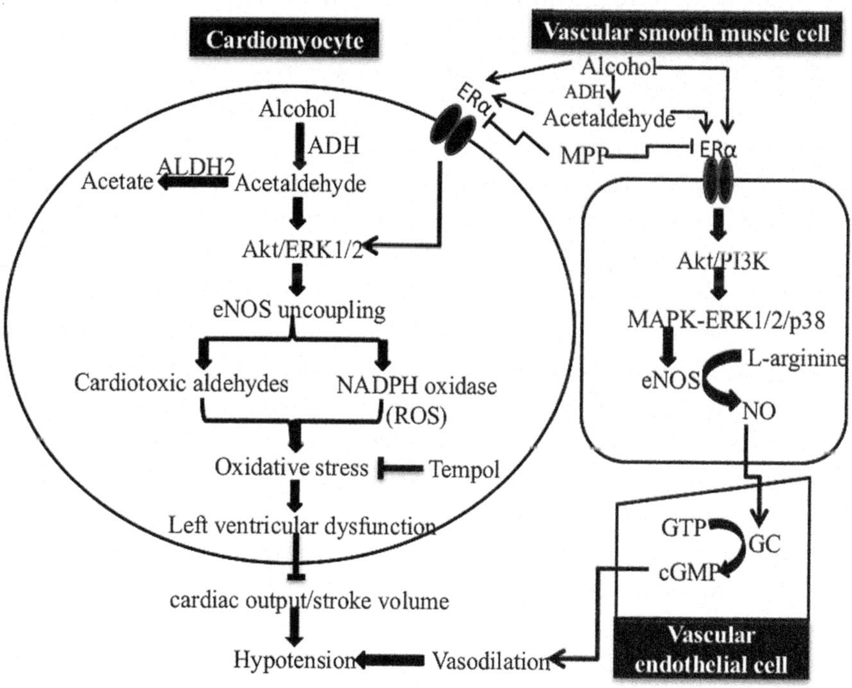

Fig. 4 Cardiac and vascular signaling events involved in the estrogen-dependent myocardial depressant and hypotensive effects of alcohol in female rats. *ADH*, alcohol dehydrogenase; *ALDH2*, alcohol dehydrogenase 2; *eNOS*, endothelial nitric oxide synthase; *ROS*, reactive oxygen species; *PI3K*, phosphoinositide 3-kinase; *GC*, guanylyl cyclase; *GTP*, guanosine triphosphate; *MAPK-ERK1/2*, mitogen-activated protein kinase-extracellular signal-regulated kinases

4.3 *Inhibition of Alcohol Metabolism Attenuates Its Cardiovascular Toxicity*

Recent pharmacological evidence further supports acetaldehyde contribution to estrogen-dependent myocardial oxidative stress and dysfunction because the latter were (1) partially reduced following ADH/CYP2E1 (4-methylpyrazole; 4-MP) or catalase (3-AT) inhibition and (2) virtually abolished following combined enzyme inhibition (4-MP plus 3-AT) [210]. The oxidative stress caused by alcohol in cardiac tissues (ROS generation) and concomitant increases in ERK1/2 phosphorylation were ameliorated in similar fashions after inhibition of the oxidative metabolism of alcohol to acetaldehyde [210]. The partial attenuation of the alcohol effects by 3-AT or 4-MP might be explained by the ability of the functional enzyme(s) to compensate for the inhibited one. Also, the elevations in blood alcohol levels that appeared after enzyme inhibition might have produced direct cardiotoxicity [13]. Alternatively, the accumulation of non-oxidative alcohol metabolites such as fatty acid ethyl ester in blood [38] and heart [119] that follows the inhibition of alcohol oxidative metabolism does not seem to contribute to the alcohol effects because the latter were virtually abolished after combined ADH/ALDH inhibition [210].

Further support for ADH- or catalase-mediated acetaldehyde production in the adverse cardiovascular effects of alcohol includes the involvement of the enhanced acetaldehyde production, via ADH, in alcohol-evoked cardiomyocyte contractile dysfunction, autophagy, mitochondrial damage, and apoptosis [86, 87, 128]. A role for a catalase-based oxidation in the estrogen-dependent cardiovascular and oxidative damage caused by alcohol in the female population is also likely because (1) estrogen enhances catalase catalytic activity in the female rat heart [21, 60, 96, 209], (2) alcohol increases cardiac catalase activity and causes myocardial oxidative stress and dysfunction in ovariectomized rats treated with the ERα agonist propylpyrazole triol [211], and (3) catalase inhibition by 3-AT attenuates the alcohol-evoked left ventricular dysfunction [210]. More pharmacologic and molecular studies are necessary to identify the precise roles of estrogen receptor subtypes in alcohol metabolism and related cardiovascular anomalies.

5 ALDH2-Mediated Cardioprotection Dampens Alcohol-Evoked Cardiotoxicity

ALDH2 is an important mitochondrial redox enzyme that combats oxidative stress via facilitating the oxidation of toxic aldehydes into less toxic acids. By reducing the cellular "aldehyde load," ALDH2 may contribute to the cardioprotective effect of alcohol and to the protection against cardiovascular disease. ALDH2*2 is an inactive variant of ALDH2 with no capacity for metabolizing acetaldehyde [30, 143]. Approximately 40% of East Asians carry at least one copy of the gene encoding the defective ALDH2*2 variant. In people who carry two copies of the defective gene (i.e., homozygous for ALDH2*2/*2), the activity of the enzyme is reduced by >95% compared with people who are homozygous for the normal, active form of the enzyme (i.e., ALDH2*1). On the other hand, in people who carry only one copy of the mutant gene and one copy of the normal gene (i.e., who are heterozygous ALDH2*1/*2), the activity of the resulting enzyme is about 40% of the normal ALDH2 [120]. ALDH2*2 carriers exhibited higher levels of troponin I and malondialdehyde and hydroxynonenal adducts after coronary artery bypass grafting [81]. Further, the intensive care unit time and postoperative hospitalization were longer in ALDH2*2 carriers [81]. In the elegant study by Chen et al. [30], the administration of Alda-1, an ALDH2 activator, to rats prior to an ischemic event increased the catalytic activity of ALDH2 by twofold and reduced infarct size by 60% through probably its inhibitory effect on the formation of cytotoxic aldehydes. Similarly, Alda-1 attenuates the cardiotoxic effect of ethanol in rats [105].

Credible evidence supports a key role for ADH/ALDH2 imbalances in the modulation of alcohol-induced cardiomyopathy. Our findings that left ventricular dysfunction induced by alcohol is coupled with elevated ADH, but not ALDH2, activity signify the involvement of myocardial acetaldehyde accumulation in the developed cardiomyopathy [96]. Similar observations are reported by others [86, 128]. Genetic manipulations showed opposite effects of ALDH2 knockout (accentuation) [132] and overexpression (inhibition) [127] of ethanol-induced cardiac depression. Based on these reports, a therapeutic potential for ALDH2 in alcoholic complications has been proposed. For example, the transfection of fetal human cardiac myocytes with ALDH2 attenuates ROS generation, apoptosis, and phosphorylation of ERK1/2 and SAPK/JNK caused by alcohol or acetaldehyde [127]. Because our previous studies implicated the same molecular stressors in cardiac [64, 96, 209, 210] and blood pressure anomalies [58, 126] caused by alcohol, the utilization of pharmacological or molecular strategies that boost ALDH2 abundance and/or activity could be viable therapeutic options for correcting alcohol-evoked cardiac adverse effects.

6 Conclusions

In this chapter we discussed mounting experimental evidence that supports scarce clinical findings on the adverse cardiac and autonomic effects of ethanol. While the mechanisms of a potential cardioprotective effect of mild/moderate alcohol consumption are beyond the scope of this book chapter, it is important to note that some recent clinical findings question this dogma. Our discussion focused on less appreciated adverse cardiac effects produced by alcohol doses (blood concentrations) that usually have minimal cardiac effects in health men and male rats. In the latter, limited metabolism of alcohol to acetaldehyde and compensatory mechanisms including increased sympathetic tone and induction of ALDH2 seem to counterbalance the adverse cardiac effects of alcohol. By marked contrast, under less studied sex (ovarian hormones/estrogen) or pathological (hypertension) conditions, the changes in these homeostatic mechanisms seem to favor the generation and accumulation of the alcohol cardiotoxic metabolite acetaldehyde and the exacerbation of the adverse cardiac effect of alcohol. Specifically, we discussed findings that implicated enhanced catalase activity in the brainstem of hypertensive rats and in the hearts of estrogen-replete rats in exaggerated neurotoxic and cardiotoxic effects of the same dose(s) of alcohol, compared to normotensive and ovarian hormone-deprived rats, respectively. Localizing microinjections of alcohol or acetaldehyde into brainstem areas, the use of selective alcohol-metabolizing enzyme inhibitors as well as MAPK and phosphatase inhibitors and other pharmacological interventions supported the premise that enhanced alcohol metabolism to acetaldehyde is pivotal for the induction of autonomic dysregulation and cardiac dysfunction. The reviewed findings implicated the Akt-ERK1/2-NOS signaling cascade in the neurotoxic and cardiotoxic effects of alcohol. Interestingly, while direct cardiac effects of alcohol or its metabolite acetaldehyde produce cardiotoxicity, pharmacological evidence supports the shift of cardiac autonomic regulation toward vagal dominance, which significantly contributes to and might explain the sex-/estrogen-dependent exaggeration of alcohol-evoked myocardial dysfunction. More studies are needed to further understand the cellular mechanisms that underlie the increased sensitivity of cardiac myocytes in the presence of estrogen and brainstem neurons under hypertension conditions to the toxic effects of alcohol and acetaldehyde. These future studies will help identify new targets for the development of potential therapeutics to mitigate the adverse autonomic and cardiac effects of alcohol in more vulnerable populations and hypertensives as well as premenopausal and post or surgical menopausal women receiving estrogen replacement therapy.

Acknowledgments The authors' reported studies cited in this book chapter were supported by NIH grants 2R01 AA07839 and 2R01 AA14441 (Abdel-Rahman) and by the Science and Technology Development Fund, Egypt, STDF Grant No. 14895 (El-Mas).

References

1. Abdel-Rahman AA, Wooles WR (1987) Ethanol-induced hypertension involves impairment of baroreceptors. Hypertension 10(1):67–73
2. Abdel-Rahman AR, Dar MS, Wooles WR (1985) Effect of chronic ethanol administration on arterial baroreceptor function and pressor and depressor responsiveness in rats. J Pharmacol Exp Ther 232(1):194–201
3. Abdel-Rahman AR, Merrill RH, Wooles WR (1987) Effect of acute ethanol administration on the baroreceptor reflex control of heart rate in normotensive human volunteers. Clin Sci (Lond) 72(1):113–122
4. Abdel-Rahman AR, Russ R, Strickland JA, Wooles WR (1987) Acute effects of ethanol on baroreceptor reflex control of heart rate and on pressor and depressor responsiveness in rats. Can J Physiol Pharmacol 65(5):834–841
5. Abilio VC, Silva RH, Carvalho RC, Grassl C, Calzavara MB, Registro S, D'Almeida V, Ribeiro Rde A, Frussa-Filho R (2004) Important role of striatal catalase in aging- and reserpine-induced oral dyskinesia. Neuropharmacology 47(2):263–272. https://doi.org/10.1016/j.neuropharm.2004.04.003

6. Ajmera VH, Terrault NA, Harrison SA (2017) Is moderate alcohol use in nonalcoholic fatty liver disease good or bad? A critical review. Hepatology 65(6):2090–2099. https://doi.org/10.1002/hep.29055
7. Altura BM, Carella A, Altura BT (1978) Acetaldehyde on vascular smooth muscle: possible role in vasodilator action of ethanol. Eur J Pharmacol 52(1):73–83
8. Amamoto K, Okamura T, Tamaki S, Kita Y, Tsujita Y, Kadowaki T, Nakamura Y, Ueshima H (2002) Epidemiologic study of the association of low-Km mitochondrial acetaldehyde dehydrogenase genotypes with blood pressure level and the prevalence of hypertension in a general population. Hypertens Res 25(6):857–864
9. Amanvermez R, Agara E (2006) Does ascorbate/L-Cys/L-Met mixture protect different parts of the rat brain against chronic alcohol toxicity? Adv Ther 23(5):705–718
10. Antunes VR, Yao ST, Pickering AE, Murphy D, Paton JF (2006) A spinal vasopressinergic mechanism mediates hyperosmolality-induced sympathoexcitation. J Physiol 576(Pt 2):569–583. https://doi.org/10.1113/jphysiol.2006.115766
11. Arkwright PD, Beilin LJ, Rouse I, Armstrong BK, Vandongen R (1982) Effects of alcohol use and other aspects of lifestyle on blood pressure levels and prevalence of hypertension in a working population. Circulation 66(1):60–66
12. Aroor AR, Shukla SD (2004) MAP kinase signaling in diverse effects of ethanol. Life Sci 74(19):2339–2364
13. Awtry EH, Philippides GJ (2010) Alcoholic and cocaine-associated cardiomyopathies. Prog Cardiovasc Dis 52(4):289–299. https://doi.org/10.1016/j.pcad.2009.11.004
14. Baros AM, Wright TM, Latham PK, Miller PM, Anton RF (2008) Alcohol consumption, %CDT, GGT and blood pressure change during alcohol treatment. Alcohol Alcohol 43(2):192–197. https://doi.org/10.1093/alcalc/agm156
15. Beilin LJ (1995) Alcohol and hypertension. Clin Exp Pharmacol Physiol 22(3):185–188
16. Bhaskaran D, Freed CR (1989) Catechol and indole metabolism in rostral ventrolateral medulla change synchronously with changing blood pressure. J Pharmacol Exp Ther 249(2):660–666
17. Brien JF, Loomis CW (1983) Pharmacology of acetaldehyde. Can J Physiol Pharmacol 61(1):1–22
18. Brooks VL, Freeman KL, O'Donaughy TL (2004) Acute and chronic increases in osmolality increase excitatory amino acid drive of the rostral ventrolateral medulla in rats. Am J Physiol Regul Integr Comp Physiol 287(6):R1359–R1368. https://doi.org/10.1152/ajpregu.00104.2004
19. Brown RA, Savage AO (1996) Effects of acute acetaldehyde, chronic ethanol, and pargyline treatment on agonist responses of the rat aorta. Toxicol Appl Pharmacol 136(1):170–178. https://doi.org/10.1006/taap.1996.0021
20. Budas GR, Disatnik MH, Chen CH, Mochly-Rosen D (2010) Activation of aldehyde dehydrogenase 2 (ALDH2) confers cardioprotection in protein kinase C epsilon (PKCvarepsilon) knockout mice. J Mol Cell Cardiol 48(4):757–764. https://doi.org/10.1016/j.yjmcc.2009.10.030
21. Campos C, Casali KR, Baraldi D, Conzatti A, Araujo AS, Khaper N, Llesuy S, Rigatto K, Bello-Klein A (2014) Efficacy of a low dose of estrogen on antioxidant defenses and heart rate variability. Oxidative Med Cell Longev 2014:218749. https://doi.org/10.1155/2014/218749
22. Campos RR, Oliveira-Sales EB, Nishi EE, Boim MA, Dolnikoff MS, Bergamaschi CT (2011) The role of oxidative stress in renovascular hypertension. Clin Exp Pharmacol Physiol 38(2):144–152. https://doi.org/10.1111/j.1440-1681.2010.05437.x
23. Cardoso LM, Colombari DS, Menani JV, Toney GM, Chianca DA Jr, Colombari E (2009) Cardiovascular responses to hydrogen peroxide into the nucleus tractus solitarius. Am J Physiol Regul Integr Comp Physiol 297(2):R462–R469. https://doi.org/10.1152/ajpregu.90796.2008

24. Cargnello M, Roux PP (2011) Activation and function of the MAPKs and their substrates, the MAPK-activated protein kinases. Microbiol Mol Biol Rev 75(1):50–83. https://doi.org/10.1128/MMBR.00031-10
25. Castillo C, Ceballos G, Rodriguez D, Villanueva C, Medina R, Lopez J, Mendez E, Castillo EF (2006) Effects of estradiol on phenylephrine contractility associated with intracellular calcium release in rat aorta. Am J Physiol Cell Physiol 291(6):C1388–C1394. https://doi.org/10.1152/ajpcell.00556.2005
26. Chalmers J, Pilowsky P (1991) Brainstem and bulbospinal neurotransmitter systems in the control of blood pressure. J Hypertens 9(8):675–694
27. Chan TC, Sutter MC (1982) The effects of chronic ethanol consumption on cardiac function in rats. Can J Physiol Pharmacol 60(6):777–782
28. Chan TC, Wall RA, Sutter MC (1985) Chronic ethanol consumption, stress, and hypertension. Hypertension 7(4):519–524
29. Chen CA, Druhan LJ, Varadharaj S, Chen YR, Zweier JL (2008) Phosphorylation of endothelial nitric-oxide synthase regulates superoxide generation from the enzyme. J Biol Chem 283(40):27038–27047. https://doi.org/10.1074/jbc.M802269200
30. Chen CH, Budas GR, Churchill EN, Disatnik MH, Hurley TD, Mochly-Rosen D (2008) Activation of aldehyde dehydrogenase-2 reduces ischemic damage to the heart. Science 321(5895):1493–1495. https://doi.org/10.1126/science.1158554
31. Chen YC, Lu RB, Peng GS, Wang MF, Wang HK, Ko HC, Chang YC, Lu JJ, Li TK, Yin SJ (1999) Alcohol metabolism and cardiovascular response in an alcoholic patient homozygous for the ALDH2*2 variant gene allele. Alcohol Clin Exp Res 23(12):1853–1860
32. Choi EY, Hwang HJ, Kim IH, Nam TJ (2009) Protective effects of a polysaccharide from Hizikia fusiformis against ethanol toxicity in rats. Food Chem Toxicol 47(1):134–139. https://doi.org/10.1016/j.fct.2008.10.026
33. Correa M, Arizzi MN, Betz A, Mingote S, Salamone JD (2003) Open field locomotor effects in rats after intraventricular injections of ethanol and the ethanol metabolites acetaldehyde and acetate. Brain Res Bull 62(3):197–202
34. Correa M, Salamone JD, Segovia KN, Pardo M, Longoni R, Spina L, Peana AT, Vinci S, Acquas E (2012) Piecing together the puzzle of acetaldehyde as a neuroactive agent. Neurosci Biobehav Rev 36(1):404–430. https://doi.org/10.1016/j.neubiorev.2011.07.009
35. Cullen N, Carlen PL (1992) Electrophysiological actions of acetate, a metabolite of ethanol, on hippocampal dentate granule neurons: interactions with adenosine. Brain Res 588(1):49–57
36. DePuy SD, Stornetta RL, Bochorishvili G, Deisseroth K, Witten I, Coates M, Guyenet PG (2013) Glutamatergic neurotransmission between the C1 neurons and the parasympathetic preganglionic neurons of the dorsal motor nucleus of the vagus. J Neurosci 33(4):1486–1497. https://doi.org/10.1523/JNEUROSCI.4269-12.2013
37. Deshpande M, Mali VR, Pan G, Xu J, Yang XP, Thandavarayan RA, Palaniyandi SS (2016) Increased 4-hydroxy-2-nonenal-induced proteasome dysfunction is correlated with cardiac damage in streptozotocin-injected rats with isoproterenol infusion. Cell Biochem Funct 34(5):334–342. https://doi.org/10.1002/cbf.3195
38. Doyle KM, Bird DA, al-Salihi S, Hallaq Y, Cluette-Brown JE, Goss KA, Laposata M (1994) Fatty acid ethyl esters are present in human serum after ethanol ingestion. J Lipid Res 35(3):428–437
39. Duan J, Esberg LB, Ye G, Borgerding AJ, Ren BH, Aberle NS, Epstein PN, Ren J (2003) Influence of gender on ethanol-induced ventricular myocyte contractile depression in transgenic mice with cardiac overexpression of alcohol dehydrogenase. Comp Biochem Physiol A Mol Integr Physiol 134(3):607–614
40. El-Bassossy HM, Awan Z, El-Mas MM (2017) Perinatal ciclosporin A exposure elicits sex-related cardiac dysfunction and inflammation in the rat progeny. Toxicol Lett 281:35–43. https://doi.org/10.1016/j.toxlet.2017.09.002

41. El-Bassossy HM, Banjar ZM, El-Mas MM (2017) The inflammatory state provokes sexual dimorphism in left ventricular and electrocardiographic effects of chronic cyclosporine in rats. Sci Rep 7:42457. https://doi.org/10.1038/srep42457
42. El-Lakany MA, Fouda MA, El-Gowelli HM, El-Gowilly SM, El-Mas MM (2018) Gonadal hormone receptors underlie the resistance of female rats to inflammatory and cardiovascular complications of endotoxemia. Eur J Pharmacol 823:41–48. https://doi.org/10.1016/j.ejphar.2018.01.051
43. el-Mas MM, Abdel-Rahman AA (1992) Role of aortic baroreceptors in ethanol-induced impairment of baroreflex control of heart rate in conscious rats. J Pharmacol Exp Ther 262(1):157–165
44. el-Mas MM, Abdel-Rahman AA (1993) Direct evidence for selective involvement of aortic baroreceptors in ethanol-induced impairment of baroreflex control of heart rate. J Pharmacol Exp Ther 264(3):1198–1205
45. el-Mas MM, Abdel-Rahman AA (1993) Role of NMDA and non-NMDA receptors in the nucleus tractus solitarius in the depressant effect of ethanol on baroreflexes. J Pharmacol Exp Ther 266(2):602–610
46. el-Mas MM, Abdel-Rahman AA (1997) Contrasting effects of urethane, ketamine, and thiopental anesthesia on ethanol-clonidine hemodynamic interaction. Alcohol Clin Exp Res 21(1):19–27
47. El-Mas MM, Abdel-Rahman AA (1998) Ovariectomy abolishes ethanol-induced impairment of baroreflex control of heart rate in conscious rats. Eur J Pharmacol 349(2–3):253–261
48. El-Mas MM, Abdel-Rahman AA (1999) Acute hemodynamic effects of ethanol in conscious spontaneously hypertensive and normotensive rats. Alcohol Clin Exp Res 23(2):285–292
49. El-Mas MM, Abdel-Rahman AA (1999) Estrogen-dependent hypotensive effects of ethanol in conscious female rats. Alcohol Clin Exp Res 23(4):624–632
50. El-Mas MM, Abdel-Rahman AA (1999) Ethanol counteraction of I1-imidazoline but not alpha-2 adrenergic receptor-mediated reduction in vascular resistance in conscious spontaneously hypertensive rats. J Pharmacol Exp Ther 288(2):455–462
51. El-Mas MM, Abdel-Rahman AA (1999) Sexually dimorphic hemodynamic effects of intragastric ethanol in conscious rats. Clin Exp Hypertens 21(8):1429–1445
52. El-Mas MM, Abdel-Rahman AA (2000) Clonidine diminishes c-jun gene expression in the cardiovascular sensitive areas of the rat brainstem. Brain Res 856(1–2):245–249
53. El-Mas MM, Abdel-Rahman AA (2000) Ovariectomy alters the chronic hemodynamic and sympathetic effects of ethanol in radiotelemetered female rats. Clin Exp Hypertens 22(1):109–126
54. El-Mas MM, Abdel-Rahman AA (2000) Radiotelemetric evaluation of hemodynamic effects of long-term ethanol in spontaneously hypertensive and Wistar-Kyoto rats. J Pharmacol Exp Ther 292(3):944–951
55. El-Mas MM, Abdel-Rahman AA (2001) An association between the estrogen-dependent hypotensive effect of ethanol and an elevated brainstem c-jun mRNA in female rats. Brain Res 912(1):79–88
56. El-Mas MM, Abdel-Rahman AA (2001) Effect of long-term ethanol feeding on brainstem alpha(2)-receptor binding in Wistar-Kyoto and spontaneously hypertensive rats. Brain Res 900(2):324–328
57. El-Mas MM, Abdel-Rahman AA (2009) Longitudinal assessment of the effects of oestrogen on blood pressure and cardiovascular autonomic activity in female rats. Clin Exp Pharmacol Physiol 36(10):1002–1009. https://doi.org/10.1111/j.1440-1681.2009.05192.x
58. El-Mas MM, Abdel-Rahman AA (2012) Enhanced catabolism to acetaldehyde in rostral ventrolateral medullary neurons accounts for the pressor effect of ethanol in spontaneously hypertensive rats. Am J Physiol Heart Circ Physiol 302(3):H837–H844. https://doi.org/10.1152/ajpheart.00958.2011

59. El-Mas MM, Abdel-Rahman AA (2012) Exacerbation of myocardial dysfunction and autonomic imbalance contributes to the estrogen-dependent chronic hypotensive effect of ethanol in female rats. Eur J Pharmacol 679(1–3):95–100. https://doi.org/10.1016/j.ejphar.2012.01.008
60. El-Mas MM, Abdel-Rahman AA (2014) Nongenomic effects of estrogen mediate the dose-related myocardial oxidative stress and dysfunction caused by acute ethanol in female rats. Am J Physiol Endocrinol Metab 306(7):E740–E747. https://doi.org/10.1152/ajpendo.00465.2013
61. El-Mas MM, Abdel-Rahman AA (2014) Ser/thr phosphatases tonically attenuate the ERK-dependent pressor effect of ethanol in the rostral ventrolateral medulla in normotensive rats. Brain Res 1577:21–28. https://doi.org/10.1016/j.brainres.2014.06.026
62. El-Mas MM, Abdel-Rahman AA (2015) Estrogen modulation of the ethanol-evoked myocardial oxidative stress and dysfunction via DAPK3/Akt/ERK activation in male rats. Toxicol Appl Pharmacol 287(3):284–292. https://doi.org/10.1016/j.taap.2015.06.015
63. El-Mas MM, El-Gowelli HM, Ghazal AR, Harraz OF, Mohy El-Din MM (2009) Facilitation of central imidazoline I(1)-site/extracellular signal-regulated kinase/p38 mitogen-activated protein kinase signalling mediates the hypotensive effect of ethanol in rats with acute renal failure. Br J Pharmacol 158(6):1629–1640. https://doi.org/10.1111/j.1476-5381.2009.00444.x
64. El-Mas MM, Fan M, Abdel-Rahman AA (2009) Facilitation of myocardial PI3K/Akt/nNOS signaling contributes to ethanol-evoked hypotension in female rats. Alcohol Clin Exp Res 33(7):1158–1168. https://doi.org/10.1111/j.1530-0277.2009.00939.x
65. El-Mas MM, Fan M, Abdel-Rahman AA (2011) Upregulation of cardiac NOS due to endotoxemia and vagal overactivity contributes to the hypotensive effect of chronic ethanol in female rats. Eur J Pharmacol 650(1):317–323. https://doi.org/10.1016/j.ejphar.2010.10.032
66. El-Mas MM, Fan M, Abdel-Rahman AA (2013) Role of rostral ventrolateral medullary ERK/JNK/p38 MAPK signaling in the pressor effects of ethanol and its oxidative product acetaldehyde. Alcohol Clin Exp Res 37(11):1827–1837. https://doi.org/10.1111/acer.12179
67. El-Mas MM, Mohy El-Din MM, Helmy MM, Omar AG (2012) Redox imbalances incite the hypertensive, baroreflex, and autonomic effects of cyclosporine in rats. Eur J Pharmacol 694(1–3):82–88. https://doi.org/10.1016/j.ejphar.2012.08.021
68. El-Mas MM, Zhang J, Abdel-Rahman AA (2006) Upregulation of vascular inducible nitric oxide synthase mediates the hypotensive effect of ethanol in conscious female rats. J Appl Physiol (1985) 100(3):1011–1018. https://doi.org/10.1152/japplphysiol.01058.2005
69. El-Yazbi AF, Ibrahim KS, El-Gowelli HM, El-Deeb NM, El-Mas MM (2017) Modulation by NADPH oxidase of the chronic cardiovascular and autonomic interaction between cyclosporine and NSAIDs in female rats. Eur J Pharmacol 806:96–104. https://doi.org/10.1016/j.ejphar.2017.04.016
70. English JM, Cobb MH (2002) Pharmacological inhibitors of MAPK pathways. Trends Pharmacol Sci 23(1):40–45
71. Enomoto N, Takase S, Yasuhara M, Takada A (1991) Acetaldehyde metabolism in different aldehyde dehydrogenase-2 genotypes. Alcohol Clin Exp Res 15(1):141–144
72. Eriksson CJ (2015) Genetic-epidemiological evidence for the role of acetaldehyde in cancers related to alcohol drinking. Adv Exp Med Biol 815:41–58. https://doi.org/10.1007/978-3-319-09614-8_3
73. Eriksson CJ, Fukunaga T, Sarkola T, Lindholm H, Ahola L (1996) Estrogen-related acetaldehyde elevation in women during alcohol intoxication. Alcohol Clin Exp Res 20(7):1192–1195
74. Farquhar WB, Wenner MM, Delaney EP, Prettyman AV, Stillabower ME (2006) Sympathetic neural responses to increased osmolality in humans. Am J Physiol Heart Circ Physiol 291(5):H2181–H2186. https://doi.org/10.1152/ajpheart.00191.2006
75. Fukumura D, Gohongi T, Kadambi A, Izumi Y, Ang J, Yun CO, Buerk DG, Huang PL, Jain RK (2001) Predominant role of endothelial nitric oxide synthase in vascular endothelial growth factor-induced angiogenesis and vascular permeability. Proc Natl Acad Sci U S A 98(5):2604–2609. https://doi.org/10.1073/pnas.041359198

76. Gaitanaki C, Konstantina S, Chrysa S, Beis I (2003) Oxidative stress stimulates multiple MAPK signalling pathways and phosphorylation of the small HSP27 in the perfused amphibian heart. J Exp Biol 206(Pt 16):2759–2769
77. Gao L, Li Y, Schultz HD, Wang WZ, Wang W, Finch M, Smith LM, Zucker IH (2010) Downregulated Kv4.3 expression in the RVLM as a potential mechanism for sympathoexcitation in rats with chronic heart failure. Am J Physiol Heart Circ Physiol 298(3):H945–H955. https://doi.org/10.1152/ajpheart.00145.2009
78. George A, Figueredo VM (2010) Alcohol and arrhythmias: a comprehensive review. J Cardiovasc Med (Hagerstown) 11(4):221–228. https://doi.org/10.2459/JCM.0b013e328334b42d
79. Ghali MGZ (2017) The brainstem network controlling blood pressure: an important role for pressor sites in the caudal medulla and cervical spinal cord. J Hypertens 35(10):1938–1947. https://doi.org/10.1097/HJH.0000000000001427
80. Goel S, Sharma A, Garg A (2018) Effect of alcohol consumption on cardiovascular health. Curr Cardiol Rep 20(4):19. https://doi.org/10.1007/s11886-018-0962-2
81. Gong D, Zhang L, Zhang Y, Wang F, Zhou X, Sun H (2018) East Asian variant of aldehyde dehydrogenase 2 is related to worse cardioprotective results after coronary artery bypass grafting. Interact Cardiovasc Thorac Surg. https://doi.org/10.1093/icvts/ivy204
82. Gordon FJ, Mark AL (1983) Impaired baroreflex control of vascular resistance in prehypertensive Dahl S rats. Am J Phys 245(2):H210–H217. https://doi.org/10.1152/ajpheart.1983.245.2.H210
83. Gordon FJ, Matsuguchi H, Mark AL (1981) Abnormal baroreflex control of heart rate in prehypertensive and hypertensive Dahl genetically salt-sensitive rats. Hypertension 3(3 Pt 2):I135–I141
84. Greenberg SS, Xie J, Wang Y, Kolls J, Shellito J, Nelson S, Summer WR (1993) Ethanol relaxes pulmonary artery by release of prostaglandin and nitric oxide. Alcohol 10(1):21–29
85. Grogan JR, Kochar MS (1994) Alcohol and hypertension. Arch Fam Med 3(2):150–154
86. Guo R, Hu N, Kandadi MR, Ren J (2012) Facilitated ethanol metabolism promotes cardiomyocyte contractile dysfunction through autophagy in murine hearts. Autophagy 8(4):593–608. https://doi.org/10.4161/auto.18997
87. Guo R, Ren J (2010) Alcohol dehydrogenase accentuates ethanol-induced myocardial dysfunction and mitochondrial damage in mice: role of mitochondrial death pathway. PLoS One 5(1):e8757. https://doi.org/10.1371/journal.pone.0008757
88. Guyenet PG, Schreihofer AM, Stornetta RL (2001) Regulation of sympathetic tone and arterial pressure by the rostral ventrolateral medulla after depletion of C1 cells in rats. Ann N Y Acad Sci 940:259–269
89. Haorah J, Ramirez SH, Floreani N, Gorantla S, Morsey B, Persidsky Y (2008) Mechanism of alcohol-induced oxidative stress and neuronal injury. Free Radic Biol Med 45(11):1542–1550. https://doi.org/10.1016/j.freeradbiomed.2008.08.030
90. Harada S, Agarwal DP, Goedde HW (1981) Aldehyde dehydrogenase deficiency as cause of facial flushing reaction to alcohol in Japanese. Lancet 2(8253):982
91. Haseba T (2014) Molecular evidences of non-ADH pathway in alcohol metabolism and Class III alcohol dehydrogenase (ADH3). Nihon Arukoru Yakubutsu Igakkai Zasshi 49(3):159–168
92. Herschman HR, Lim RW, Brankow DW, Fujiki H (1989) The tumor promoters 12-O-tetradecanoylphorbol-13-acetate and okadaic acid differ in toxicity, mitogenic activity and induction of gene expression. Carcinogenesis 10(8):1495–1498
93. Hipolito L, Sanchez MJ, Polache A, Granero L (2007) Brain metabolism of ethanol and alcoholism: an update. Curr Drug Metab 8(7):716–727
94. Hurtado R, Celani M, Geber S (2016) Effect of short-term estrogen therapy on endothelial function: a double-blinded, randomized, controlled trial. Climacteric 19(5):448–451. https://doi.org/10.1080/13697137.2016.1201809

95. Husain K, Ansari RA, Ferder L (2014) Alcohol-induced hypertension: mechanism and prevention. World J Cardiol 6(5):245–252. https://doi.org/10.4330/wjc.v6.i5.245
96. Ibrahim BM, Fan M, Abdel-Rahman AA (2014) Oxidative stress and autonomic dysregulation contribute to the acute time-dependent myocardial depressant effect of ethanol in conscious female rats. Alcohol Clin Exp Res 38(5):1205–1215. https://doi.org/10.1111/acer.12363
97. Ibrahim KS, El-Yazbi AF, El-Gowelli HM, El-Mas MM (2017) Opposite modulatory effects of selective and non-selective cyclooxygenase inhibition on cardiovascular and autonomic consequences of cyclosporine in female rats. Basic Clin Pharmacol Toxicol 120(6):571–581. https://doi.org/10.1111/bcpt.12754
98. Ireland MA, Vandongen R, Davidson L, Beilin LJ, Rouse IL (1984) Acute effects of moderate alcohol consumption on blood pressure and plasma catecholamines. Clin Sci (Lond) 66(6):643–648
99. Itoh T, Matsumoto M, Nakamura M, Okada A, Shirahashi N, Hougaku H, Hashimoto H, Sakaguchi M, Handa N, Takeshita T, Morimoto K, Hori M (1997) Effects of daily alcohol intake on the blood pressure differ depending on an individual's sensitivity to alcohol: oriental flushing as a sign to stop drinking for health reasons. J Hypertens 15(11):1211–1217
100. Iwai N, Tago N, Yasui N, Kokubo Y, Inamoto N, Tomoike H, Shioji K (2004) Genetic analysis of 22 candidate genes for hypertension in the Japanese population. J Hypertens 22(6):1119–1126
101. Jayakumar AR, Panickar KS, Murthy Ch R, Norenberg MD (2006) Oxidative stress and mitogen-activated protein kinase phosphorylation mediate ammonia-induced cell swelling and glutamate uptake inhibition in cultured astrocytes. J Neurosci 26(18):4774–4784. https://doi.org/10.1523/JNEUROSCI.0120-06.2006
102. Jeavons CM, Zeiner AR (1984) Effects of elevated female sex steroids on ethanol and acetaldehyde metabolism in humans. Alcohol Clin Exp Res 8(4):352–358
103. Ji A, Lou P, Dong Z, Xu C, Zhang P, Chang G, Li T (2018) The prevalence of alcohol dependence and its association with hypertension: a population-based cross-sectional study4 in Xuzhou city, China. BMC Public Health 18(1):364. https://doi.org/10.1186/s12889-018-5276-1
104. Kandadi MR, Hu N, Ren J (2013) ULK1 plays a critical role in AMPK-mediated myocardial autophagy and contractile dysfunction following acute alcohol challenge. Curr Pharm Des 19(27):4874–4887
105. Kang PF, Wu WJ, Tang Y, Xuan L, Guan SD, Tang B, Zhang H, Gao Q, Wang HJ (2016) Activation of ALDH2 with low concentration of ethanol attenuates myocardial ischemia/reperfusion injury in diabetes rat model. Oxidative Med Cell Longev 2016:6190504. https://doi.org/10.1155/2016/6190504
106. Kannel WB (1975) Role of blood pressure in cardiovascular disease: the Framingham Study. Angiology 26(1 Pt. 1):1–14. https://doi.org/10.1177/000331977502600101
107. Karahanian E, Quintanilla ME, Tampier L, Rivera-Meza M, Bustamante D, Gonzalez-Lira V, Morales P, Herrera-Marschitz M, Israel Y (2011) Ethanol as a prodrug: brain metabolism of ethanol mediates its reinforcing effects. Alcohol Clin Exp Res 35(4):606–612. https://doi.org/10.1111/j.1530-0277.2011.01439.x
108. Karuppiah K, Druhan LJ, Chen CA, Smith T, Zweier JL, Sessa WC, Cardouncl AJ (2011) Suppression of eNOS-derived superoxide by caveolin-1: a biopterin-dependent mechanism. Am J Physiol Heart Circ Physiol 301(3):H903–H911. https://doi.org/10.1152/ajpheart.00936.2010
109. Katahira K, Mikami H, Otsuka A, Moriguchi A, Kohara K, Higashimori K, Okuda N, Nagano M, Morishita R, Ogihara T (1994) Differential control of vascular tone and heart rate by different amino acid neurotransmitters in the rostral ventrolateral medulla of the rat. Clin Exp Pharmacol Physiol 21(7):545–556

110. Kawano Y, Abe H, Kojima S, Ashida T, Yoshida K, Imanishi M, Yoshimi H, Kimura G, Kuramochi M, Omae T (1992) Acute depressor effect of alcohol in patients with essential hypertension. Hypertension 20(2):219–226
111. Kelbaek H, Gjorup T, Brynjolf I, Christensen NJ, Godtfredsen J (1985) Acute effects of alcohol on left ventricular function in healthy subjects at rest and during upright exercise. Am J Cardiol 55(1):164–167
112. Kim KH, Young BD, Bender JR (2014) Endothelial estrogen receptor isoforms and cardiovascular disease. Mol Cell Endocrinol 389(1–2):65–70. https://doi.org/10.1016/j.mce.2014.02.001
113. Kinoshita H, Jessop DS, Finn DP, Coventry TL, Roberts DJ, Ameno K, Jiri I, Harbuz MS (2001) Acetaldehyde, a metabolite of ethanol, activates the hypothalamic-pituitary-adrenal axis in the rat. Alcohol Alcohol 36(1):59–64
114. Kishi T, Hirooka Y, Konno S, Ogawa K, Sunagawa K (2010) Angiotensin II type 1 receptor-activated caspase-3 through ras/mitogen-activated protein kinase/extracellular signal-regulated kinase in the rostral ventrolateral medulla is involved in sympathoexcitation in stroke-prone spontaneously hypertensive rats. Hypertension 55(2):291–297. https://doi.org/10.1161/HYPERTENSIONAHA.109.138636
115. Klatsky AL (1990) Blood pressure and alcohol intake. In: J L B (ed) Hypertension: pathophysiology, diagnosis, and management. Raven Press Ltd, New York, pp 277–294
116. Krenz M, Korthuis RJ (2012) Moderate ethanol ingestion and cardiovascular protection: from epidemiologic associations to cellular mechanisms. J Mol Cell Cardiol 52(1):93–104. https://doi.org/10.1016/j.yjmcc.2011.10.011
117. Kucerova J, Strbak V (2001) The osmotic component of ethanol and urea action is critical for their immediate stimulation of thyrotropin-releasing hormone (TRH) release from rat brain septum. Physiol Res 50(3):309–314
118. Lange LG, Sobel BE (1983) Mitochondrial dysfunction induced by fatty acid ethyl esters, myocardial metabolites of ethanol. J Clin Invest 72(2):724–731. https://doi.org/10.1172/JCI111022
119. Laposata EA, Lange LG (1986) Presence of nonoxidative ethanol metabolism in human organs commonly damaged by ethanol abuse. Science 231(4737):497–499
120. Larson HN, Weiner H, Hurley TD (2005) Disruption of the coenzyme binding site and dimer interface revealed in the crystal structure of mitochondrial aldehyde dehydrogenase "Asian" variant. J Biol Chem 280(34):30550–30556. https://doi.org/10.1074/jbc.M502345200
121. Lauar MR, Colombari DS, De Paula PM, Colombari E, Cardoso LM, De Luca LA Jr, Menani JV (2010) Inhibition of central angiotensin II-induced pressor responses by hydrogen peroxide. Neuroscience 171(2):524–530. https://doi.org/10.1016/j.neuroscience.2010.08.048
122. Leal S, Ricardo Jorge DO, Joana B, Maria SS, Isabel SS (2017) Heavy alcohol consumption effects on blood pressure and on kidney structure persist after long-term withdrawal. Kidney Blood Press Res 42(4):664–675. https://doi.org/10.1159/000482022
123. Lee RD, An SM, Kim SS, Rhee GS, Kwack SJ, Seok JH, Chae SY, Park CH, Choi YW, Kim HS, Cho HY, Lee BM, Park KL (2005) Neurotoxic effects of alcohol and acetaldehyde during embryonic development. J Toxicol Environ Health A 68(23–24):2147–2162. https://doi.org/10.1080/15287390500177255
124. Leibing E, Meyer T (2016) Enzymes and signal pathways in the pathogenesis of alcoholic cardiomyopathy. Herz 41(6):478–483. https://doi.org/10.1007/s00059-016-4459-8
125. Lekontseva O, Chakrabarti S, Jiang Y, Cheung CC, Davidge ST (2011) Role of neuronal nitric-oxide synthase in estrogen-induced relaxation in rat resistance arteries. J Pharmacol Exp Ther 339(2):367–375. https://doi.org/10.1124/jpet.111.183798
126. Li G, Wang X, Abdel-Rahman AA (2005) Brainstem norepinephrine neurons mediate ethanol-evoked pressor response but not baroreflex dysfunction. Alcohol Clin Exp Res 29(4):639–647

127. Li SY, Li Q, Shen JJ, Dong F, Sigmon VK, Liu Y, Ren J (2006) Attenuation of acetaldehyde-induced cell injury by overexpression of aldehyde dehydrogenase-2 (ALDH2) transgene in human cardiac myocytes: role of MAP kinase signaling. J Mol Cell Cardiol 40(2):283–294. https://doi.org/10.1016/j.yjmcc.2005.11.006
128. Li SY, Ren J (2008) Cardiac overexpression of alcohol dehydrogenase exacerbates chronic ethanol ingestion-induced myocardial dysfunction and hypertrophy: role of insulin signaling and ER stress. J Mol Cell Cardiol 44(6):992–1001. https://doi.org/10.1016/j.yjmcc.2008.02.276
129. Liang Q, Carlson EC, Borgerding AJ, Epstein PN (1999) A transgenic model of acetaldehyde overproduction accelerates alcohol cardiomyopathy. J Pharmacol Exp Ther 291(2):766–772
130. Lieber CS (1997) Ethanol metabolism, cirrhosis and alcoholism. Clin Chim Acta 257(1):59–84
131. Lohmeier TE, Iliescu R (2015) The baroreflex as a long-term controller of arterial pressure. Physiology (Bethesda) 30(2):148–158. https://doi.org/10.1152/physiol.00035.2014
132. Ma H, Yu L, Byra EA, Hu N, Kitagawa K, Nakayama KI, Kawamoto T, Ren J (2010) Aldehyde dehydrogenase 2 knockout accentuates ethanol-induced cardiac depression: role of protein phosphatases. J Mol Cell Cardiol 49(2):322–329. https://doi.org/10.1016/j.yjmcc.2010.03.017
133. Magner PO, Ethier JH, Kamel KS, Halperin ML (1991) Interpretation of the urine osmolality: the role of ethanol and the rate of excretion of osmoles. Clin Invest Med 14(4):355–358
134. Manzo-Avalos S, Saavedra-Molina A (2010) Cellular and mitochondrial effects of alcohol consumption. Int J Environ Res Public Health 7(12):4281–4304. https://doi.org/10.3390/ijerph7124281
135. Mao L, Abdel-Rahman AA (1995) Blockade of L-glutamate receptors in the rostral ventrolateral medulla contributes to ethanol-evoked impairment of baroreflexes in conscious rats. Brain Res Bull 37(5):513–521
136. Mao L, Li G, Abdel-Rahman AA (2003) Effect of ethanol on reductions in norepinephrine electrochemical signal in the rostral ventrolateral medulla and hypotension elicited by I1-receptor activation in spontaneously hypertensive rats. Alcohol Clin Exp Res 27(9):1471–1480. https://doi.org/10.1097/01.ALC.0000086062.95225.0C
137. Martin SR, Emanuel K, Sears CE, Zhang YH, Casadei B (2006) Are myocardial eNOS and nNOS involved in the beta-adrenergic and muscarinic regulation of inotropy? A systematic investigation. Cardiovasc Res 70(1):97–106. https://doi.org/10.1016/j.cardiores.2006.02.002
138. Matyas C, Varga ZV, Mukhopadhyay P, Paloczi J, Lajtos T, Erdelyi K, Nemeth BT, Nan M, Hasko G, Gao B, Pacher P (2016) Chronic plus binge ethanol feeding induces myocardial oxidative stress, mitochondrial and cardiovascular dysfunction, and steatosis. Am J Physiol Heart Circ Physiol 310(11):H1658–H1670. https://doi.org/10.1152/ajpheart.00214.2016
139. McBride WJ, Li TK, Deitrich RA, Zimatkin S, Smith BR, Rodd-Henricks ZA (2002) Involvement of acetaldehyde in alcohol addiction. Alcohol Clin Exp Res 26(1):114–119
140. Meriin AB, Yaglom JA, Gabai VL, Zon L, Ganiatsas S, Mosser DD, Zon L, Sherman MY (1999) Protein-damaging stresses activate c-Jun N-terminal kinase via inhibition of its dephosphorylation: a novel pathway controlled by HSP72. Mol Cell Biol 19(4):2547–2555
141. Meyer MR, Haas E, Barton M (2006) Gender differences of cardiovascular disease: new perspectives for estrogen receptor signaling. Hypertension 47(6):1019–1026. https://doi.org/10.1161/01.HYP.0000223064.62762.0b
142. Meyer MR, Prossnitz ER, Barton M (2011) The G protein-coupled estrogen receptor GPER/GPR30 as a regulator of cardiovascular function. Vasc Pharmacol 55(1–3):17–25. https://doi.org/10.1016/j.vph.2011.06.003
143. Mochly-Rosen D, Zakhari S (2010) Focus on: the cardiovascular system: what did we learn from the French (Paradox)? Alcohol Res Health 33(1–2):76–86
144. Morales JA, Ram JL, Song J, Brown RA (1997) Acetaldehyde inhibits current through voltage-dependent calcium channels. Toxicol Appl Pharmacol 143(1):70–74. https://doi.org/10.1006/taap.1996.8072

145. Nakao LS, Kadiiska MB, Mason RP, Grijalba MT, Augusto O (2000) Metabolism of acetaldehyde to methyl and acetyl radicals: in vitro and in vivo electron paramagnetic resonance spin-trapping studies. Free Radic Biol Med 29(8):721–729
146. Nasser SA, Elmallah AI, Sabra R, Khedr MM, El-Din MM, El-Mas MM (2014) Blockade of endothelin ET(A), but not thromboxane, receptors offsets the cyclosporine-evoked hypertension and interrelated baroreflex and vascular dysfunctions. Eur J Pharmacol 727:52–59. https://doi.org/10.1016/j.ejphar.2014.01.034
147. Nieminen MT, Salaspuro M (2018) Local acetaldehyde-an essential role in alcohol-related upper gastrointestinal tract carcinogenesis. Cancers (Basel) 10(1). https://doi.org/10.3390/cancers10010011
148. Nishimura FT, Fukunaga T, Kajiura H, Umeno K, Takakura H, Ono T, Nishijo H (2002) Effects of aldehyde dehydrogenase-2 genotype on cardiovascular and endocrine responses to alcohol in young Japanese subjects. Auton Neurosci 102(1–2):60–70
149. Nuutinen H, Lindros KO, Salaspuro M (1983) Determinants of blood acetaldehyde level during ethanol oxidation in chronic alcoholics. Alcohol Clin Exp Res 7(2):163–168
150. O'Keefe EL, DiNicolantonio JJ, O'Keefe JH, Lavie CJ (2018) Alcohol and CV health: Jekyll and Hyde J-curves. Prog Cardiovasc Dis 61(1):68–75. https://doi.org/10.1016/j.pcad.2018.02.001
151. Oba T, Maeno Y, Ishida K (2005) Differential contribution of clinical amounts of acetaldehyde to skeletal and cardiac muscle dysfunction in alcoholic myopathy. Curr Pharm Des 11(6):791–780
152. Oh YI, Kim JH, Kang CW (2008) Effects of ethanol on insulin-like growth factor-I system in primary cultured rat hepatocytes: implications of JNK1/2 and alcoholdehydrogenase. World J Gastroenterol 14(27):4324–4331
153. Pastor R, Aragon CM (2008) Ethanol injected into the hypothalamic arcuate nucleus induces behavioral stimulation in rats: an effect prevented by catalase inhibition and naltrexone. Behav Pharmacol 19(7):698–705. https://doi.org/10.1097/FBP.0b013e328315ecd7
154. Peana AT, Sanchez-Catalan MJ, Hipolito L, Rosas M, Porru S, Bennardini F, Romualdi P, Caputi FF, Candeletti S, Polache A, Granero L, Acquas E (2017) Mystic acetaldehyde: the never-ending story on alcoholism. Front Behav Neurosci 11:81. https://doi.org/10.3389/fnbeh.2017.00081
155. Perissinotto E, Buja A, Maggi S, Enzi G, Manzato E, Scafato E, Mastrangelo G, Frigo AC, Coin A, Crepaldi G, Sergi G, Group IW (2010) Alcohol consumption and cardiovascular risk factors in older lifelong wine drinkers: the Italian Longitudinal Study on Aging. Nutr Metab Cardiovasc Dis 20(9):647–655. https://doi.org/10.1016/j.numecd.2009.05.014
156. Perry PJ, Doroudgar S, Van Dyke P (2017) Ethanol forensic toxicology. J Am Acad Psychiatry Law 45(4):429–438
157. Pfeiffer D, Jurisch D, Neef M, Hagendorff A (2016) Alcohol and arrhythmias. Herz 41(6):498–502. https://doi.org/10.1007/s00059-016-4463-z
158. Polizio AH, Pena C (2005) Effects of angiotensin II type 1 receptor blockade on the oxidative stress in spontaneously hypertensive rat tissues. Regul Pept 128(1):1–5. https://doi.org/10.1016/j.regpep.2004.12.004
159. Potter JF, Beevers DG (1984) Pressor effect of alcohol in hypertension. Lancet 1(8369):119–122
160. Pragst F, Auwaerter V, Sporkert F, Spiegel K (2001) Analysis of fatty acid ethyl esters in hair as possible markers of chronically elevated alcohol consumption by headspace solid-phase microextraction (HS-SPME) and gas chromatography-mass spectrometry (GC-MS). Forensic Sci Int 121(1–2):76–88
161. Puddey IB, Beilin LJ, Vandongen R, Rouse IL, Rogers P (1985) Evidence for a direct effect of alcohol consumption on blood pressure in normotensive men. A randomized controlled trial. Hypertension 7(5):707–713

162. Quertemont E (2003) Discriminative stimulus effects of ethanol with a conditioned taste aversion procedure: lack of acetaldehyde substitution. Behav Pharmacol 14(4):343–350. https://doi.org/10.1097/01.fbp.0000082130.08343.47
163. Quertemont E (2004) Genetic polymorphism in ethanol metabolism: acetaldehyde contribution to alcohol abuse and alcoholism. Mol Psychiatry 9(6):570–581. https://doi.org/10.1038/sj.mp.4001497
164. Quertemont E, Tambour S, Tirelli E (2005) The role of acetaldehyde in the neurobehavioral effects of ethanol: a comprehensive review of animal studies. Prog Neurobiol 75(4):247–274. https://doi.org/10.1016/j.pneurobio.2005.03.003
165. Ramchandani VA, Bosron WF, Li TK (2001) Research advances in ethanol metabolism. Pathol Biol (Paris) 49(9):676–682
166. Ratna A, Mandrekar P (2017) Alcohol and cancer: mechanisms and therapies. Biomol Ther 7(3). https://doi.org/10.3390/biom7030061
167. Rehm J, Hasan OSM, Imtiaz S, Neufeld M (2017) Quantifying the contribution of alcohol to cardiomyopathy: a systematic review. Alcohol 61:9–15. https://doi.org/10.1016/j.alcohol.2017.01.011
168. Rekik M, El-Mas MM, Mustafa JS, Abdel-Rahman AA (2002) Role of endothelial adenosine receptor-mediated vasorelaxation in ethanol-induced hypotension in hypertensive rats. Eur J Pharmacol 452(2):205–214
169. Ren J, Davidoff AJ, Brown RA (1997) Acetaldehyde depresses shortening and intracellular Ca2+ transients in adult rat ventricular myocytes. Cell Mol Biol (Noisy-le-Grand) 43(6):825–834
170. Ren J, Wang GJ, Petrovski P, Ren BH, Brown RA (2002) Influence of hypertension on acetaldehyde-induced vasorelaxation in rat thoracic aorta. Pharmacol Res 45(3):195–199. https://doi.org/10.1006/phrs.2002.0947
171. Ren J, Wold LE (2008) Mechanisms of alcoholic heart disease. Ther Adv Cardiovasc Dis 2(6):497–506. https://doi.org/10.1177/1753944708095137
172. Reslan OM, Yin Z, do Nascimento GR, Khalil RA (2013) Subtype-specific estrogen receptor-mediated vasodilator activity in the cephalic, thoracic, and abdominal vasculature of female rat. J Cardiovasc Pharmacol 62(1):26–40. https://doi.org/10.1097/FJC.0b013e31828bc88a
173. Robinson DL, Lara JA, Brunner LJ, Gonzales RA (2000) Quantification of ethanol concentrations in the extracellular fluid of the rat brain: in vivo calibration of microdialysis probes. J Neurochem 75(4):1685–1693
174. Roth Flach RJ, Bennett AM (2010) Mitogen-activated protein kinase phosphatase-1 – a potential therapeutic target in metabolic disease. Expert Opin Ther Targets 14(12):1323–1332. https://doi.org/10.1517/14728222.2010.528395
175. Russ R, Abdel-Rahman AR, Wooles WR (1991) Role of the sympathetic nervous system in ethanol-induced hypertension in rats. Alcohol 8(4):301–307
176. Salaspuro M (2017) Key role of local acetaldehyde in upper GI tract carcinogenesis. Best Pract Res Clin Gastroenterol 31(5):491–499. https://doi.org/10.1016/j.bpg.2017.09.016
177. Sallam MY, El-Gowilly SM, Abdel-Galil AG, El-Mas MM (2016) Modulation by central MAPKs/PI3K/sGc of the TNF-alpha/iNOS-dependent hypotension and compromised cardiac autonomic control in endotoxic rats. J Cardiovasc Pharmacol 68(2):171–181. https://doi.org/10.1097/FJC.0000000000000400
178. Sapkota M, Wyatt TA (2015) Alcohol, aldehydes, adducts and airways. Biomol Ther 5(4):2987–3008. https://doi.org/10.3390/biom5042987
179. Schoppet M, Maisch B (2001) Alcohol and the heart. Herz 26(5):345–352
180. Seitz HK, Mueller S (2015) Alcohol and cancer: an overview with special emphasis on the role of acetaldehyde and cytochrome P450 2E1. Adv Exp Med Biol 815:59–70. https://doi.org/10.1007/978-3-319-09614-8_4
181. Seliskar M, Rozman D (2007) Mammalian cytochromes P450 – importance of tissue specificity. Biochim Biophys Acta 1770(3):458–466. https://doi.org/10.1016/j.bbagen.2006.09.016

182. Seyedabadi M, Goodchild AK, Pilowsky PM (2001) Differential role of kinases in brain stem of hypertensive and normotensive rats. Hypertension 38(5):1087–1092
183. Silverstein PS, Kumar S, Kumar A (2014) HIV-1, HCV and alcohol in the CNS: potential interactions and effects on neuroinflammation. Curr HIV Res 12(4):282–292
184. Skotzko CE, Vrinceanu A, Krueger L, Freudenberger R (2009) Alcohol use and congestive heart failure: incidence, importance, and approaches to improved history taking. Heart Fail Rev 14(1):51–55. https://doi.org/10.1007/s10741-007-9048-8
185. Song BJ, Akbar M, Jo I, Hardwick JP, Abdelmegeed MA (2015) Translational implications of the alcohol-metabolizing enzymes, including cytochrome P450-2E1, in alcoholic and nonalcoholic liver disease. Adv Pharmacol 74:303–372. https://doi.org/10.1016/bs.apha.2015.04.002
186. Stampfer MJ, Colditz GA, Willett WC, Speizer FE, Hennekens CH (1988) A prospective study of moderate alcohol consumption and the risk of coronary disease and stroke in women. N Engl J Med 319(5):267–273. https://doi.org/10.1056/NEJM198808043190503
187. Steiner JL, Lang CH (2017) Alcoholic cardiomyopathy: disrupted protein balance and impaired cardiomyocyte contractility. Alcohol Clin Exp Res 41(8):1392–1401. https://doi.org/10.1111/acer.13405
188. Steiner JL, Lang CH (2017) Etiology of alcoholic cardiomyopathy: mitochondria, oxidative stress and apoptosis. Int J Biochem Cell Biol 89:125–135. https://doi.org/10.1016/j.biocel.2017.06.009
189. Stickel F, Moreno C, Hampe J, Morgan MY (2017) The genetics of alcohol dependence and alcohol-related liver disease. J Hepatol 66(1):195–211. https://doi.org/10.1016/j.jhep.2016.08.011
190. Stocker SD, Osborn JL, Carmichael SP (2008) Forebrain osmotic regulation of the sympathetic nervous system. Clin Exp Pharmacol Physiol 35(5–6):695–700. https://doi.org/10.1111/j.1440-1681.2007.04835.x
191. Sudhir K, Elser MD, Jennings GL, Komesaroff PA (1997) Estrogen supplementation decreases norepinephrine-induced vasoconstriction and total body norepinephrine spillover in perimenopausal women. Hypertension 30(6):1538–1543
192. Swingle M, Ni L, Honkanen RE (2007) Small-molecule inhibitors of ser/thr protein phosphatases: specificity, use and common forms of abuse. Methods Mol Biol 365:23–38. https://doi.org/10.1385/1-59745-267-X:23
193. Takagi S, Baba S, Iwai N, Fukuda M, Katsuya T, Higaki J, Mannami T, Ogata J, Goto Y, Ogihara T (2001) The aldehyde dehydrogenase 2 gene is a risk factor for hypertension in Japanese but does not alter the sensitivity to pressor effects of alcohol: the Suita study. Hypertens Res 24(4):365–370
194. Tibirica E, Mermet C, Feldman J, Gonon F, Bousquet P (1989) Correlation between the inhibitory effect on catecholaminergic ventrolateral medullary neurons and the hypotension evoked by clonidine: a voltammetric approach. J Pharmacol Exp Ther 250(2):642–647
195. Turlapaty PD, Altura BT, Altura BM (1979) Ethanol reduces Ca2+ concentrations in arterial and venous smooth muscle. Experientia 35(5):639–640
196. Umoh NA, Walker RK, Al-Rubaiee M, Jeffress MA, Haddad GE (2014) Acute alcohol modulates cardiac function as PI3K/Akt regulates oxidative stress. Alcohol Clin Exp Res 38(7):1847–1864. https://doi.org/10.1111/acer.12459
197. Urbano-Marquez A, Estruch R, Fernandez-Sola J, Nicolas JM, Pare JC, Rubin E (1995) The greater risk of alcoholic cardiomyopathy and myopathy in women compared with men. JAMA 274(2):149–154
198. Varga K, Kunos G (1990) Ethanol inhibition of baroreflex bradycardia: role of brainstem GABA receptors. Br J Pharmacol 101(4):773–775
199. Varga K, Kunos G (1992) Inhibition of baroreflex bradycardia by ethanol involves both GABAA and GABAB receptors in the brainstem of the rat. Eur J Pharmacol 214(2–3):223–232

200. Wakabayashi I (2010) Associations between alcohol drinking and multiple risk factors for atherosclerosis in smokers and nonsmokers. Angiology 61(5):495–503. https://doi.org/10.1177/0003319709358694
201. Wang Q, Brunner HR, Burnier M (2004) Determination of cardiac contractility in awake unsedated mice with a fluid-filled catheter. Am J Physiol Heart Circ Physiol 286(2):H806–H814. https://doi.org/10.1152/ajpheart.00291.2003
202. Wang S, Ren J (2018) Role of autophagy and regulatory mechanisms in alcoholic cardiomyopathy. Biochim Biophys Acta 1864(6. Pt A):2003–2009. https://doi.org/10.1016/j.bbadis.2018.03.016
203. Wang X, Li G, Abdel-Rahman AA (2005) Site-dependent inhibition of neuronal c-jun in the brainstem elicited by imidazoline I1 receptor activation: role in rilmenidine-evoked hypotension. Eur J Pharmacol 514(2–3):191–199. https://doi.org/10.1016/j.ejphar.2005.03.021
204. Wang Y, Zeigler MM, Lam GK, Hunter MG, Eubank TD, Khramtsov VV, Tridandapani S, Sen CK, Marsh CB (2007) The role of the NADPH oxidase complex, p38 MAPK, and Akt in regulating human monocyte/macrophage survival. Am J Respir Cell Mol Biol 36(1):68–77. https://doi.org/10.1165/rcmb.2006-0165OC
205. Watanabe A, Kobayashi M, Hobara N, Nakatsukasa H, Nagashima H, Fujimoto A (1985) A report of unusually high blood ethanol and acetaldehyde levels in two surviving patients. Alcohol Clin Exp Res 9(1):14–16
206. Yan X, Wu L, Lin Q, Dai X, Hu H, Wang K, Zhang C, Shao M, Cai L, Tan Y (2017) From the cover: alcohol inhibition of the enzymatic activity of glyceraldehyde 3-phosphate dehydrogenase impairs cardiac glucose utilization, contributing to alcoholic cardiomyopathy. Toxicol Sci 159(2):392–401. https://doi.org/10.1093/toxsci/kfx140
207. Yang ZW, Wang J, Zheng T, Altura BT, Altura BM (2001) Ethanol-induced contractions in cerebral arteries: role of tyrosine and mitogen-activated protein kinases. Stroke 32(1):249–257
208. Yang ZW, Wang J, Zheng T, Altura BT, Altura BM (2002) Roles of tyrosine kinase-, 1-phosphatidylinositol 3-kinase-, and mitogen-activated protein kinase-signaling pathways in ethanol-induced contractions of rat aortic smooth muscle: possible relation to alcohol-induced hypertension. Alcohol 28(1):17–28
209. Yao F, Abdel-Rahman AA (2016) Estrogen receptor ERalpha plays a major role in ethanol-evoked myocardial oxidative stress and dysfunction in conscious female rats. Alcohol 50:27–35. https://doi.org/10.1016/j.alcohol.2015.11.002
210. Yao F, Abdel-Rahman AA (2017) Combined catalase and ADH inhibition ameliorates ethanol-induced myocardial dysfunction despite causing oxidative stress in conscious female rats. Alcohol Clin Exp Res 41(9):1541–1550. https://doi.org/10.1111/acer.13442
211. Yao F, Abdel-Rahman AA (2017) Estrogen receptors alpha and beta play major roles in ethanol-evoked myocardial oxidative stress and dysfunction in conscious ovariectomized rats. Alcohol Clin Exp Res 41(2):279–290. https://doi.org/10.1111/acer.13290
212. Yedlapati SH, Mendu A, Stewart SH (2016) Alcohol-related diagnoses and increased mortality in acute myocardial infarction patients: an analysis of the nationwide inpatient sample. J Hosp Med 11(8):563–567. https://doi.org/10.1002/jhm.2584
213. Yeligar SM, Machida K, Tsukamoto H, Kalra VK (2009) Ethanol augments RANTES/CCL5 expression in rat liver sinusoidal endothelial cells and human endothelial cells via activation of NF-kappa B, HIF-1 alpha, and AP-1. J Immunol 183(9):5964–5976. https://doi.org/10.4049/jimmunol.0901564
214. Yoshizawa S, Matsushima R, Watanabe MF, Harada K, Ichihara A, Carmichael WW, Fujiki H (1990) Inhibition of protein phosphatases by microcystins and nodularin associated with hepatotoxicity. J Cancer Res Clin Oncol 116(6):609–614

215. Yu Y, Zhong MK, Li J, Sun XL, Xie GQ, Wang W, Zhu GQ (2007) Endogenous hydrogen peroxide in paraventricular nucleus mediating cardiac sympathetic afferent reflex and regulating sympathetic activity. Pflugers Arch 454(4):551–557. https://doi.org/10.1007/s00424-007-0256-9
216. Zagon A, Smith AD (1993) Monosynaptic projections from the rostral ventrolateral medulla oblongata to identified sympathetic preganglionic neurons. Neuroscience 54(3):729–743
217. Zhang X, Li SY, Brown RA, Ren J (2004) Ethanol and acetaldehyde in alcoholic cardiomyopathy: from bad to ugly en route to oxidative stress. Alcohol 32(3):175–186. https://doi.org/10.1016/j.alcohol.2004.01.005
218. Zhao X, Park J, Ho D, Gao S, Yan L, Ge H, Iismaa S, Lin L, Tian B, Vatner DE, Graham RM, Vatner SF (2012) Cardiomyocyte overexpression of the alpha1A-adrenergic receptor in the rat phenocopies second but not first window preconditioning. Am J Physiol Heart Circ Physiol 302(8):H1614–H1624. https://doi.org/10.1152/ajpheart.01072.2011
219. Zilkens RR, Burke V, Hodgson JM, Barden A, Beilin LJ, Puddey IB (2005) Red wine and beer elevate blood pressure in normotensive men. Hypertension 45(5):874–879. https://doi.org/10.1161/01.HYP.0000164639.83623.76
220. Zimatkin SM (1991) Histochemical study of aldehyde dehydrogenase in the rat CNS. J Neurochem 56(1):1–11
221. Zimatkin SM, Buben AL (2007) Ethanol oxidation in the living brain. Alcohol Alcohol 42(6):529–532. https://doi.org/10.1093/alcalc/agm059
222. Zimatkin SM, Lindros KO (1996) Distribution of catalase in rat brain: aminergic neurons as possible targets for ethanol effects. Alcohol Alcohol 31(2):167–174
223. Zimatkin SM, Pronko SP, Vasiliou V, Gonzalez FJ, Deitrich RA (2006) Enzymatic mechanisms of ethanol oxidation in the brain. Alcohol Clin Exp Res 30(9):1500–1505. https://doi.org/10.1111/j.1530-0277.2006.00181.x
224. Zimatkin SM, Rout UK, Koivusalo M, Buhler R, Lindros KO (1992) Regional distribution of low-Km mitochondrial aldehyde dehydrogenase in the rat central nervous system. Alcohol Clin Exp Res 16(6):1162–1167

Environmental Aldehyde Sources and the Health Implications of Exposure

Pritam Sinharoy, Stacy L. McAllister, Megana Vasu, and Eric R. Gross

Abstract Aldehydes, which are present within the air as well as food and beverage sources, are highly reactive molecules that can be cytotoxic, mutagenic, and carcinogenic. To prevent harm from reactive aldehyde exposure, the enzyme aldehyde dehydrogenase 2 (ALDH2) metabolizes reactive aldehydes to a less toxic form. However, the genetic variant of ALDH2, ALDH2*2, significantly reduces the ability to metabolize reactive aldehydes in humans. Therefore, frequent environmental aldehyde exposure, coupled with inefficient aldehyde metabolism, could potentially lead to an increased health risk for diseases such as cancer or cardiovascular disease.

Here, we discuss the environmental sources of reactive aldehydes and the potential health implications particularly for those with an ALDH2*2 genetic variant. We also suggest when considering the ALDH2*2 genetic variant the safety limits of reactive aldehyde exposure may have to be reevaluated. Moreover, the ALDH2*2 genetic variant can also be used as an example for how to implement precision medicine in the field of environmental health sciences.

Keywords Reactive aldehyde · Aldehyde dehydrogenase 2 · Cigarette · Alcohol · 4-HNE · ALDH2*2

1 Introduction

Aldehydes are highly reactive electrophiles abundant within our environment. Many sources of aldehydes exist and are present within the air we inhale in addition to the products we use and consume. Exposure to these aldehydes occurs outdoors and indoors, including within the workplace. Lifestyle choices such as tobacco cigarettes, e-cigarettes, and alcohol also expose people to aldehydes. Aldehydes are also present in foods, nonalcoholic beverages, cosmetics, and hand sanitizers (Fig. 1).

P. Sinharoy · S. L. McAllister · M. Vasu · E. R. Gross (✉)
Department of Anesthesiology, Perioperative and Pain Medicine,
Stanford University School of Medicine, Stanford, CA, USA
e-mail: ergross@stanford.edu

J. Ren et al. (eds.), *Aldehyde Dehydrogenases*, Advances in Experimental Medicine and Biology 1193, https://doi.org/10.1007/978-981-13-6260-6_2

Fig. 1 Exogenous sources of reactive aldehyde exposure. These include air pollution produced by industrial power plants and automobiles, alcoholic beverages, tobacco products including cigarettes and e-cigarettes, fried foods, cosmetics, and lacquers in paints. As pictured, the ALDH2 enzyme reduces an aldehyde to a less harmful acid. However, for those with an ALDH2*2 variant, the efficiency of this metabolism is reduced by >60%

Together, exposure to these aldehyde sources, coupled with genetic differences which reduce aldehyde metabolism, may influence the risk of developing diseases such as cancer and cardiovascular disease.

Environmental aldehyde sources include those such as acrolein, acetaldehyde, and formaldehyde. Acrolein is considered a human carcinogen that irritates the upper respiratory tract when inhaled. Acetaldehyde and formaldehyde are also classified as Group 1 carcinogens by the International Agency for Research on Cancer (IARC) [1]. These toxic properties of aldehydes are due to their electrophilic nature, which can modify DNA and proteins, thus the label "reactive" aldehydes [2, 3]. For example, acrolein reacts with proteins specifically on histidine, cysteine, and lysine amino acids by Michael addition or Schiff base formation [4]. Acrolein can also block protein sulfhydryl group formation on lysine residues and result in impaired protein function [5]. Overall, these effects from exogenous sources of reactive aldehydes can be cytotoxic, mutagenic, and carcinogenic.

A critical enzyme responsible for reducing aldehydes to less reactive forms is aldehyde dehydrogenase 2 (ALDH2). ALDH2, primarily known for metabolizing acetaldehyde (produced as an intermediate in the metabolism of alcohol to acetic acid), is also important in the detoxification of other reactive aldehydes such as formaldehyde and acrolein [6, 7]. However, individuals with an ALDH2 genetic variant have significantly reduced enzymatic activity to metabolize these aldehydes.

In homozygotes for the ALDH2 variant (ALDH2*2/*2), the metabolism of aldehydes is severely limited (~96% activity loss), whereas in heterozygotes (ALDH2*1/*2) the enzymatic activity is reduced by 60–80% relative to those individuals that are without the variant (ALDH2*1/*1) [2, 8]. Throughout this rest of this chapter, we will denote the ALDH2 variant as ALDH2*2.

With exposure, the limited ability to metabolize reactive aldehydes for those with the ALDH2*2 variant can potentially pose an increased risk for developing diseases such as cancer or cardiovascular disease. However, some of these environmental exposures could be easily modifiable. Here we discuss the sources of reactive aldehydes within the environment and their effects on human health, particularly for those with a limited ability to metabolize reactive aldehydes.

2 Outdoor, Indoor, and Occupational Aldehyde Exposure

In this section, we will discuss the outdoor, indoor, and occupational exposure of reactive aldehydes and how these exposures may potentially impact people with an ALDH2*2 genetic variant. These exposures include air pollution from automobiles and industrial waste, indoor sources such as paint, and occupational sources such as working with aldehydes and surgical smoke.

2.1 Outdoor Exposures

2.1.1 Air Pollution

Air pollution is a significant concern for Asian countries but also a concern for the United States. According to the World Health Organization (WHO), air pollution is the greatest environmental risk to human health [9]. Sources of air pollution can be natural or anthropogenic and include biomass and fossil fuel combustion, automobile exhaust, and industrial power plant and wood burning fumes [10–12]. Typical city or urban air pollutant exposure consists of aldehydes, ketones, hydrocarbons, and particulate matter. The studies below detail how aldehydes contribute to air pollution.

At six locations in Japan including in urban cities (Sanda, Nishiwaki, Toyooka, and Sumoto), at roadside (Ashiya), and at an industrial site (Takasago), researchers measured 101 volatile organic compound (VOC) concentrations, including acetaldehyde and formaldehyde [13]. Relative to carcinogenic effect, hazardousness of these compounds was evaluated by calculating the factor called excess cancer incidence. For formaldehyde at all sites, the excess cancer incidence exceeded 10^{-5} per μg m^{-3}. This is a level of concern for a carcinogenic effect since the United States Environmental Protection Agency quantitatively estimates that there is a carcinogenic risk for formaldehyde at or above 1.3×10^{-5} per μg m^{-3} [13]. The authors concluded that elevated VOCs, such as formaldehyde, were partially attributed to higher traffic levels generating greater amounts of automobile exhaust in the designated areas [12, 13]. Overall, these studies emphasized the potential harm one exogenous source of pollution, automobile exhaust, can potentially have on changing the risk for carcinogenesis.

In Asia, ambient carbonyl compound concentrations were assessed in seven major cities (Beijing, Chengdu, Guangzhou, Shanghai, Wuhan, Xiamen, and Yantai) and two rural areas (Qinghai Lake, Qinghai, and Lhasa, Tibet) during both the summer and winter seasons. These average concentrations showed seasonal variability dependent on temperature. Among the nine sampling sites measured, the average concentrations of propionaldehyde, formaldehyde, and acetaldehyde were found to be 0.25, 5.07, and 1.91 ppb by volume in the summer and 0.17, 2.04, and 1.42 ppb by volume during the winter, respectively [14]. This study also found the relatively high levels of acetaldehyde emissions were a direct result of increased use of ethanol-blended gasoline (e-gasoline) in vehicles during the summer season.

In Beijing, ambient air aldehyde contents were also measured, and the concentrations of formaldehyde, acetaldehyde, and acrolein were found to be $29.3 \pm 15.1\ \mu g/m^3$, $27.1 \pm 15.7\ \mu g/m^3$, and $2.3 \pm 1.0\ \mu g/m^3$. These levels are considered at the high end of concentration ranges measured in cities around the world [15]. For example, in Savannah, Georgia, the relative levels of formaldehyde and acetaldehyde were $2.0\ \mu g/m^3$ and $2.3\ \mu g/m^3$, compared to Beijing concentrations of $29.3 \pm 15.1\ \mu g/m^3$ and $27.1 \pm 15.7\ \mu g/m^3$, ~14 times lower than that of Beijing [16].

Air pollution-mediated aldehyde exposure is highly concerning for those with an ALDH2*2 variant especially living in major East Asian cities. Air pollution continues to be a significant source of environmental reactive aldehyde exposure throughout East Asia and in particular China. Therefore, the overall risk for exposure for individuals with an ALDH2*2 variant, particularly with the increased migration to urban cities, is of potential concern. Below, we will discuss in more detail automobile exhaust and industrial waste which also contribute to outdoor levels of reactive aldehydes.

2.1.2 Automobiles

Automobile exhaust includes the reactive aldehydes formaldehyde, acetaldehyde, and acrolein [26]. These aldehydes are produced during the burning of fossil fuels and constitute ~1–2% of the volatile organic compounds produced from vehicle exhaust [13]. In one study, diesel engine exhaust gas emission was monitored using three types of diesel fuel (red diesel, biofuel, and gas oil). In all types of diesel fuel, the most abundant reactive aldehydes were formaldehyde, acetaldehyde, and acrolein that were all within a range of 1000–2000 ppb [17]. Biofuel diesel had the greatest levels of acetaldehyde when compared to red diesel and gas oil.

With their increasing popularity, reformulated automotive fuels are an additional source of aldehyde emission. Depending on whether ethanol or methanol is added to the automotive fuels, an increased amount of acetaldehyde or formaldehyde is emitted in automobile exhaust. In a study comparing ethanol-blended fuel exhaust with gasoline exhaust, ethanol-blended exhaust emits predominantly acetaldehyde (1.2–12 g/kWh), whereas gasoline exhaust emits predominantly formaldehyde (0.74–2.3 g/kWh) [18]. With increased ethanol blending, acetaldehyde emissions increased so that pure ethanol used as fuel had acetaldehyde emissions 35–44 times higher than gasoline. Ethanol also influenced formaldehyde emissions so that addition of 50% ethanol to gasoline resulted in a 30–50% increase in formaldehyde emissions [18].

The increased popularity of gasoline-electric, hybrid cars and electric cars facilitates a reduction in automobile emission of acetaldehyde and formaldehyde that may potentially assist in alleviating the elevated levels measured in major cities described above. For ALDH2*2 variant, vehicle selection could also potentially reduce direct reactive aldehyde exposure while driving an automobile.

2.1.3 Industrial Waste

During industrial manufacturing, aldehydes are released as a by-product into the atmosphere. Significant sources of reactive aldehydes produced during the manufacturing process include formaldehyde, butyraldehyde, acetaldehyde, and acrolein. Formaldehyde and butyraldehyde are used as precursors for resin and plasticizer (softener) production, respectively [19]. The reactivity and availability of butyraldehyde makes it a popular material in plasticizer production and other industrial materials [20]. Formaldehyde is used extensively in commercial processes such as in the synthetic resin industry, due to its high chemical reactivity and thermal stability. In Hong Kong, an industrial workplace pilot study found formaldehyde as the most abundant carbonyl compound among all workplace air samples, accounting for 22.0–44.0% of total carbonyls measured on a molar basis. When formaldehyde levels were measured in a paint and wax manufacturing plant and a food-processing factory, formaldehyde levels exceeded the WHO air quality guideline of 81.8 ppb [20].

During the manufacturing processes, combustion emissions are of optimal importance. Combustion sources include power plant petrochemical, diesel-fueled engines, and polyethylene plastic, which release formaldehyde, acetaldehyde, and acrolein [21, 22]. Manufacturing-based acrolein emissions also include volatilization from treated waters and contaminated waste streams, formation as a photooxidation product of various hydrocarbon pollutants, and use in petroleum operations [23–25].

Industrial waste carbonyl contents should be closely monitored particularly in East Asian countries where there is a high prevalence of people with the ALDH2*2 variant. Most standards in regard to what is considered a safe level of industrial product exposure are based upon the ability to efficiently metabolize by-products from industrial manufacturing such as aldehydes to less toxic forms. Therefore, reconsidering the standards for what are the safe levels of exposure for these aldehydes in the industrial workplace or from industrial waste may need to be considered for those who have an ALDH2*2 genetic variant.

2.2 Indoor Exposures

Within a building, aldehydes can infiltrate from external sources described above or can also be present in the air as a result of generation indoors. In six different New Jersey housing sites, levels of indoor formaldehyde, acetaldehyde, and seven other aldehydes were measured. The combined concentration of these nine aldehydes measured indoors was 63 ± 22 ppb compared to outdoor measured values of 19 ± 11 ppb. Of the nine aldehydes assessed, formaldehyde was the most abundant at 55 ppb with acetaldehyde the second most abundant at 3 ppb [26].

In newly installed and painted buildings, higher acetaldehyde levels were found indoors relative to outdoors, suggesting paint and/or lacquers are a significant acetaldehyde source which declines with building age [27]. Dry lacquers which contain aldehydes such as propanol, n-nonanal, and n-hexanal are also used to paint radiators [28]. This is concerning because after a radiator is painted, the heat generated during radiator use causes dry lacquer combustion and aldehydes to be aerosolized into the air. Other lesser-known indoor aldehyde sources include wood burning fireplaces, gas stoves, gas heaters, and synthetic carpets, which can emit formaldehyde and acrolein [11, 26].

2.3 Occupational Exposure

Persons working in aldehyde production industries or as laboratory technicians, healthcare professionals, or funeral home employees may be exposed to higher levels of aldehydes, especially when working with formaldehyde, acrolein, and acetaldehyde [29, 30].

Occupational exposure to these reactive aldehydes may occur via inhalation of reactive aldehyde vapor or direct skin exposure. For people working directly with formaldehyde in an industrial setting, it was found that ~3.5% of workers were exposed to formaldehyde air concentrations greater than 3 ppm, well above the workplace formaldehyde exposure limit set by OSHA at ~0.75 ppm over an 8-hour workday. In this same study, less than 12% of workers were exposed to concentrations greater than 1 ppm; however, over 88% of workers were exposed to concentrations of 0.5 ppm or higher, nearing the OSHA limit [31].

For healthcare professionals, additional sources of reactive aldehyde exposure include electrocautery smoke and widely used hospital sanitizers and disinfectants. During surgery, electrocautery surgical smoke contains formaldehyde, acetone, benzene, and acrylamide [32]. Hospital disinfectants also contain reactive aldehyde sources including formaldehyde and ortho-phthalaldehyde (OPA) [33]. Laboratory researchers can also be exposed to high levels of aldehydes. For example, in a cancer research institute study, formaldehyde exposure levels for laboratory workers ranged from 4.9 to 268.7 μg/m^3 [29].

In humans, evidence suggests when using liver mitochondrial fractions that formaldehyde metabolism is ~3× slower for those with an ALDH2*2 variant compared to those that do not [34]. This suggests that exposure to formaldehyde could perhaps be more toxic and potentially carcinogenic for those with an ALDH2*2 variant since formaldehyde cannot be metabolized as efficiently to the less reactive formic acid. Further, exposure of ALDH2 knockout mice to 500 ppm of inhalational acetaldehyde leads to higher erosion and degeneration of the respiratory epithelium (55.6% vs. 22.2%) and dorsal skin (77.8% vs. 0.0%) and hemorrhaging of the nasal cavity for the ALDH2 knockout mice when compared to the wild-type ALDH2 mice [35]. This suggests that workers with the ALDH2*2 variant are potentially more susceptible to harm from inhalational exposure to reactive aldehydes in the workplace. Therefore, standard limits in regard to workplace exposure may have to be re-examined with the ALDH2*2 genetic variant in mind.

Overall, these studies suggest that outdoor, indoor, and occupational sources of reactive aldehydes are prevalent and of potential concern. Elevated exposure to these aldehyde sources could pose an environmental risk for human health especially for those with an ALDH2*2 variant, and the thresholds considered safe for exposure in these environments need to be potentially re-examined.

3 Lifestyle Choices

Lifestyle choices contribute to an increased risk of aldehyde exposure including drinking alcohol, smoking tobacco products, consuming certain dietary sources, and using particular cosmetic products including perfumes. Here, we will discuss specific lifestyle choices that are sources of aldehyde exposure and their implication on human health, especially in the context of people with an ALDH2*2 variant who cannot catabolize aldehydes efficiently.

3.1 Alcohol

A well-known source of reactive aldehyde exposure is alcohol. Upon ingestion, alcohol is converted to acetaldehyde by the enzyme alcohol dehydrogenase (ADH). Acetaldehyde is then metabolized by ALDH2 to the less reactive acetic acid. Within the cell, ALDH2 has the highest enzymatic affinity for acetaldehyde [36, 37]. The biochemical effect of acetaldehyde after alcohol consumption, which includes DNA- and protein-induced adduct formation, has been detailed in recent reviews [38, 39].

Individuals who are heterozygotes for the ALDH2*2 variant result in an accumulation of acetaldehyde which is ~5-fold higher compared to the wild-type ALDH2 enzyme after alcohol consumption [40]. This acetaldehyde accumulation produces the phenotype of facial flushing and elevated heart rate seen in those with an

ALDH2*2 variant. Those homozygous for the ALDH2*2 variant have an aversion to alcohol consumption since the ability to metabolize acetaldehyde is <4% compared to the wild-type ALDH2 enzyme. Thus, the ALDH2*2 variant is considered protective from developing alcoholism, supported since the occurrence of alcoholism is reduced for those individuals with an ALDH2*2 variant [41].

Although the ALDH2*2 variant may protect from alcoholism, individuals who are heterozygotes for the ALDH2*2 variant, ~560 million people worldwide, still consume alcohol. For example, in Korea, where ~20% of the population carries an ALDH2*2 variant, alcohol consumption is one of the highest per capita in the world [42].

In particular, the reduced metabolism of acetaldehyde resulting from an ALDH2*2 variant after alcohol consumption is linked to a risk for developing esophageal cancer and cardiovascular disease. In regard to esophageal cancer, a seminal study by Yokoyama was the first to link an increased incidence of esophageal cancer caused by alcohol consumption to those with an ALDH2*2 genetic variant [43]. The odds of developing esophageal cancer are also dose-dependently related to the amount of alcohol consumed. Even a moderate consumption of alcohol (9–17.9 units/week) for those with an ALDH2*2 variant increases the odds ratio to 40 for esophageal cancer [43, 44]. The importance of relaying this significant health risk to the public in addition to healthcare professionals is critical and is stressed in several studies [45, 46]. Further, the impact of the ALDH2 enzyme on aldehyde metabolism and alcohol-induced cardiomyopathy was recently reviewed [47].

3.2 *Cigarettes and E-cigarettes*

Smoking imposes a tremendous economic burden accounting for more than $170 billion each year in healthcare costs for adults in the United States and more than $422 billion each year worldwide [48, 49]. To date, smoking remains the leading cause of preventable death in the United States, accounting for one out of five deaths annually [50, 51]. The typical components of a tobacco cigarette are filters (designed to trap smoke), tipping paper (which wraps around the filter), tobacco fillers, cigarette paper (for holding the tobacco fillers), and adhesives. In contrast to cigarettes, the primary components of an e-cigarette are an e-liquid cartridge, an atomizer/heating element, a microprocessor, and a battery (Fig. 2). This is important to note since even though the structural compositions of cigarettes and e-cigarettes are quite different, both of these ways designed to deliver nicotine produce reactive aldehydes.

The smoke from a cigarette is composed of more than 4700 different chemical compounds. These include aldehydes such as formaldehyde, propionaldehyde, butyraldehyde, acetaldehyde, acrolein, and crotonaldehyde [52, 53]. When compared to non-smokers, a person smoking one cigarette produces a 3.5-fold increase in aldehyde levels in the saliva. After smoking ten cigarettes or more a day, a two-fold increase in these compounds was found in the saliva [54]. As these compounds when smoking are inhaled into the lungs, exhaled breath condensate of cigarette

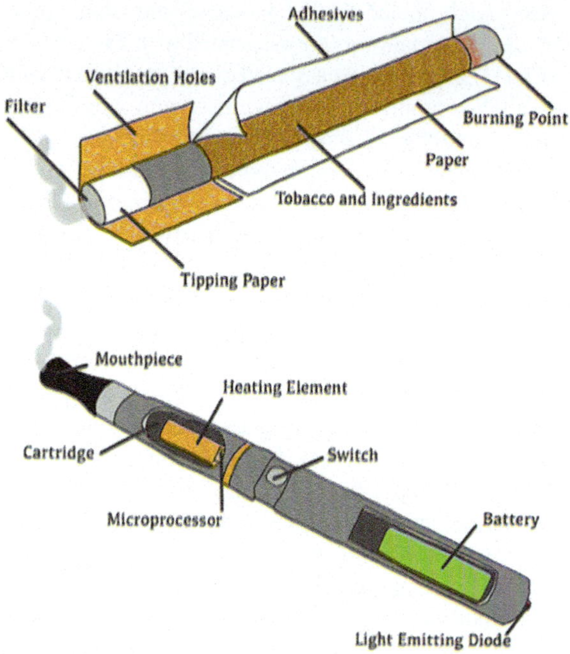

Fig. 2 Schematic illustration describing cigarette and e-cigarette components. Although both are largely different in regard to appearance and the materials used for composition, both cigarette and e-cigarette aerosol contain reactive aldehydes

smokers also contains higher levels of endogenous aldehydes including malondialdehyde (57 ± 2 nmol/L), hexanal (64 ± 4 nmol/L), and heptanal (27 ± 4 nmol/L) compared to non-smokers (18 ± 6 nmol/L, 14 ± 4 nmol/L, and 19 ± 1 nmol/L, respectively) [54].

Epidemiological evidence also suggests that smoking is a key factor in developing cancer and cardiopulmonary disease [55]. Particularly reactive aldehydes in cigarette smoke, especially acetaldehyde and acrolein, are directly linked to a higher risk for developing lung, oral, and gastrointestinal cancer [38]. Moreover, aldehyde levels in cigarette and e-cigarette smoke are responsible for 33% of deaths from cardiovascular diseases and 20% of deaths from ischemic heart disease [51]. Aldehydes derived from cigarette smoke affect heart contractile function and damage blood vessel structure and function by affecting vascular endothelial and epithelial cells [56].

Further in regard to the ALDH2*2 variant, a study consisting of 410 Japanese patients who underwent coronary angiography between 2010 and 2016 demonstrated that smokers had a significantly higher prevalence of coronary spastic angina (CSA) (63.1% vs. 48.9%) than non-smokers [57]. Patients with the ALDH2*2 variant also had a higher risk for CSA when compared to non-smokers with a wild-type ALDH2 enzyme (58.5% vs. 40.8%), whereas the ALDH2*2 variant exaggerated the risk in smokers (82.5% vs. 56.6%) when compared to wild-type ALDH2.

Cigarette smoke exposure is also the primary risk factor for chronic obstructive pulmonary disease (COPD) with a general underlying hypothesis that reactive oxygen and reactive aldehydes contribute to the development of COPD [54, 58]. A recent study including 967 Japanese individuals revealed that the presence of an ALDH2*2 variant was not associated with the development of COPD but having an ALDH2*2 variant was associated with significantly lower lung functioning parameters when compared to individuals with wild-type ALDH2 [59]. Moreover, this study found that 16 weeks of cigarette smoke exposure increased the lung volume and resulted in mild emphysema for wild-type ALDH2 mice, whereas ALDH2*2 variant mice were resistant to emphysema development in response to cigarette smoke exposure. Although the link between cigarette smoking and COPD is clearly established, these data suggest how differences in the metabolism of reactive aldehydes may affect COPD requires further study.

E-cigarette aerosol, similar to cigarettes, also contains reactive aldehydes including acrolein, acetaldehyde, and formaldehyde [60–62]. Moreover, a recent report suggests that when measuring reactive aldehyde content across 66 different e-cigarette liquids, acetaldehyde concentrations are generally ~2 fold higher than formaldehyde or acrolein [63]. However, the quantity of these three reactive aldehydes produced is variable and dependent upon the battery voltage, type of e-cigarette vaporizer, and amount and type of liquid within the e-cigarette cartridge. For example, changing the battery voltage from 3.7 to 4.8 V increases the combined reactive aldehyde production of acetaldehyde, formaldehyde, and acrolein by 3.5-fold. Additionally, a double-coil vaporizer when compared to a single-coil vaporizer will decrease resistance and increase the power. Therefore, higher-power e-cigarettes (9.1 W) can increase acetaldehyde levels ~65-fold, formaldehyde levels ~49-fold, and acrolein levels ~3-fold in the e-cigarette aerosol when compared to a lower-power e-cigarette (4.6 W) [60]. Clearly, e-cigarette research is a new and emerging field, and how e-cigarette aldehyde exposure in combination with genetic differences in reactive aldehyde metabolism may affect the risk of developing diseases such as cardiovascular disease or cancer is largely unknown.

3.3 Beverage and Food Sources

Aldehydes are abundantly found in several dietary sources. Besides alcoholic beverages, methanol and ethanol are natural constituents of fruit juice, which are enzymatically metabolized by the enzyme alcohol dehydrogenase (ADH) into formaldehyde (range from 3.7 to 60 ppm) and acetaldehyde (ranging from 0.0005 to 230 ppm), respectively [64, 65]. Aldehydes are natural components of fruits, vegetables, spices, and nuts. For example, peas contain traces of acetaldehyde, whereas cinnamon contains cinnamaldehyde. Almonds and cherries contain benzaldehyde, whereas anisaldehyde and salicylaldehyde appear in anise and vanilla extracts [66]. Moreover, fermentation is a widely known process used for millennia for preserving food and beverages. According to the WHO, fermented and processed foods including cheese, yogurt, meat, kefir, kimchi, and tofu also contain traceable amounts of formaldehyde (5.7–20 ppm) and acetaldehyde (0.2–0.6 ppm) [64].

Fats are also a significant source of aldehydes, and when cooked, over 20 different aldehydes are produced [66]. Aldehydes are formed when frying food or heating oils to cook food. These aldehydes are mainly generated from the thermal oxidation of the polyunsaturated triacylglycerols [67]. Cooking oil heated at a temperature of 180 °C produces high amounts of aerosolized acrolein (canola oil 53.5 ± 3.9 mg/h and safflower oil 57.3 ± 6.7 mg/h) which are typically inhaled while standing over cooking food [68]. When soybean oil is used to cook deep-fried potatoes, 4-hydroxynonenal (4-HNE) is a major polar lipophilic compound in the thermally oxidized frying oil [69]. Additional studies further validate that deep-frying food, especially at high temperatures and for prolonged periods of time, generate reactive aldehydes [70, 71]. For example, in Taiwan restaurant exhaust streams, 18 carbonyl species were measured. Formaldehyde, acetaldehyde, acetone, and butyraldehyde contributed 55.01–94.52% of total carbonyls in the dining areas for the restaurants measured [72].

3.4 *Cosmetics*

Cosmetic products including perfumes, deodorants, skin care products, and nail polish removers are common in daily lifestyle. According to the American Cancer Society, professional hair treatment products such as keratin smoothening products contain formaldehyde or formaldehyde-releasing agents which can raise indoor formaldehyde concentrations to a hazardous level (range 0.33–1.88 ppm) in hair salons [73]. Various other aldehydes found in perfumes and deodorants are the most common cause of hypersensitivity to fragrances and allergic reactions [74]. For example, cinnamaldehyde, a well-known chemical irritant and sensitizing agent, is a major component of the "fragrance mix" and commonly found in deodorants [75]. In 1991, when the EPA tested 31 fragrance products, among the 20 most commonly found toxic chemicals were propylene glycol, acetone, benzaldehyde, ethanol, and benzyl alcohol.

Aldehydes are also present in hand sanitizers where a hand sanitizer may contain 60–90% ethanol. During topical application of ethanol and aldehydes, the skin is the most susceptible organ to ethanol- and aldehyde-induced damage. Both ethanol and aldehydes are highly diffusible through the lipophilic layer of the outer skin and can covalently modify the epidermis to form an immunogenic antigen. The antigen-carrying cell can migrate to the lymph nodes, where the antigen is recognized by specific T cells that can proliferate and disseminate throughout the body via the blood, causing an inflammatory skin response [76]. Although there is a lack of evidence linking topical ethanol application to skin cancer, several studies confirm that when taken orally, ethanol increases the risk for skin cancer and other types of cancer [77]. Lastly, considering the susceptible nature of ALDH2*2 variant to ethanol and subsequently generated acetaldehyde-induced health disorders, it is advisable to avoid usage of ethanol-based cosmetic products.

Taken together, these studies suggest alcohol intake, use of tobacco products, fried foods, and cosmetic products are sources of reactive aldehydes. Reactive aldehyde exposure associated with these lifestyle choices increases the chances of developing cancer and cardiovascular diseases [51] specifically for those with an ALDH2*2 variant [78].

4 Impact on Humans with Reduced Reactive Aldehyde Metabolism

Reactive aldehyde exposure damages the electron transport chain, leading to an overproduction of mitochondrial reactive oxygen species (Fig. 3). This in turn affects cellular properties contributing to various diseases [79]. The health implications of reactive aldehyde exposure from these environmental sources listed above are even more concerning for those ~560 million people worldwide who cannot efficiently metabolize these reactive aldehydes due to ALDH2*2 genetic variant. Therefore, people with an ALDH2*2 variant are vulnerable to reactive aldehyde exposure-induced oxidative stress and associated pathophysiological conditions.

Extensive research over the past several years has recognized the ALDH2*2 variant as a major contributor to several human diseases including cardiovascular disorders, stroke, and cancer [2, 38, 80]. For example, a recent meta-analysis of nine case-control studies concluded that the ALDH2*2 variant is associated with an increased risk for developing coronary heart diseases, including myocardial infarction [81]. Therefore, those with an ALDH2*2 variant exposed to exogenous reactive aldehyde sources as discussed above are more likely to develop medical complications as a result of limited reactive aldehyde metabolism.

Since ALDH2*2 individuals are more susceptible to reactive aldehyde-associated health complications, it is important that information regarding the health risks associated with reactive aldehyde exposure be disseminated widely to the public. Many exogenous reactive aldehyde sources, such as alcohol consumption or cigarette use, are avoidable or modifiable. In situations where exposure cannot be avoided, safety precautions can be taken to limit or minimize exposure.

5 Conclusion

Over the past several years, extensive research has advanced the understanding of the relationship between exogenous reactive aldehyde sources and the harmful effects of their exposure. Whether outdoors, indoors, or at the place of occupational employment, the potential for reactive aldehyde exposure and the resultant negative health consequences are particularly important to consider for those with an ALDH2*2 variant.

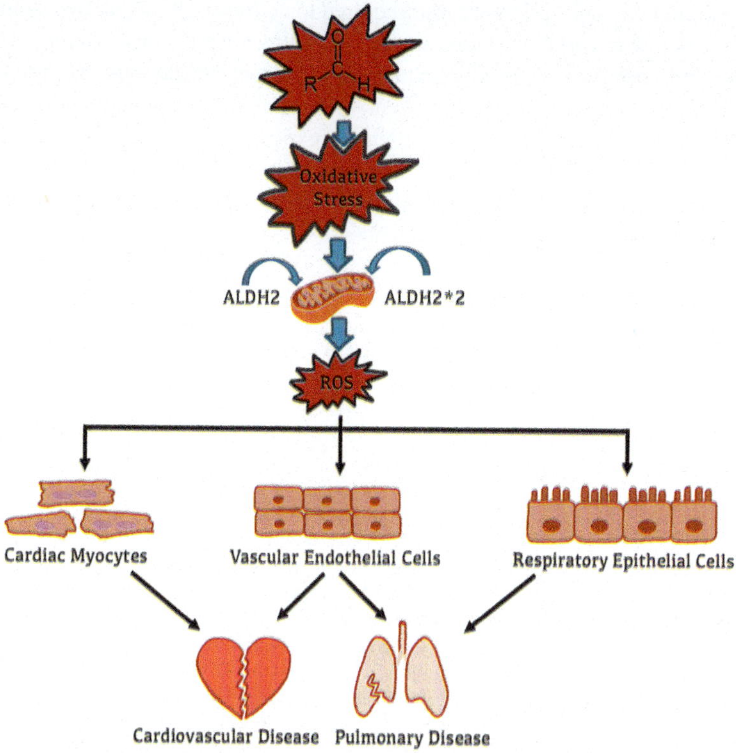

Fig. 3 Schematic representation of how reactive aldehyde exposure harms the cardiopulmonary system. ALDH2 metabolizes toxic reactive aldehydes into a less toxic form. For those with an ALDH2*2 variant, exposure to exogenous reactive aldehydes increases intracellular oxidative stress and impairs mitochondrial function due to an inefficient metabolism of reactive aldehydes. This is followed by an increase in mitochondrial reactive oxygen species (ROS) production. ROS directly affects the structure and function of cardiac myocytes, vascular endothelial cells, and pulmonary epithelial cells, resulting in cellular dysfunction that can ultimately lead to cardiopulmonary disease

At the forefront of protection from environmental sources of reactive aldehydes is public awareness. Unless knowledge is properly disseminated, individuals will not be cognizant of the risks associated with reactive aldehyde exposure or capable of taking the necessary steps to minimize exposure. Precautionary measures and lifestyle modifications to reduce reactive aldehyde exposure should be a focus of public health to ultimately reduce the potential risk for developing cancer or cardiovascular disease.

There are also several key questions that remain unanswered. More research is needed to understand how susceptible those with an ALDH2*2 variant are to reactive aldehydes from environmental sources. In addition, it is also important to further understand how much the relative risk for cancer or cardiovascular disease is increased with reactive aldehyde exposure for those with an ALDH2*2 genetic variant. Basic and clinical studies are also required to determine how reactive aldehydes produced with the workplace environment, including hospitals, may potentially and more preferentially affect those with an ALDH2*2 variant. New therapeutic strategies should also be encouraged to prevent, minimize, and treat health-related disorders arising as a result of reactive aldehyde exposure. Together, considering the genetics of reactive aldehyde metabolism with respect to environmental sources of aldehyde exposure can lead to developing a basis for a precision medicine platform in the field of environmental health sciences.

Acknowledgments PSR and SM received career development awards from TRDRP (27FT-0037) and NICHD K99 (HD093858), respectively. All additional work was supported by a TRDRP High Impact Pilot Award (26IP-0038), and NHLBI (HL144388) to ERG.

References

1. O'Brien PJ, Siraki AG, Shangari N (2005) Aldehyde sources, metabolism, molecular toxicity mechanisms, and possible effects on human health. Crit Rev Toxicol 35(7):609–662
2. Chen CH, Ferreira JC, Gross ER, Mochly-Rosen D (2014) Targeting aldehyde dehydrogenase 2: new therapeutic opportunities. Physiol Rev 94(1):1–34
3. Matsuda T, Yabushita H, Kanaly RA, Shibutani S, Yokoyama A (2006) Increased DNA damage in ALDH2-deficient alcoholics. Chem Res Toxicol 19(10):1374–1378
4. Cai J, Bhatnagar A, Pierce WM Jr (2009) Protein modification by acrolein: formation and stability of cysteine adducts. Chem Res Toxicol 22(4):708–716
5. Stevens JF, Maier CS (2008) Acrolein: sources, metabolism, and biomolecular interactions relevant to human health and disease. Mol Nutr Food Res 52(1):7–25
6. Perez-Miller S, Younus H, Vanam R, Chen CH, Mochly-Rosen D, Hurley TD (2010) Alda-1 is an agonist and chemical chaperone for the common human aldehyde dehydrogenase 2 variant. Nat Struct Mol Biol 17(2):159–164
7. Zambelli VO, Gross ER, Chen CH, Gutierrez VP, Cury Y, Mochly-Rosen D (2014) Aldehyde dehydrogenase-2 regulates nociception in rodent models of acute inflammatory pain. Sci Transl Med 6(251):251ra118
8. Yoshida A, Huang IY, Ikawa M (1984) Molecular abnormality of an inactive aldehyde dehydrogenase variant commonly found in Orientals. Proc Natl Acad Sci U S A 81(1):258–261
9. World Health Organization (2018) Air pollution. http://www.who.int/airpollution/en/
10. Schauer JJ, Kleeman MJ, Cass GR, Simoneit BR (2001) Measurement of emissions from air pollution sources. 3. C1-C29 organic compounds from fireplace combustion of wood. Environ Sci Technol 35(9):1716–1728
11. Zhang J, Smith KR (1999) Emissions of carbonyl compounds from various cookstoves in China. Environ Sci Technol 33(14):2311–2320
12. Lipari F, Dasch JM, Scruggs WF (1984) Aldehyde emissions from wood-burning fireplaces. Environ Sci Technol 18(5):326–330
13. Okada Y, Nakagoshi A, Tsurukawa M, Matsumura C, Eiho J, Nakano T (2012) Environmental risk assessment and concentration trend of atmospheric volatile organic compounds in Hyogo Prefecture. Jpn Environ Sci Pollut Res Int 19(1):201–213
14. Ho KF, Ho SSH, Huang RJ, Dai WT, Cao JJ, Tian L et al (2015) Spatiotemporal distribution of carbonyl compounds in China. Environ Pollut 197:316–324
15. Altemose B, Gong J, Zhu T, Hu M, Zhang L, Cheng H et al (2015) Aldehydes in relation to air pollution sources: a case study around the Beijing Olympics. Atmos Environ (1994) 109:61–69

16. MacIntosh DL, Zimmer-Dauphinee SA, Manning RO, Williams PL (2000) Aldehyde concentrations in ambient air of coastal Georgia, U.S.A. Environ Monit Assess 63(3):409–429
17. Smith D, Španěl P, Dabill D, Cocker J, Rajan B (2004) On-line analysis of diesel engine exhaust gases by selected ion flow tube mass spectrometry. Rapid Commun Mass Spectrom 18(23):2830–2838
18. Magnusson R, Nilsson C, Andersson B (2002) Emissions of aldehydes and ketones from a two-stroke engine using ethanol and ethanol-blended gasoline as fuel. Environ Sci Technol 36(8):1656–1664
19. Franz AW, Kronemayer H, Pfeiffer D, Pilz RD, Reuss G, Disteldorf W, Gamer AO, Hilt A (2016) Formaldehyde. In: Ullmann's encyclopedia of industrial chemistry. Wiley-VCH Verlag GmbH & Co. KGaA
20. Ho SS, Ip HS, Ho KF, Ng LP, Chan CS, Dai WT et al (2013) Hazardous airborne carbonyls emissions in industrial workplaces in China. J Air Waste Manag Assoc 63(7):864–877
21. Hodgkin JH, Galbraith MN, Chong YK (1982) Combustion products from burning polyethylene. J Macromol Sci Part A Chem 17(1):35–44
22. Jonsson A, Persson KA, Grigoriadis V (1985) Measurements of some low molecular-weight oxygenated, aromatic, and chlorinated hydrocarbons in ambient air and in vehicle emissions. Environ Int 11(2):383–392
23. Eisler R (1994) Acrolein hazards to fish, wildlife, and invertebrates: a synoptic review. U.S. Dept. of the Interior, National Biological Survey, Washington, DC. Report. Laurel, MD: 1994 Report 28; Biological Report 23
24. Ghilarducci DP, Tjeerdema RS (1995) Fate and effects of acrolein. Rev Environ Contam Toxicol 144:95–146
25. Faroon O, Roney N, Taylor J, Ashizawa A, Lumpkin MH, Plewak DJ (2008) Acrolein environmental levels and potential for human exposure. Toxicol Ind Health 24(8):543–564
26. Zhang J, Lioy PJ, He Q (1994) Characteristics of aldehydes: concentrations, sources, and exposures for indoor and outdoor residential microenvironments. Environ Sci Technol 28(1):146–152
27. Crump DR, Gardiner D (1989) Sources and concentrations of aldehydes and ketones in indoor environments in the UK. Environ Int 15(1):455–462
28. Ullrich D, Nagel R, Seifert B (1982) The effect of lacquer coatings on indoor air quality using as example radiator lacquers. Schriftenr Ver Wasser Boden Lufthyg 53:283–298
29. Pala M, Ugolini D, Ceppi M, Rizzo F, Maiorana L, Bolognesi C et al (2008) Occupational exposure to formaldehyde and biological monitoring of Research Institute workers. Cancer Detect Prev 32(2):121–126
30. Scarselli A, Corfiati M, Di Marzio D, Iavicoli S (2017) National estimates of exposure to formaldehyde in Italian workplaces. Ann Work Expo Health 61(1):33–43
31. Siegel DM, Frankos VH, Schneiderman MA (1983) Formaldehyde risk assessment for occupationally exposed workers. RTP 3(4):355–371
32. Krones CJ, Conze J, Hoelzl F, Stumpf M, Klinge U, Möller M et al (2007) Chemical composition of surgical smoke produced by electrocautery, harmonic scalpel and argon beaming – a short study. Eur Surg 39(2):118–121
33. Mwanga HH, Baatjies R, Jeebhay MF (2018) 646 exposure to aldehydes among health care workers in a large tertiary hospital in Cape Town, South Africa. Occup Environ Med 75(Suppl 2):A337
34. Wang RS, Nakajima T, Kawamoto T, Honma T (2002) Effects of aldehyde dehydrogenase-2 genetic polymorphisms on metabolism of structurally different aldehydes in human liver. Drug Metab Dispos 30(1):69–73
35. Oyama T, Isse T, Ogawa M, Muto M, Uchiyama I, Kawamoto T (2007) Susceptibility to inhalation toxicity of acetaldehyde in Aldh2 knockout mice. Front Biosci 12:1927–1934
36. Parrilla R, Okawa K, Lindros KO, Zimmerman UJ, Kobayashi K, Williamson JR (1974) Functional compartmentation of acetaldehyde oxidation in rat liver. J Biol Chem 249(15):4926–4933

37. Eriksson CJ, Marselos M, Koivula T (1975) Role of cytosolic rat liver aldehyde dehydrogenase in the oxidation of acetaldehyde during ethanol metabolism in vivo. Biochem J 152(3):709–712
38. Gross ER, Zambelli VO, Small BA, Ferreira JC, Chen CH, Mochly-Rosen D (2015) A personalized medicine approach for Asian Americans with the aldehyde dehydrogenase 2*2 variant. Annu Rev Pharmacol Toxicol 55:107–127
39. Heymann HM, Gardner AM, Gross ER (2018) Aldehyde-induced DNA and protein adducts as biomarker tools for alcohol use disorder. Trends Mol Med 24(2):144–155
40. Yokoyama A, Tsutsumi E, Imazeki H, Suwa Y, Nakamura C, Mizukami T et al (2008) Salivary acetaldehyde concentration according to alcoholic beverage consumed and aldehyde dehydrogenase-2 genotype. Alcohol Clin Exp Res 32(9):1607–1614
41. Harada S, Agarwal DP, Goedde HW (1985) Aldehyde dehydrogenase polymorphism and alcohol metabolism in alcoholics. Alcohol (Fayetteville, NY) 2(3):391–392
42. Monzavi SM, Afshari R, Rehman N (2015) Alcohol related disorders in Asia Pacific region: prevalence, health consequences and impacts on the nations. Asia Pac J Med Toxicol 4(1):1–8
43. Yokoyama A, Omori T (2005) Genetic polymorphisms of alcohol and aldehyde dehydrogenases and risk for esophageal and head and neck cancers. Alcohol (Fayetteville, NY) 35(3):175–185
44. Yokoyama A, Omori T, Yokoyama T (2010) Alcohol and aldehyde dehydrogenase polymorphisms and a new strategy for prevention and screening for cancer in the upper aerodigestive tract in East Asians. Keio J Med 59(4):115–130
45. Brooks PJ, Enoch MA, Goldman D, Li TK, Yokoyama A (2009) The alcohol flushing response: an unrecognized risk factor for esophageal cancer from alcohol consumption. PLoS Med 6(3):e50
46. McAllister SL, Sun K, Gross ER (2016) Developing precision medicine for people of East Asian descent. J Biomed Sci 23:80
47. Zhang Y, Ren J (2011) ALDH2 in alcoholic heart diseases: molecular mechanism and clinical implications. Pharmacol Ther 132(1):86–95
48. Xu X, Bishop EE, Kennedy SM, Simpson SA, Pechacek TF (2015) Annual healthcare spending attributable to cigarette smoking: an update. Am J Prev Med 48(3):326–333
49. Goodchild M, Nargis N, Tursan d'Espaignet E (2018) Global economic cost of smoking-attributable diseases. Tob Control 27:58–64. https://doi.org/10.1136/tobaccocontrol-2016-053305
50. Federal Trade Commission (2017) United States Federal Trade Commission Cigarette Report for 2015. Published online www.ftc.gov
51. CDC (2010) How tobacco smoke causes disease: the biology and behavioral basis for smoking-attributable disease: a report of the surgeon general. Centers for Disease Control and Prevention (US), Atlanta
52. Eschner MS, Selmani I, Groger TM, Zimmermann R (2011) Online comprehensive two-dimensional characterization of puff-by-puff resolved cigarette smoke by hyphenation of fast gas chromatography to single-photon ionization time-of-flight mass spectrometry: quantification of hazardous volatile organic compounds. Anal Chem 83(17):6619–6627
53. Facchinetti F, Amadei F, Geppetti P, Tarantini F, Di Serio C, Dragotto A et al (2007) Alpha,beta-unsaturated aldehydes in cigarette smoke release inflammatory mediators from human macrophages. Am J Respir Cell Mol Biol 37(5):617–623
54. Corradi M, Rubinstein I, Andreoli R, Manini P, Caglieri A, Poli D et al (2003) Aldehydes in exhaled breath condensate of patients with chronic obstructive pulmonary disease. Am J Respir Crit Care Med 167(10):1380–1386
55. Vineis P, Alavanja M, Buffler P, Fontham E, Franceschi S, Gao YT et al (2004) Tobacco and cancer: recent epidemiological evidence. JNCI 96(2):99–106
56. DelloStritto DJ, Sinharoy P, Connell PJ, Fahmy JN, Cappelli HC, Thodeti CK et al (2016) 4-Hydroxynonenal dependent alteration of TRPV1-mediated coronary microvascular signaling. Free Radic Biol Med 101:10–19

57. Mizuno Y, Hokimoto S, Harada E, Kinoshita K, Yoshimura M, Yasue H (2017) Variant aldehyde dehydrogenase 2 (ALDH2*2) in East Asians interactively exacerbates tobacco smoking risk for coronary spasm—possible role of reactive aldehydes. Circ J 81(1):96–102
58. Kirkham PA, Barnes PJ (2013) Oxidative stress in COPD. Chest 144(1):266–273
59. Kuroda A, Hegab AE, Jingtao G, Yamashita S, Hizawa N, Sakamoto T et al (2017) Effects of the common polymorphism in the human aldehyde dehydrogenase 2 (ALDH2) gene on the lung. Respir Res 18(1):69
60. Ogunwale MA, Li M, Ramakrishnam Raju MV, Chen Y, Nantz MH, Conklin DJ et al (2017) Aldehyde detection in electronic cigarette aerosols. ACS Omega 2(3):1207–1214
61. Sinharoy P, McAllister S, Gross ER (2018) E-cigarette vapor elevates heart rate in mice with limited reactive aldehyde metabolism. FASEB J 32(1 suppl):848.16–.16
62. Wang P, Chen W, Liao J, Matsuo T, Ito K, Fowles J et al (2017) A device-independent evaluation of carbonyl emissions from heated electronic cigarette solvents. PLoS One 12(1):e0169811
63. Fagan P, Pokhrel P, Herzog TA, Moolchan ET, Cassel KD, Franke AA et al (2017) Sugar and aldehyde content in flavored electronic cigarette liquids. Nicotine Tob Res 20(8):985–992
64. Feron VJ, Til HP, de Vrijer F, Woutersen RA, Cassee FR, van Bladeren PJ (1991) Aldehydes: occurrence, carcinogenic potential, mechanism of action and risk assessment. Mutat Res 259(3–4):363–385
65. Kavet R, Nauss KM (1990) The toxicity of inhaled methanol vapors. Crit Rev Toxicol 21(1):21–50
66. Guillén-Sans R, Guzmán-Chozas M (1995) Aldehydes in food and its relation with the TBA test for rancidity. Lipid/Fett 97(7–8):285–286
67. Ramirez MR, Estevez M, Morcuende D, Cava R (2004) Effect of the type of frying culinary fat on volatile compounds isolated in fried pork loin chops by using SPME-GC-MS. J Agric Food Chem 52(25):7637–7643
68. Katragadda HR, Fullana A, Sidhu S, Carbonell-Barrachina ÁA (2010) Emissions of volatile aldehydes from heated cooking oils. Food Chem 120(1):59–65
69. Seppanen CM, Csallany AS (2004) Incorporation of the toxic aldehyde 4-hydroxy-2-trans-nonenal into food fried in thermally oxidized soybean oil. J Am Oil Chem Soc 81(12):1137–1141
70. Guillén MD, Uriarte PS (2012) Aldehydes contained in edible oils of a very different nature after prolonged heating at frying temperature: presence of toxic oxygenated α,β unsaturated aldehydes. Food Chem 131(3):915–926
71. Katsuta I, Shimizu M, Yamaguchi T, Nakajima Y (2008) Emission of volatile aldehydes from DAG-rich and TAG-rich oils with different degrees of unsaturation during deep-frying. J Am Oil Chem Soc 85(6):513–519
72. Cheng J-H, Lee Y-S, Chen K-S (2016) Carbonyl compounds in dining areas, kitchens and exhaust streams in restaurants with varying cooking methods in Kaohsiung, Taiwan. J Environ Sci 41:218–226
73. Golden R, Valentini M (2014) Formaldehyde and methylene glycol equivalence: critical assessment of chemical and toxicological aspects. Regul Toxicol Pharmacol 69(2):178–186
74. Patlewicz GY, Wright ZM, Basketter DA, Pease CK, Lepoittevin JP, Arnau EG (2002) Structure-activity relationships for selected fragrance allergens. Contact Dermatitis 47(4):219–226
75. Smith CK, Moore CA, Elahi EN, Smart AT, Hotchkiss SA (2000) Human skin absorption and metabolism of the contact allergens, cinnamic aldehyde, and cinnamic alcohol. Toxicol Appl Pharmacol 168(3):189–199
76. Johansen JD (2003) Fragrance contact allergy: a clinical review. Am J Clin Dermatol 4(11):789–798
77. Yen H, Dhana A, Okhovat J-P, Qureshi A, Keum N, Cho E (2017) Alcohol intake and risk of nonmelanoma skin cancer: a systematic review and dose-response meta-analysis. Br J Dermatol 177(3):696–707

78. Chen C-H, Sun L, Mochly-Rosen D (2010) Mitochondrial aldehyde dehydrogenase and cardiac diseases. Cardiovasc Res 88(1):51–57
79. Manzo-Avalos S, Saavedra-Molina A (2010) Cellular and mitochondrial effects of alcohol consumption. Int J Environ Res Public Health 7(12):4281–4304
80. Ohsawa I, Nishimaki K, Murakami Y, Suzuki Y, Ishikawa M, Ohta S (2008) Age-dependent neurodegeneration accompanying memory loss in transgenic mice defective in mitochondrial aldehyde dehydrogenase 2 activity. J Neurosci 28(24):6239–6249
81. Wang Q, Zhou S, Wang L, Lei M, Wang Y, Miao C et al (2013) ALDH2 rs671 polymorphism and coronary heart disease risk among Asian populations: a meta-analysis and meta-regression. DNA Cell Biol 32(7):393–399

ALDH2 and Cardiovascular Disease

Che-Hong Chen, Julio C. B. Ferreira, and Daria Mochly-Rosen

Abstract Aldehyde dehydrogenase 2 (ALDH2) is a non-cytochrome P450 mitochondrial aldehyde oxidizing enzyme. It is best known for its role in the metabolism of acetaldehyde, a common metabolite from alcohol drinking. More evidences have been accumulated in recent years to indicate a greater role of ALDH2 in the metabolism of other endogenous and exogenous aldehydes, especially lipid peroxidation-derived reactive aldehyde under oxidative stress. Many cardiovascular diseases are associated with oxidative stress and mitochondria dysfunction. Considering that an estimated 560 million East Asians carry a common ALDH2 deficient variant which causes the well-known alcohol flushing syndrome due to acetaldehyde accumulation, the importance of understanding the role of ALDH2 in these diseases should be highlighted. There are several unfavorable cardiovascular conditions that are associated with ALDH2 deficiency. This chapter reviews the function of ALDH2 in various pathological conditions of the heart in relation to aldehyde toxicity. It also highlights the importance and clinical implications of interaction between ALDH2 deficiency and alcohol drinking on cardiovascular disease among the East Asians.

1 Introduction

The effect of alcohol and cardiovascular disease has been controversial and remains an active area of basic research, epidemiology, and public health discussion in recent years. Current recommended alcohol drinking guidelines vary greatly from country to country globally [29], highlighting the need for clarification with respect to the risk of alcohol consumption and cardiovascular diseases versus other health

C.-H. Chen · D. Mochly-Rosen (✉)
Department of Chemical and Systems Biology, Stanford University, School of Medicine, Stanford, CA, USA
e-mail: mochly@stanford.edu

J. C. B. Ferreira
Department of Anatomy, Institute of Biomedical Sciences, University of Sao Paulo, Sao Paulo, Brazil

J. Ren et al. (eds.), *Aldehyde Dehydrogenases*, Advances in Experimental Medicine and Biology 1193, https://doi.org/10.1007/978-981-13-6260-6_3

risks, such as cancer, central nervous system (CNS) diseases, liver diseases, or metabolic syndromes. In the United States, based on the 2015–2020 Dietary Guidelines for Americans, moderate alcohol drinking is recommended. This moderate alcohol consumption is defined as up to two drinks per day for men and one drink per day for women with one drink defined as 14 grams of pure alcohol [28]. Moderate alcohol drinking has been associated with reduced risks for coronary heart disease and mortality due to heart disease, decreased risk of ischemic stroke, and decreased risk of diabetes [22]. However, in a very recent study on risk assessment for alcohol consumption, based on data collected from nearly 600,000 alcohol drinkers from 19 high-income countries, it was shown that alcohol consumption greater than ~1 drink per day (7.1 standard US drinks per week or 100 g per week) is already positively associated with increased risks for coronary disease (except for myocardial infarction), heart failure, fatal hypertensive disease, fatal aortic aneurysm, and stroke. In comparison to individuals consuming less than 100 g per week, those who reported drinking between 100 and 200 g per week (7.1–14.2 drinks per week), between 200 and 350 g per week (14.2–25 drinks per week), or > 350 g per week (>25 drinks per week) had reduced life expectancy of approximately 6 months, 1–2 years, or 4–5 years at age 40 years, respectively [69]. Twenty percent of the above alcohol-related survival difference could be attributed to excessive death from cardiovascular disease. The opposite effect of alcohol on myocardial infarction and other cardiovascular disease indicates a need for research to understand the complex molecular mechanisms of ethanol and its metabolites on different aspects of heart function.

Ethanol is metabolized by two dehydrogenase enzymes in the liver as the main organ of detoxification. Alcohol dehydrogenase (ADH) catalyzes the first step of ethanol oxidation to acetaldehyde, and the mitochondrial aldehyde dehydrogenase 2 (ALDH2) further oxidizes acetaldehyde to acetate for excretion to the blood and final conversion to CO_2 [74]. In addition to liver, ALDH2 is also expressed in multiple human tissues and organs including the heart, lung, brain, muscle, kidney, adipose, bone marrow, and skin [22]. ALDH2 is the most efficient enzyme in metabolizing ethanol-derived acetaldehyde, but it can also remove other reactive, toxic aldehydes derived from membrane lipid peroxidation under ischemic or oxidative stress. These well-known endogenous reactive aldehydes include 4-hydroxynonenal (4HNE), malondialdehyde (MDA), 3,4-dihydroxyphenylacetaldehyde (DOPAL), and acrolein [2, 72].

A common ALDH2 deficient variant [ALDH2*2 allele, rs671(A), or E487K single amino acid substitution] exists in an estimated 560 million East Asians (or about 8% of the world population). Individuals carrying the ALDH2*2 variant have very low ALDH2 enzyme activity and cannot detoxify acetaldehyde efficiently, which leads to rapid accumulation of acetaldehyde and the well-known Asian alcohol facial flushing reaction following consumption of as little as one drink (14 g ethanol) of any alcoholic beverage [7]. The ALDH2*2 mutation was thought to originate from one single founder in Southeast China nearly 2000–3000 years ago and to quickly spread to other East Asian countries, such as Japan, Korea, Taiwan, and Singapore [36]. ALDH2 deficiency is therefore very common among East

Asians. The accumulation of acetaldehyde also leads to cardiac palpitation, nausea, and headache. Such unpleasant physiological reactions have been correlated with less alcohol consumption and lower rate of alcohol addiction among the ALDH2*2 carriers [71]. Nevertheless, acetaldehyde toxicity in ALDH2*2 subjects is a well-established cause of upper aerodigestive track cancers (oral cavity, pharynx, larynx, esophagus) and other cancers of the colorectum and liver and female breast cancer [3, 7]. In 2007, the World Health Organization (WHO) classified alcoholic beverages as a Group 1 carcinogen, especially for individuals with the ALDH2*2 variants, due to acetaldehyde toxicity [3]. The ALDH2*2 variant has also been shown to be a risk factor for coronary artery disease [70], alcohol-induced cardiac dysfunction [25], stroke [63], insensitivity to nitroglycerin treatment, [33] and many other pathological conditions among the East Asians [10].

We first reported a critical role of ALDH2 in cardioprotection against ischemia-reperfusion (I/R) injuries [9]. It was demonstrated that direct activation of the ALDH2 enzyme activity, by an ALDH2-specific small molecule activator, Alda-1 (Aldehyde dehydrogenase activator-1), significantly reduced I/R damage in myocardial infarction [9]. Alda-1 is both an enzyme agonist and a molecular chaperon which facilitates proper protein folding of ALDH2; it interacted with ALDH2 at the catalytic tunnel of the tetrameric enzyme to accelerate catalysis of the substrate and was also able to restore the structural defect of the ALDH2*2 enzyme caused by the E487K mutation [56]. Alda-1 was therefore an effective activator for both the ALDH2 wild-type and ALDH2*2 mutant enzymes [9, 56]. Since the identification of this new role of ALDH2 in cardioprotection, many comprehensive reviews have been published on the subject of ALDH2 and cardiovascular disease [8, 11, 17, 37, 43, 54, 55, 70, 76, 77]. This current review will focus on more recent advances in this expanding field of heart disease.

2 Myocardial Infarction and Heart Failure

Despite the discovery of effective pharmacological and non-pharmacological interventions capable of preventing deaths during acute myocardial infarction, these patients remain at high risk to develop pathological cardiac remodeling and heart failure in the following years. Therefore, myocardial infarction and heart failure are leading causes of morbidity and death worldwide. In this section, we describe the critical role of aldehydes in myocardial infarction, other cardiomyopathies, and heart failure.

A significant amount of evidence has recently emerged to indicate a greater role of endogenous and exogenous aldehydes in the establishment and progression of myocardial infarction and heart failure [10]. During myocardial infarction or heart failure, excessive production of reactive oxygen species exacerbates the peroxidation of mitochondrial polyunsaturated fatty acids, which subsequently generates toxic secondary aldehydes such as malondialdehyde and 4-hydroxy-nonenal (4HNE) [59]. 4HNE is a highly reactive aldehyde that readily interacts with proteins and forms adducts via Michael addition. Excessive 4HNE adduct formation has a negative impact on heart physiology by impairing mitochondrial bioenergetics [19], protein quality control [16], and contractility properties [18, 19]. Moreover, exposure of isolated adult rat cardiomyocytes to 4HNE is sufficient to impair calcium handling with impact on contractility properties [5, 51].

Elevated cardiac aldehyde levels, such as 4HNE, is also associated with the pathophysiology of different cardiomyopathies as seen in rodent models and in humans [18, 42, 50]. The progression of myocardial infarction into heart failure is characterized by reduced ALDH2 activity and accumulation of cardiac 4HNE protein adducts [18, 19]. Animals with elevated 4HNE levels due to genetic disruption of ALDH2 activity or overexpression of alcohol dehydrogenase are more susceptible to develop cardiomyopathy and heart failure [9, 20, 30]. From a clinical perspective, 4HNE protein adducts accumulate in cardiac biopsies from patients with hypertrophic and dilated cardiomyopathies and heart failure compared with control donor hearts. Of interest, improved outcome in heart failure patients treated with beta-blocker (carvedilol) is associated with a prominent reduction in the levels of cardiac 4HNE adducts [50]. Finally, circulating 4HNE levels increase according to the progression of myocardial infarction and heart failure both in rat models of the disease and in humans, therefore suggesting its value as a relevant biomarker of chronic cardiac diseases [18, 42].

The cardioprotective effects of ALDH2 against alcoholic and ischemic cardiomyopathies as well as against heart failure have been demonstrated over the past decade. Overexpression of ALDH2 attenuates infarct size and cardiac dysfunctional properties in an in vivo mouse model of acute myocardial infarction [38]. [These mice are also resistant to acute ethanol toxicity [39, 40].] Conversely, ALDH2 knockout mice present elevated cardiac damage during acute myocardial infarction, which is associated with increased oxidative stress and elevated 4HNE protein adducts [38, 67]. Moreover, the absence of ALDH2 aggravates ethanol-mediated cardiac toxicity in rodents [39]. The molecular mechanisms associated with aldehyde cardiotoxicity include hyperphosphorylation of GSK-3beta, ASK-1, GATA4, and CREB, which often lead to bioenergetics dysfunction, oxidative stress, and cell death [15, 57, 77]. The cardioprotection against acute myocardial infarction seen in ALDH2 overexpression appears to be related to activation of Akt and AMPK mTOR signaling cascades [38].

Considering the critical role of 4HNE and other reactive aldehydes, the ALDH2-selective activator, Alda-1, has been used to counteract the deleterious effect of aldehydic load in the establishment and progression of chronic cardiac diseases, such as myocardial infarction and heart failure. Acute Alda-1 treatment is sufficient to neutralize the negative impact of 4HNE on mitochondrial bioenergetics and cardiomyocyte contractility properties [18, 19]. These experiments were performed in isolated cardiac mitochondria and isolated adult cardiomyocytes from rats, respectively.

ALDH2 activation using Alda-1 also protects against aldehydic stress when tested in intact organs and in whole organisms. As an example, Alda-1 administration is sufficient to increase cardiac ALDH2 activity by twofold and protect against acute myocardial infarction in both rats and mice [9, 66]. This cardioprotection induced by Alda-1 is associated with reduced levels of cardiac 4-HNE protein adducts and acetaldehyde and better mitochondrial metabolism and redox status during acute myocardial infarction [9, 19, 66]. Indeed, ALDH2 activity has a tight inverse correlation with cardiac damage during acute myocardial infarction [9, 66]. The benefits of ALDH2 activation during acute myocardial infarction are not only restricted to cardiomyocytes. Alda-1 antagonizes the impact of aldehydes on cardiac mast cell degranulation [32] and fibroblast proliferation [19], which are well-known players involved in cardiac degeneration/remodeling during myocardial infarction and heart failure (Fig. 1).

Sustained ALDH2 activation also protects against ischemic cardiomyopathy and heart failure in rats. During these pathological processes, the aldehydic load contributes to mitochondrial dysfunction, reactive oxygen species production, and impaired calcium handling, which result in pathological cardiac hypertrophy and reduced contractility properties [18]. Sustained Alda-1 treatment of infarcted animals for 6 weeks, starting at day 1 after surgery, is sufficient to reduce cardiac aldehydic levels, protect mitochondrial metabolism, and improve cardiac ejection fraction, therefore preventing the progression to heart failure [18].

We also found that continuous Alda-1 treatment is also important to improve heart failure outcome in rodent models of heart failure [19]. Chronic administration of Alda-1 using osmotic pumps, starting 4 weeks after myocardial infarction, at a time when heart failure is present, improves cardiomyocyte shortening, cardiac function, left ventricular compliance, and diastolic function under basal conditions, and after isoproterenol stimulation [19]. Importantly, sustained Alda-1 treatment promotes a cardiac anti-remodeling effect and malfunction by suppressing myocardial hypertrophy and fibrosis. Alda-1 reverses the accumulation of 4HNE-protein

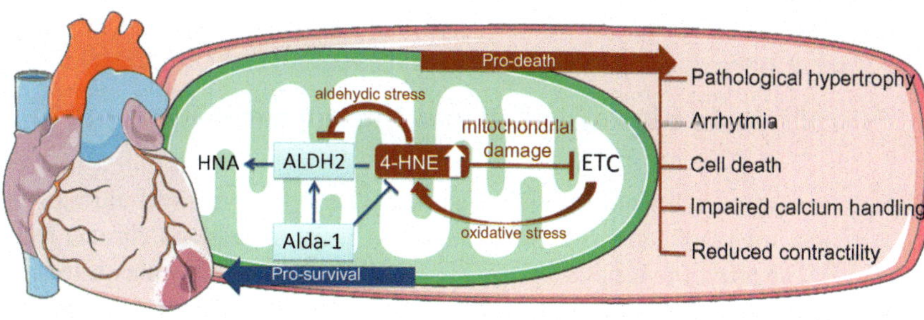

Fig. 1 Proposed model for aldehydic stress during cardiac diseases. Excessive mitochondrial lipid peroxidation and consequent 4-HNE generation can yield mitochondrial ALDH2 inactivation, energy metabolism dysfunction, and uncontrolled oxidative stress, which results in further 4-HNE generation during acute and chronic cardiac diseases, including myocardial infarction and heart failure. Impaired 4-HNE metabolism and consequent accumulation of free 4-HNE favors 4-HNE-protein modification with a negative impact in cardiomyocyte contractility properties, cell death, and cardiac remodeling. Pharmacological ALDH2 activation using Alda-1 improves the outcome of cardiac diseases through increased clearance of 4-HNE. Heart, cell, and mitochondrion images by smart.servier.com are licensed under creativecommons.org/licenses/by/3.0/

adducts and protein carbonylation in hearts subjected to myocardial infarction. This benefit is associated with improved mitochondrial function, including elevated mitochondrial respiratory control ratios and reduced H_2O_2 release. This selective ALDH2 activation decreased mitochondrial Ca2+-induced permeability transition opening and cytochrome c release in failing hearts [19].

Further supporting a mitochondrial mechanism for ALDH2 activation, Alda-1 treatment preserves mitochondrial function following in vitro increase in aldehydic load. Notably, sustained activation of ALDH2 using Alda-1 improves the clinical outcome following myocardial infarction and heart failure in rodents, through decreasing cardiac reactive aldehydes and improving mitochondrial bioenergetics [19]. Finally, we also found that Alda-1 treatment in vitro reverses ALDH2 inactivation in ventricular specimens of patients with end-stage heart failure and protected human pluripotent stem cell-derived cardiomyocytes against ischemic damage [19]. Together, these data show that increased aldehydic load causes cardiotoxicity and drugs that reduce aldehyde toxicity and ameliorate mitochondrial dysfunction by targeting mitochondrial ALDH2 may prevent or reduce the progression of a number of pathologies associated with chronic cardiac diseases, such as myocardial infarction and heart failure. One exception of the beneficial role of ALDH2 seems to be in the aging heart. Autophagy is an important mechanism of protein quality control in aging and adaptive cardiac remodeling. Zhang et al. reported that autophagy is regulated by ALDH2-mediated interaction between Bcl-2 and Beclin-1 and AMPK phosphorylation in the cardiac aging process [78]. In transgenic mice, global overexpression of ALDH2 led to accentuated suppression of autophagy and a 7.7% shortened life span as compared with the wild-type ALDH2 between young mice and old mice. Furthermore, in an examination among healthy elderly population, there was positive correlation between cardiac function and geometry and the ALDH2*2 genotype indicating a more subdued cardiac remodeling induced by aging in the ALDH2*2 heterozygous and homozygous elderly individual [78].

3 Nitroglycerin Bioconversion and Nitroglycerin Tolerance

Nitroglycerin (glyceryl trinitrate, GTN) is one of the most prescribed nitric oxide (NO) donors for vasodilation for the treatment of acute angina and chronic heart failure [6]. The effect of vasodilation by GTN is rapid, which usually occurs in minutes. However, it is also well-known that prolonged use of GTN often leads to its ineffectiveness as patients develop tolerance to GTN [14, 45]. The mechanism of GTN bioconversion and GTN tolerance is still not completely understood. Chen and Stamler first [13] demonstrated that ALDH2, via its reductase activity, was the enzyme responsible for the bioconversion of GTN to NO which results in cGMP-mediated vasodilation. It was further revealed that biotransformation of GTN by ALDH2 also led to direct oxidation of cysteine 302, the critical catalytic amino acid of ALDH2, which leads to enzyme inactivation and may explain the clinical observation of GTN intolerance in patients [4, 13, 45, 49, 68]. In an ischemia-reperfusion animal model, Sun et al. showed that administration of Alda-1, the ALDH2 activator, before sustained treatment of nitroglycerin protected against cardiac injury from GTN-induced ALDH2 inactivation [62]. Dithiothreitol (DTT) was effective in reactivating oxidized ALDH2, but neither cysteine nor the most abundant intracellular thiol donor, glutathione, was effective in protecting ALDH2 against GTN inactivation [12]. It was postulated that there existed a sink of potent endogenous DTT-like reductant as a thiol donor for reactivation of ALDH2 and sustained bioconversion from GTN to NO in initial hours of GTN treatment, and it was the depletion of this thiol donor that led to GTN tolerance [12]. However, the search for such a potent cellular thiol donor as an effective agent to overcome ALDH2-medicated GTN intolerance in vivo remains elusive. Recently, using a fluorescent NO sensor in vascular smooth muscle cells with ALDH2 C301S/C303S mutant construct, Opelt et al. showed that despite a slow recovery of ALDH2 inactivation, vascular smooth muscle cells were capable of providing a burst of low but sufficient physiological level of NO to achieve the initial vessel relaxation [53]. This observation implies that a robust strong reactivating reductant may not exist in cell and low level of ALDH2 activity is sufficient to achieve vasodilation and explain GTN intolerance simultaneously.

With the involvement of ALDH2 in GTN biotransformation, enzyme inactivation, and intolerance, it is well-justified to evaluate the cardiovascular outcome of GTN treatment in ALDH2*2 patients with angina. Previous clinical studies confirmed a marked decrease in nitroglycerin efficacy in Chinese patients carrying the mutant ALDH2*2 (n = 33) relatively to carriers of the wild-type enzyme (n = 43) [33]. It was shown that GTN biotransformation using purified human fetal liver wild-type ALDH2 was 10 times more efficient than corresponding ALDH2*2 human fetal liver enzyme [33]. Furthermore, at high dose of GTN, a significantly greater blood flow was observed in ALDH2 wild-type and in ALDH2*2 human subjects [33, 41]. However, the effect of GTN bioconversion and intolerance with regard to ALDH2 in clinics with ALDH2*2 heterozygous and homozygous genotypes remains inconclusive. A more recent study by Miura et al. showed that there was no difference in the vasodilatory effect of sublingual GTN tablets administered to a small group of healthy Japanese with all three ALDH2 genotypes [47]. In contrast to previous studies that used indirect assessments of changes in systemic circulation, blood flow, or pain relief, this study measured directly GTN-mediated vessel dilation diameters. However, the study was conducted with a very small number of individuals of six ALDH2*1/*1, seven ALDH2*1/*2, and seven ALDH2*2/*2 [47]. In another larger randomized, open-label, crossover trial with 117 healthy Japanese (47 ALDH2*1/*1, 48 ALDH2*1/*2, and 22 ALDH2*2/*2), it was shown that there was no difference in the amplitude of vasodilation induced by GTN among the three different ALDH2 genotypes [60]. But the ALDH2*1/*2 and ALDH2*2/*2 individuals had a much slower response time, especially among the ALDH2*2/*2 homozygotes [60]. These studies indicate the need of further clarification on the role of ALDH2 in GTN bioconversion and the phenomenon of tolerance. Considering the wide use of GTN for angina and heart failure treatment, more observations on GTN efficacy among individuals with different ALDH2 genotypes are also needed to determine if any adjustment in clinic practice is warranted.

4 Arrhythmia

Excessive alcohol drinking can induce irregular heart rhythms in healthy individuals. The phenomenon is often associated with binge drinking, common during the holiday season, and is known as the "holiday heart syndrome" [46]. The most common arrhythmia is atrial fibrillation (AF), which is characterized by rapid and irregular heartbeats of the atria. AF can increase the risks of heart failure, dementia, and stroke [48]. Acetaldehyde is a potent inducer of arrhythmia. Gallardo-Carpentier et al. showed that acetaldehyde as low as 30 uM was sufficient to illicit arrhythmia caused by delayed after-depolarization and triggered activity in canine Purkinje fibers [58]. Acetaldehyde and reactive aldehydes generated under ischemia-reperfusion or hypoxia can cause arrhythmia via cardiac mast cell degranulation and renin release [32]. Acetaldehyde also triggers release of norepinephrine in cardiac sympathetic nerve endings and arrhythmia in myocardial ischemia-reperfusion [58]. It is therefore conceivable that accelerated removal of toxic aldehyde can confer a beneficial effect both in preventing cell death in I/R and arrhythmia, induced by the release of hormones and neurotransmitters. It was, indeed, demonstrated that activation of ALDH2, either by its activator, Alda-1, or via ε-Protein Kinase C (εPKC)-medicated ALDH2 signaling pathway, was sufficient to prevent ischemia-reperfusion-induced renin release from the mast cells or norepinephrine release from the neurons to confer cardioprotection against reperfusion arrhythmia [1, 32, 44, 58]. More recently, the role of ALDH2 in regulating the renin/angiotensin/norepinephrine arrhythmia was further shown to be mediated through the G_i-coupled S1P receptor, and sphingosine-1-phosphate, a sphingolipid produced in the I/R heart, was sufficient to achieve inhibition of I/R-induced cardiac arrhythmia [44]. As expected, in the ALDH2*2 knock-in mouse model, such protective mechanism against arrhythmia was lost in either isolated hearts or cultured mast cells [44].

The relationship between alcoholic beverage drinking and AF has been studied recently in human subjects carrying different genotypes of ethanol metabolizing genes. Among a group of 281 Taiwanese patients who had AF and received catheter ablation, the recurrence of paroxysmal AF was associated with both the amount of alcohol consumed and the alcohol dehydrogenase-1B (ADH1B) genotype. Nondrinkers and light drinkers consistently had better sinus rhythm maintenance rate than moderate or former drinkers, and the ADH1B*1/*2 and *2/*2 genotype had a great hazard ratio than the ADH1B*1*1 genotype [24]. In another study involving 577 Japanese AF patients and 1935 non-AF control subjects, the allele frequency of ALDH2*2 was significantly lower in the AF patient group than in the non-AF control group [52]. On the other hand, in subset analyses among the 182 patients with lone AF and 914 controls, the ADH1B*1 allele was significantly higher in the AF patient group than in the non-AF control group. It was postulated that the protective effect against AF observed in the Japanese patients may be due to either a prolonged metabolic conversion of ethanol to acetaldehyde or a lower alcohol consumption due to the influence of the ALDH1B and ALDH2*2 genotypes

[52]. Whether low level of acetaldehyde, like ethanol, can mediate a "preconditioning" effect against AF remains to be explored.

5 Coronary Artery Disease

The ALDH2*2 variant has consistently been associated with higher risks of coronary artery disease (CAD) in several GWAS studies and meta-analyses in East Asian populations [21, 23, 34, 64, 75, 79]. Hypertension is one of the leading risk factors for CAD. However, the relationship between hypertension and CAD among ALDH2*2 carriers seems to be more complicated with confounding factors that are known to be influenced by the ALDH2*2 variants, such as alcohol consumption [27], serum triglyceride level [65], cholesterol level [31], blood pressure [35], and metabolic syndrome [80]. These reports highlighted the complex interaction between ALDH2*2, alcohol consumption, and cardiovascular disease. Yet, no information regarding the amount of alcohol consumption were included in these studies. Since ALDH2 genotypes greatly influence the amount of alcohol consumption and the amount of alcohol consumption is also strongly correlated with coronary heart disease, it is necessary to include both the amount of alcohol consumption and the ALDH2 genotypes information when conducting evaluation of the effect of ALDH2 and cardiovascular outcome in East Asians. Using comprehensive echocardiograph and two-dimensional speckle tracking, Hung et al. examined the effect of light-to-moderate alcohol drinking on cardiac function among nearly 4000 asymptomatic Taiwanese participants and found that even with low-to-moderate dose of habitual alcohol consumption, both left ventricular and left atrial functions were significantly compromised in an alcohol dose-dependent manner [26]. In addition, light-to-moderate drinking was also associated with higher incidences of hypertension; higher fasting glucose, HbA1c, and triglyceride; and lower HDL [26]. In another controlled large-scale epidemiological study involving ~24,000 South Koreans, ALDH2 genotypes were closely correlated with the amount of alcohol consumption and indexes of liver enzyme. Furthermore, increased alcohol consumption was associated with increased coronary artery calcification [73]. In a most detailed heart function evaluation, combining both the amount of alcohol consumption and ALDH2 genotyping information from ~1600 participants, it was confirmed that even light-to-moderate alcohol drinking could confer significant subclinical adverse effects on cardiac systolic functions [25]. Furthermore, this adverse effect was worse in subjects carrying the common ALDH2*2 variant, with most pronounced cardiac dysfunction and structural deformation in the subgroup of drinkers carrying the common ALDH2*2 and ADH1B*2 fast alcohol metabolizing allele [25]. A comprehensive review summarizing the effect of alcohol consumption, ALDH2 genotype, and cardiovascular disease in Korea has been published recently [61] emphasizing the need to separate the effect of alcohol consumption based on ALDH2 genotypes.

6 Conclusion and Perspective

ALDH2 deficiency is one of the most common enzymopathies in human affecting 560 million East Asians, or 8% of the world population. This East Asian-specific ALDH2 deficiency is caused by a single amino acid substitution, which results in the production of a dysfunctional enzyme. ALDH2 is the main enzyme for detoxification of ethanol-derived acetaldehyde. ALDH2 also plays a key role in the metabolism of lipid peroxidation-derived reactive aldehydes generated under acute or chronic oxidative stress such as myocardial infarction, cardiomyopathy, and heart failure. Individuals carrying the inactive ALDH2*2 variant are susceptible to higher risks of many cardiovascular diseases. Strong evidences have demonstrated the mechanistic effects of aldehydic toxicity in carriers of the ALDH2*2 variant. It is therefore clear that modulation of ALDH2 activity can be a new and attractive therapeutic strategy for treating cardiovascular disease. A small molecule ALDH2 activator, Alda-1, has been discovered and demonstrated to be effective in enhancing the activity of both the ALDH2*1 and ALDH2*2 enzymes. Positive results have been observed by acute and chronic treatments of Alda-1 in several proof of concept animal models related to cardiovascular conditions, such as myocardial infraction, cardiomyopathy, heart failure, and protection against nitroglycerin-induced ALDH2 inactivation under ischemia-reperfusion. In cardiomyocytes derived from ALDH2*2 patient pluripotent stem cells, Alda-1 also increased cell survival under ischemia. Considering that 8% of the world population carry the ALDH2*2 genotype and that accumulating molecular and epidemiological data have revealed a strong negative impact of this genotype in cardiovascular disease, a personalized precision medicine approach should be considered for individuals born with this common enzyme deficiency. This includes preventive measures, clinical research, and translational development targeting ALDH2. Current US moderate alcohol drinking guideline recommending two drinks per day for men and one drink per day for women for cardioprotection has not taken into account for the East Asian-specific differences in body weight or the variant ALDH2 genotype. More recent studies challenge the notion of the beneficial effect of moderate ethanol consumption on the heart, especially among the East Asians. New alcohol drinking guideline incorporating genetic information and alcohol education program providing evidences on harmful alcohol drinking behavior which causes cardiovascular disease are therefore needed for public health policy and awareness especially for the East Asian descendants and in countries that have high prevalence of the ALDH2*2 variant.

Acknowledgment Che-Hong Chen is supported by R37 MERIT award NIH AAA11147 to Daria Mochly-Rosen, Department of Chemical and Systems Biology, Stanford University, School of Medicine, and Julio C.B. Ferreira is supported by FAPESP 2017/16694-2.

References

1. Aldi S, Takano K, Tomita K, Koda K, Chan NY, Marino A, Salazar-Rodriguez M, Thurmond RL, Levi R (2014) Histamine H4-receptors inhibit mast cell renin release in ischemia/reperfusion via protein kinase C epsilon-dependent aldehyde dehydrogenase type-2 activation. J Pharmacol Exp Ther 349:508–517
2. Anderson EJ, Katunga LA, Willis MS (2012) Mitochondria as a source and target of lipid peroxidation products in healthy and diseased heart. Clin Exp Pharmacol Physiol 39:179–193
3. Baan R, Straif K, Grosse Y, Secretan B, El Ghissassi F, Bouvard V, Altieri A, Cogliano V, WHO International Agency for Research on Cancer Monograph Working Group (2007) Carcinogenicity of alcoholic beverages. Lancet Oncol 8:292–293
4. Beretta M, Sottler A, Schmidt K, Mayer B, Gorren AC (2008) Partially irreversible inactivation of mitochondrial aldehyde dehydrogenase by nitroglycerin. J Biol Chem 283:30735–30744
5. Bhatnagar A (1995) Electrophysiological effects of 4-hydroxynonenal, an aldehydic product of lipid peroxidation, on isolated rat ventricular myocytes. Circ Res 76:293–304
6. Boden WE, Finn AV, Patel D, Peacock WF, Thadani U, Zimmerman FH (2012) Nitrates as an integral part of optimal medical therapy and cardiac rehabilitation for stable angina: review of current concepts and therapeutics. Clin Cardiol 35:263–271
7. Brooks PJ, Enoch MA, Goldman D, Li TK, Yokoyama A (2009) The alcohol flushing response: an unrecognized risk factor for esophageal cancer from alcohol consumption. PLoS Med 6:e50
8. Budas GR, Disatnik MH, Mochly-Rosen D (2009) Aldehyde dehydrogenase 2 in cardiac protection: a new therapeutic target? Trends Cardiovasc Med 19:158–164
9. Chen CH, Budas GR, Churchill EN, Disatnik MH, Hurley TD, Mochly-Rosen D (2008) Activation of aldehyde dehydrogenase-2 reduces ischemic damage to the heart. Science 321:1493–1495
10. Chen CH, Ferreira JC, Gross ER, Mochly-Rosen D (2014) Targeting aldehyde dehydrogenase 2: new therapeutic opportunities. Physiol Rev 94:1–34
11. Chen CH, Sun L, Mochly-Rosen D (2010) Mitochondrial aldehyde dehydrogenase and cardiac diseases. Cardiovasc Res 88:51–57
12. Chen Z, Stamler JS (2006) Bioactivation of nitroglycerin by the mitochondrial aldehyde dehydrogenase. Trends Cardiovasc Med 16:259–265
13. Chen Z, Zhang J, Stamler JS (2002) Identification of the enzymatic mechanism of nitroglycerin bioactivation. Proc Natl Acad Sci U S A 99:8306–8311
14. Daiber A, Oelze M, Wenzel P, Wickramanayake JM, Schuhmacher S, Jansen T, Lackner KJ, Torzewski M, Munzel T (2009) Nitrate tolerance as a model of vascular dysfunction: roles for mitochondrial aldehyde dehydrogenase and mitochondrial oxidative stress. Pharmacol Rep 61:33–48
15. Doser TA, Turdi S, Thomas DP, Epstein PN, Li SY, Ren J (2009) Transgenic overexpression of aldehyde dehydrogenase-2 rescues chronic alcohol intake-induced myocardial hypertrophy and contractile dysfunction. Circulation 119:1941–1949
16. Ferreira JC, Boer BN, Grinberg M, Brum PC, Mochly-Rosen D (2012) Protein quality control disruption by PKCbetaII in heart failure; rescue by the selective PKCbetaII inhibitor, betaIIV5-3. PLoS One 7:e33175
17. Ferreira JC, Mochly-Rosen D (2012) Nitroglycerin use in myocardial infarction patients. Circ J 76:15–21
18. Gomes KM, Bechara LR, Lima VM, Ribeiro MA, Campos JC, Dourado PM, Kowaltowski AJ, Mochly-Rosen D, Ferreira JC (2015) Aldehydic load and aldehyde dehydrogenase 2 profile during the progression of post-myocardial infarction cardiomyopathy: benefits of Alda-1. Int J Cardiol 179:129–138
19. Gomes KM, Campos JC, Bechara LR, Queliconi B, Lima VM, Disatnik MH, Magno P, Chen CH, Brum PC, Kowaltowski AJ, Mochly-Rosen D, Ferreira JC (2014) Aldehyde dehydrogenase 2 activation in heart failure restores mitochondrial function and improves ventricular function and remodelling. Cardiovasc Res 103:498–508

20. Gong D, Zhang Y, Zhang H, Gu H, Jiang Q, Hu S (2012) Aldehyde dehydrogenase-2 activation during cardioplegic arrest enhances the cardioprotection against myocardial ischemia-reperfusion injury. Cardiovasc Toxicol 12:350–358
21. Gu J, Li L (2014) Reply to: ALDH2 Glu504Lys polymorphism and susceptibility to coronary artery disease and myocardial infarction in East Asians: a meta-analysis. Arch Med Res 45:281
22. Guo X, Disatnik MH, Monbureau M, Shamloo M, Mochly-Rosen D, Qi X (2013) Inhibition of mitochondrial fragmentation diminishes Huntington's disease-associated neurodegeneration. J Clin Invest 123:5371–5388
23. Han H, Wang H, Yin Z, Jiang H, Fang M, Han J (2013) Association of genetic polymorphisms in ADH and ALDH2 with risk of coronary artery disease and myocardial infarction: a meta-analysis. Gene 526:134–141
24. Hu YF, Chang YT, Lin YJ, Chang SL, Lo LW, Huang YH, Liu TT, Chen CH, Tuan TC, Chao TF, Chung FP, Liao JN, Te ALD, Huang CF, Chen SA (2017) The roles of alcohol dehydrogenase in patients with atrial fibrillation. Pacing Clin Electrophysiol 40:1446–1453
25. Hung CL, Chang SC, Chang SH, Chi PC, Lai YJ, Wang SW, Wu YJ, Yeh HI, Lin SJ, Chen CH, Mochly-Rosen D, Wang LY, MAGNET Study Investigator (2017) Genetic polymorphisms of alcohol metabolizing enzymes and alcohol consumption are associated with asymptomatic cardiac remodeling and subclinical systolic dysfunction in large community-dwelling Asians. Alcohol Alcohol 52:638–646
26. Hung CL, Goncalves A, Lai YJ, Lai YH, Sung KT, Lo CI, Liu CC, Kuo JY, Hou CJ, Chao TF, Bulwer BE, Lin SJ, Yeh HI, Lam CS (2016) Light to moderate habitual alcohol consumption is associated with subclinical ventricular and left atrial mechanical dysfunction in an asymptomatic population: dose-response and propensity analysis. J Am Soc Echocardiogr 29:1043–1051. e4
27. Jorgenson E, Thai KK, Hoffmann TJ, Sakoda LC, Kvale MN, Banda Y, Schaefer C, Risch N, Mertens J, Weisner C, Choquet H (2017) Genetic contributors to variation in alcohol consumption vary by race/ethnicity in a large multi-ethnic genome-wide association study. Mol Psychiatry 22:1359–1367
28. Joshi AU, Saw NL, Vogel H, Cunnigham AD, Shamloo M, Mochly-Rosen D (2018) Inhibition of Drp1/Fis1 interaction slows progression of amyotrophic lateral sclerosis. EMBO Mol Med 10:e8166
29. Kalinowski A, Humphreys K (2016) Governmental standard drink definitions and low-risk alcohol consumption guidelines in 37 countries. Addiction 111:1293–1298
30. Karliner JS (2009) Lessons from the besotted heart. J Am Coll Cardiol 54:2197–2198
31. Kato N, Takeuchi F, Tabara Y, Kelly TN, Go MJ, Sim X, Tay WT, Chen CH, Zhang Y, Yamamoto K, Katsuya T, Yokota M, Kim YJ, Ong RT, Nabika T, Gu D, Chang LC, Kokubo Y, Huang W, Ohnaka K, Yamori Y, Nakashima E, Jaquish CE, Lee JY, Seielstad M, Isono M, Hixson JE, Chen YT, Miki T, Zhou X, Sugiyama T, Jeon JP, Liu JJ, Takayanagi R, Kim SS, Aung T, Sung YJ, Zhang X, Wong TY, Han BG, Kobayashi S, Ogihara T, Zhu D, Iwai N, Wu JY, Teo YY, Tai ES, Cho YS, He J (2011) Meta-analysis of genome-wide association studies identifies common variants associated with blood pressure variation in East Asians. Nat Genet 43:531–538
32. Koda K, Salazar Rodriguez M, Corti F, Chan NY, Estephan R, Silver RB, Mochly-Rosen D, Levi R (2010) Aldehyde dehydrogenase activation prevents reperfusion arrhythmias by inhibiting local renin release from cardiac mast cells. Circulation 122:771–781
33. Li Y, Zhang D, Jin W, Shao C, Yan P, Xu C, Sheng H, Liu Y, Yu J, Xie Y, Zhao Y, Lu D, Nebert DW, Harrison DC, Huang W, Jin L (2006) Mitochondrial aldehyde dehydrogenase-2 (ALDH2) Glu504Lys polymorphism contributes to the variation in efficacy of sublingual nitroglycerin. J Clin Invest 116:506–511
34. Li YY, Wang H, Wu JJ, Kim HJ, Yang XX, Geng HY, Gong G (2018) ALDH2 gene G487A polymorphism and coronary artery disease: a meta-analysis including 5644 participants. J Cell Mol Med 22:1666–1674

35. Lu X, Wang L, Lin X, Huang J, Charles Gu C, He M, Shen H, He J, Zhu J, Li H, Hixson JE, Wu T, Dai J, Lu L, Shen C, Chen S, He L, Mo Z, Hao Y, Mo X, Yang X, Li J, Cao J, Chen J, Fan Z, Li Y, Zhao L, Li H, Lu F, Yao C, Yu L, Xu L, Mu J, Wu X, Deng Y, Hu D, Zhang W, Ji X, Guo D, Guo Z, Zhou Z, Yang Z, Wang R, Yang J, Zhou X, Yan W, Sun N, Gao P, Gu D (2015) Genome-wide association study in Chinese identifies novel loci for blood pressure and hypertension. Hum Mol Genet 24:865–874
36. Luo HR, Wu GS, Pakstis AJ, Tong L, Oota H, Kidd KK, Zhang YP (2009) Origin and dispersal of atypical aldehyde dehydrogenase ALDH2487Lys. Gene 435:96–103
37. Luo XJ, Liu B, Ma QL, Peng J (2014) Mitochondrial aldehyde dehydrogenase, a potential drug target for protection of heart and brain from ischemia/reperfusion injury. Curr Drug Targets 15:948–955
38. Ma H, Guo R, Yu L, Zhang Y, Ren J (2011) Aldehyde dehydrogenase 2 (ALDH2) rescues myocardial ischaemia/reperfusion injury: role of autophagy paradox and toxic aldehyde. Eur Heart J 32:1025–1038
39. Ma H, Li J, Gao F, Ren J (2009) Aldehyde dehydrogenase 2 ameliorates acute cardiac toxicity of ethanol: role of protein phosphatase and forkhead transcription factor. J Am Coll Cardiol 54:2187–2196
40. Ma H, Yu L, Byra EA, Hu N, Kitagawa K, Nakayama KI, Kawamoto T, Ren J (2010) Aldehyde dehydrogenase 2 knockout accentuates ethanol-induced cardiac depression: role of protein phosphatases. J Mol Cell Cardiol 49:322–329
41. Mackenzie IS, Maki-Petaja KM, Mceniery CM, Bao YP, Wallace SM, Cheriyan J, Monteith S, Brown MJ, Wilkinson IB (2005) Aldehyde dehydrogenase 2 plays a role in the bioactivation of nitroglycerin in humans. Arterioscler Thromb Vasc Biol 25:1891–1895
42. Mak S, Lehotay DC, Yazdanpanah M, Azevedo ER, Liu PP, Newton GE (2000) Unsaturated aldehydes including 4-OH-nonenal are elevated in patients with congestive heart failure. J Card Fail 6:108–114
43. Mali VR, Palaniyandi SS (2014) Regulation and therapeutic strategies of 4-hydroxy-2-nonenal metabolism in heart disease. Free Radic Res 48:251–263
44. Marino A, Sakamoto T, Robador PA, Tomita K, Levi R (2017) S1P receptor 1-mediated anti-renin-angiotensin system cardioprotection: pivotal role of mast cell aldehyde dehydrogenase type 2. J Pharmacol Exp Ther 362:230–242
45. Mayer B, Beretta M (2008) The enigma of nitroglycerin bioactivation and nitrate tolerance: news, views and troubles. Br J Pharmacol 155:170–184
46. Menz V, Grimm W, Hoffmann J, Maisch B (1996) Alcohol and rhythm disturbance: the holiday heart syndrome. Herz 21:227–231
47. Miura T, Nishinaka T, Terada T, Yonezawa K (2017) Vasodilatory effect of nitroglycerin in Japanese subjects with different aldehyde dehydrogenase 2 (ALDH2) genotypes. Chem Biol Interact 276:40–45
48. Munger TM, Wu LQ, Shen WK (2014) Atrial fibrillation. J Biomed Res 28:1–17
49. Munzel T, Daiber A, Gori T (2011) Nitrate therapy: new aspects concerning molecular action and tolerance. Circulation 123:2132–2144
50. Nakamura K, Kusano K, Nakamura Y, Kakishita M, Ohta K, Nagase S, Yamamoto M, Miyaji K, Saito H, Morita H, Emori T, Matsubara H, Toyokuni S, Ohe T (2002) Carvedilol decreases elevated oxidative stress in human failing myocardium. Circulation 105:2867–2871
51. Nakamura K, Miura D, Kusano KF, Fujimoto Y, Sumita-Yoshikawa W, Fuke S, Nishii N, Nagase S, Hata Y, Morita H, Matsubara H, Ohe T, Ito H (2009) 4-hydroxy-2-nonenal induces calcium overload via the generation of reactive oxygen species in isolated rat cardiac myocytes. J Card Fail 15:709–716
52. Nakano Y, Ochi H, Onohara Y, Sairaku A, Tokuyama T, Matsumura H, Tomomori S, Amioka M, Hironomobe N, Motoda C, Oda N, Chayama K, Chen CH, Gross ER, Mochly-Rosen D, Kihara Y (2016) Genetic variations of aldehyde dehydrogenase 2 and alcohol dehydrogenase 1B are associated with the etiology of atrial fibrillation in Japanese. J Biomed Sci 23:89

53. Opelt M, Wolkart G, Eroglu E, Waldeck-Weiermair M, Malli R, Graier WF, Kollau A, Fassett JT, Schrammel A, Mayer B, Gorren ACF (2018) Sustained formation of nitroglycerin-derived nitric oxide by aldehyde dehydrogenase-2 in vascular smooth muscle without added reductants: implications for the development of nitrate tolerance. Mol Pharmacol 93:335–343
54. Pang J, Wang J, Zhang Y, Xu F, Chen Y (2017) Targeting acetaldehyde dehydrogenase 2 (ALDH2) in heart failure—recent insights and perspectives. Biochim Biophys Acta 1863:1933–1941
55. Pang JJ, Barton LA, Chen YG, Ren J (2015) Mitochondrial aldehyde dehydrogenase in myocardial ischemia-reperfusion injury: from bench to bedside. Sheng Li Xue Bao 67:535–544
56. Perez-Miller S, Younus H, Vanam R, Chen CH, Mochly-Rosen D, Hurley TD (2010) Alda-1 is an agonist and chemical chaperone for the common human aldehyde dehydrogenase 2 variant. Nat Struct Mol Biol 17:159–164
57. Ren J (2007) Acetaldehyde and alcoholic cardiomyopathy: lessons from the ADH and ALDH2 transgenic models. Novartis Found Symp 285:69–76; discussion 76–9, 198–9
58. Robador PA, Seyedi N, Chan NY, Koda K, Levi R (2012) Aldehyde dehydrogenase type 2 activation by adenosine and histamine inhibits ischemic norepinephrine release in cardiac sympathetic neurons: mediation by protein kinase Cepsilon. J Pharmacol Exp Ther 343:97–105
59. Roede JR, Jones DP (2010) Reactive species and mitochondrial dysfunction: mechanistic significance of 4-hydroxynonenal. Environ Mol Mutagen 51:380–390
60. Sakata S, Yoshihara T, Arima H, Shiraishi F, Oniki H, Takahashi-Yanaga F, Matsumura K, Sasaguri T (2011) Differential effects of organic nitrates on arterial diameter among healthy Japanese participants with different mitochondrial aldehyde dehydrogenase 2 genotypes: randomised crossover trial. BMJ Open 1:e000133
61. Shin MJ, Cho Y, Davey Smith G (2017) Alcohol consumption, aldehyde dehydrogenase 2 gene polymorphisms, and cardiovascular health in Korea. Yonsei Med J 58:689–696
62. Sun L, Ferreira JC, Mochly-Rosen D (2011) ALDH2 activator inhibits increased myocardial infarction injury by nitroglycerin tolerance. Sci Transl Med 3:107ra111
63. Sung YF, Lu CC, Lee JT, Hung YJ, Hu CJ, Jeng JS, Chiou HY, Peng GS (2016) Homozygous ALDH2*2 is an independent risk factor for ischemic stroke in Taiwanese men. Stroke 47:2174–2179
64. Takeuchi F, Yokota M, Yamamoto K, Nakashima E, Katsuya T, Asano H, Isono M, Nabika T, Sugiyama T, Fujioka A, Awata N, Ohnaka K, Nakatochi M, Kitajima H, Rakugi H, Nakamura J, Ohkubo T, Imai Y, Shimamoto K, Yamori Y, Yamaguchi S, Kobayashi S, Takayanagi R, Ogihara T, Kato N (2012) Genome-wide association study of coronary artery disease in the Japanese. Eur J Hum Genet 20:333–340
65. Tan A, Sun J, Xia N, Qin X, Hu Y, Zhang S, Tao S, Gao Y, Yang X, Zhang H, Kim ST, Peng T, Lin X, Li L, Mo L, Liang Z, Shi D, Huang Z, Huang X, Liu M, Ding Q, Trent JM, Zheng SL, Mo Z, Xu J (2012) A genome-wide association and gene-environment interaction study for serum triglycerides levels in a healthy Chinese male population. Hum Mol Genet 21:1658–1664
66. Ueta CB, Campos JC, Albuquerque RPE, Lima VM, Disatnik MH, Sanchez AB, Chen CH, Medeiros MHG, Yang W, Mochly-Rosen D, Ferreira JCB (2018) Cardioprotection induced by a brief exposure to acetaldehyde: role of aldehyde dehydrogenase 2. Cardiovasc Res 114(7):1006–1015
67. Wenzel P, Muller J, Zurmeyer S, Schuhmacher S, Schulz E, Oelze M, Pautz A, Kawamoto T, Wojnowski L, Kleinert H, Munzel T, Daiber A (2008) ALDH-2 deficiency increases cardiovascular oxidative stress—evidence for indirect antioxidative properties. Biochem Biophys Res Commun 367:137–143
68. Wenzl MV, Beretta M, Griesberger M, Russwurm M, Koesling D, Schmidt K, Mayer B, Gorren AC (2011) Site-directed mutagenesis of aldehyde dehydrogenase-2 suggests three distinct pathways of nitroglycerin biotransformation. Mol Pharmacol 80:258–266

69. Wood AM, Kaptoge S, Butterworth AS, Willeit P, Warnakula S, Bolton T, Paige E, Paul DS, Sweeting M, Burgess S, Bell S, Astle W, Stevens D, Koulman A, Selmer RM, Verschuren WMM, Sato S, Njolstad I, Woodward M, Salomaa V, Nordestgaard BG, Yeap BB, Fletcher A, Melander O, Kuller LH, Balkau B, Marmot M, Koenig W, Casiglia E, Cooper C, Arndt V, Franco OH, Wennberg P, Gallacher J, De La Camara AG, Volzke H, Dahm CC, Dale CE, Bergmann MM, Crespo CJ, Van Der Schouw YT, Kaaks R, Simons LA, Lagiou P, Schoufour JD, Boer JMA, Key TJ, Rodriguez B, Moreno-Iribas C, Davidson KW, Taylor JO, Sacerdote C, Wallace RB, Quiros JR, Tumino R, Blazer DG 2nd, Linneberg A, Daimon M, Panico S, Howard B, Skeie G, Strandberg T, Weiderpass E, Nietert PJ, Psaty BM, Kromhout D, Salamanca-Fernandez E, Kiechl S, Krumholz HM, Grioni S, Palli D, Huerta JM, Price J, Sundstrom J, Arriola L, Arima H, Travis RC, Panagiotakos DB, Karakatsani A, Trichopoulou A, Kuhn T, Grobbee DE, Barrett-Connor E, Van Schoor N, Boeing H, Overvad K, Kauhanen J, Wareham N, Langenberg C, Forouhi N, Wennberg M, Despres JP, Cushman M, Cooper JA, Rodriguez CJ, Sakurai M, Shaw JE, Knuiman M, Voortman T, Meisinger C et al (2018) Risk thresholds for alcohol consumption: combined analysis of individual-participant data for 599 912 current drinkers in 83 prospective studies. Lancet 391:1513–1523
70. Xu F, Sun Y, Shang R, Li M, Cui L, Cui Z, Chen Y (2014) The Glu504Lys polymorphism of aldehyde dehydrogenase 2 contributes to development of coronary artery disease. Tohoku J Exp Med 234:143–150
71. Yoshida A (1992) Molecular genetics of human aldehyde dehydrogenase. Pharmacogenetics 2:139–147
72. Yoval-Sanchez B, Rodriguez-Zavala JS (2012) Differences in susceptibility to inactivation of human aldehyde dehydrogenases by lipid peroxidation byproducts. Chem Res Toxicol 25:722–729
73. Yun KE, Chang Y, Yun SC, Davey Smith G, Ryu S, Cho SI, Chung EC, Shin H, Khang YH (2017) Alcohol and coronary artery calcification: an investigation using alcohol flushing as an instrumental variable. Int J Epidemiol 46:950–962
74. Zakhari S (2006) Overview: how is alcohol metabolized by the body? Alcohol Res Health 29:245–254
75. Zhang LL, Wang YQ, Fu B, Zhao SL, Kui Y (2015) Aldehyde dehydrogenase 2 (ALDH2) polymorphism gene and coronary artery disease risk: a meta-analysis. Genet Mol Res 14:18503–18514
76. Zhang R, Wang J, Xue M, Xu F, Chen Y (2017a) ALDH2—the genetic polymorphism and enzymatic activity regulation: their epidemiologic and clinical implications. Curr Drug Targets 18:1810–1816
77. Zhang Y, Ren J (2011) ALDH2 in alcoholic heart diseases: molecular mechanism and clinical implications. Pharmacol Ther 132:86–95
78. Zhang Y, Wang C, Zhou J, Sun A, Hueckstaedt LK, Ge J, Ren J (2017b) Complex inhibition of autophagy by mitochondrial aldehyde dehydrogenase shortens lifespan and exacerbates cardiac aging. Biochim Biophys Acta 1863:1919–1932
79. Zhao J, You L, Wang DW, Cui W (2016) Impacts of common variants in ALDH2 on coronary artery disease patients. Gene 585:104–109
80. Zhu Y, Zhang D, Zhou D, Li Z, Li Z, Fang L, Yang M, Shan Z, Li H, Chen J, Zhou X, Ye W, Yu S, Li H, Cai L, Liu C, Zhang J, Wang L, Lai Y, Ruan L, Sun Z, Zhang S, Wang H, Liu Y, Xu Y, Ling J, Xu C, Zhang Y, Lv D, Yuan Z, Zhang J, Zhang Y, Shi Y, Lai M (2017) Susceptibility loci for metabolic syndrome and metabolic components identified in Han Chinese: a multi-stage genome-wide association study. J Cell Mol Med 21:1106–1116

The Bidirectional Effect of Defective *ALDH2* Polymorphism and Disease Prevention

Akiko Matsumoto

Abstract Despite the role of aldehyde dehydrogenase 2 (ALDH2) in the detoxification of endogenous aldehydes, the defective polymorphism (rs671), which is highly prevalent among East Asians, does not show a serious phenotype, such as congenital abnormality. However, unfavorable and favorable impacts of the variant allele, *ALDH2*2*, on various disease risks have been reported. The underlying mechanisms are often complicated due to the compensatory aldehyde detoxification systems. As the phenotypes emerge due to overlapping environmental factors (e.g., alcohol intake and tobacco smoke) or individual vulnerabilities (e.g., aging and apolipoprotein E ε4 allele), polymorphism is therefore considered to be important in the field of preventative medicine. For example, it is important to recognize that *ALDH2*2* carriers are at a high risk of alcohol drinking-related cancers; however, their drinking habit has less adverse effects on physiological indices, such as blood pressure, body mass index, levels of lipids, and hepatic deviation enzymes in the blood, than in non-*ALDH2*2* carriers. Therefore, opportunities to reconsider their excessive drinking habit before adverse events occur can be missed. To perform effective disease prevention, the effects of *ALDH2*2* on various diseases and the biological mechanisms should be clarified.

Keywords ALDH2 · Gene polymorphism · rs671 · Preventive medicine · Alcohol

1 Introduction

Among the 19 known human aldehyde dehydrogenase (ALDH) isozymes [44, 69], ALDH2 is a widely distributed [26, 28, 110, 147] and highly expressed ALDH isozyme [13, 108, 110]. While many isozymes show obvious congenital anomalies with their gene mutation [69] (e.g., Sjogren's syndrome with *ALDH3A2* [34] and pyridoxine-dependent epilepsy with *ALDH7A1* [79]), no serious phenotype is observed with a single dysfunction of ALDH2 [77]. However, when observing

A. Matsumoto (✉)
Department of Social Medicine, Saga University School of Medicine, Saga, Japan
e-mail: matsumoa@cc.saga-u.ac.jp

J. Ren et al. (eds.), *Aldehyde Dehydrogenases*, Advances in Experimental Medicine and Biology 1193, https://doi.org/10.1007/978-981-13-6260-6_4

relatively mild health effects, it shows complex and diverse phenotypes. In most cases, these health effects become conspicuous when ALDH2 deficiency overlaps with other factors, such as lifestyle, aging, environmental chemical exposure, and other gene mutations, and this interaction increases the importance of *ALDH2* polymorphism in disease prevention [38, 71, 72, 76].

In the present report, the basic details and importance of the *ALDH2* polymorphism are summarized.

2 Defective Polymorphism of *ALDH2*

2.1 ALDH2 and ALDH2*2 *(rs671)*

The *ALDH2* gene is located on the long arm of chromosome 12 (12q 24.2) (43,099 bp, 13 exons) and encodes 517 amino acids to form the 56-kDa ALDH2 protein (a mature body consists of 500 amino acids), which is distributed in the mitochondrial matrix of various cells [13, 26, 28, 108, 110, 147]. ALDH2 is a tetrameric protein in which dimers are paired, with each dimer consisting of two monomers facing each other and sandwiching their functional sites (the central region of catalysis) [58, 109] (Fig. 1). The best known substrate is acetaldehyde, which can be generated by drinking alcohol. ALDH2 shows the highest affinity for acetaldehyde

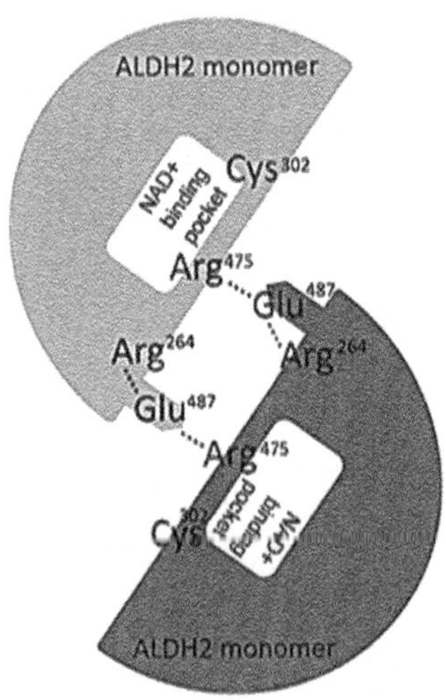

Fig. 1 A schematic illustration of the structure of ALDH2 dimer [71]
The function site of ALDH2 protein has a binding pocket for coenzyme NAD^+ and Cys302, which interacts with substrates. A pair of monomers forms a dimeric ALDH2. Glu487 (Glu504 in the precursor form, 517 amino acids) unites Arg264 of the same monomer with Arg475 of the partner monomer with hydrogen bonds (dotted line), which is critical for the assembly of both monomers

among enzymes that metabolize acetaldehyde [69, 138]. It also plays an important role in the metabolism of endogenous aldehydes [25, 57], such as the representative lipid peroxide 4-hydroxynonenal (4HNE), the glycation end product methylglyoxal [91], and the metabolite of lipids and polyamines acrolein [97] and formaldehyde, which is derived from folic acid, catecholamines, and foods [8, 10]. These cytotoxic agents [20, 45, 140, 145] are all known to be substrates of ALDH2 [6, 43, 143].

According to the database of the National Center for Biotechnology Information, more than 5000 polymorphisms are reported for *ALDH2*, including 220 missense variants. Although several correlations have been reported between *ALDH2* polymorphisms and diseases, e.g., rs886205 [32], rs737280 [22], and rs10744777 [90], an obvious phenotypic difference is only observed for rs671, in which the guanine at base 42421 in exon 12 is replaced with an adenine, and the glutamic acid at position 487 in mature ALDH2 or at position 504 in precursor ALDH2 is replaced with a lysine (Glu487Lys or Glu504Lys). Generally, the wild and variant alleles are called *ALDH2*1* and *ALDH2*2*, respectively, resulting in three types of *ALDH2* gene: wild-type homozygote (*ALDH2*1/*1*), heterozygote (*ALDH2*1/*2*), and variant-type homozygote (*ALDH2*2/*2*). *ALDH2*2* is largely restricted to Mongoloid populations. The prevalence is particularly high in China, Japan, Korea, Mongolia, and the Indochina Peninsula (approximately 50%). The variant is extremely rare in Caucasians, Africans (and their descendants), Papua New Guineans, Australian Aborigines, and Araucanians in South Chile [27, 64].

Glu487 is an important amino acid that forms a dimer cross-link [21] (Fig. 1), which greatly affects the enzyme activity. In *ALDH2*1/*2* carriers, variant monomers and wild-type monomers are produced in a ratio of 1:1 and randomly combine to form two sets of dimers (tetramer). When a variant monomer and wild monomer form a dimer, the three-dimensional structure collapses, and the enzyme activity disappears. Therefore, when a tetramer includes one variant monomer out of four monomers, the enzyme activity decreases to 50% [21]. Furthermore, the half-life of the tetramer shortens to about 50% [128]. As such, all tetramer combinations have particular existence rates, activity levels, and half-lives, and the total enzyme activity in *ALDH2*1/*2* carriers is theoretically calculated to be 15.6% of that in *ALDH2*1/*1* carriers [71] (Fig. 2). Similar numbers are reported in actual measurements [51, 55].

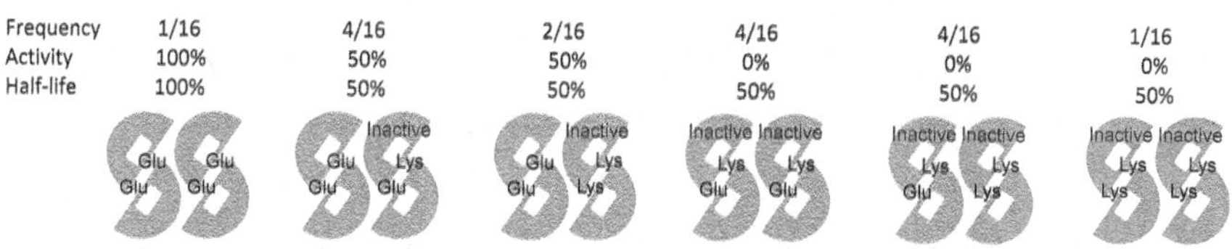

Fig. 2 Schematic illustration of ALDH2 protein derived from *ALDH2*1/*2*
Because human ALDH2 is tetrameric protein, *ALDH2*1/*2* carriers have six patterns of protein combination. Theoretically expected frequency of appearance, enzymatic activity, and half-life of each combination are shown at the top of the figure. Based on the theory that a heterodimeric ALDH2 is inactive and that a tetramer that includes more than one variant monomer has a shortened half-life (50%), the total ALDH2 activity in *ALDH2*1/*2* carriers is calculated to be 15.6% of that in *ALDH2*1/*1* carriers (1/16 × 100% + 4/16 × 50% × 50% + 2/16 × 50% × 50% + 4/16 × 0% × 50% + 4/16 × 0% × 50% + 1/16 × 0% × 50%)

2.2 *Prediction of* ALDH2*2

*ALDH2*2* possession is highly predictable without a DNA analysis [95, 137]. In a study on Japanese men asking the following questions, "Did you tend to flush in the face immediately after drinking a glass of beer for the first couple of years after you started drinking?" [137], the sensitivity and specificity were calculated to be 90% and 95%, respectively ("yes" to predict *ALDH2*2* possession). Such flushing questionnaires are suggested to be more reliable than ethanol patch tests, which examine the flushing reaction on the skin due to acetaldehyde accumulation [132].

However, the genetic polymorphism of alcohol dehydrogenase 1B (ADH1B), which oxidizes ethanol to acetaldehyde, interferes with the prediction of *ALDH2*2*. The variant ADH1B (rs1229984), which is hyperactive, is also frequently observed in Mongoloid populations [27]; indeed, the possession rate in Japanese is reported to be over 90% [15, 35, 36, 78]. Ethanol oxidation, or acetaldehyde production, after ethanol intake in wild-type *ADH1B* carriers (minority for Mongoloid populations) is slower than that in variant *ADH1B* carriers [98]; therefore, alcohol-induced discomfort, including flushing, is ameliorated in wild-type *ADH1B* carriers [33]. The detection rate of *ALDH2*2* by ethanol patch test is reported to be extremely lower in a population with *ALDH2*1/*2* and wild-type *ADH1B* (13%) than in that with *ALDH2*1/*2* and variant *ADH1B* (65–73%) [35].

2.3 *Effect of* ALDH2*2 *on Drinking Behavior*

*ALDH2*2* reduces an individual's likelihood of becoming a regular alcohol drinker, especially for *ALDH2*2/*2* carriers, who are estimated to comprise less than 10% of noncarrying drinkers in males, while the rate is even lower in females [78, 89]. The average amount of alcohol that is consumed by *ALDH2*1/*2*-carrying drinkers is considered to be 50–60% of that for *ALDH2*1/*1*-carrying drinkers, while for *ALDH2*2/*2*-carrying drinkers, the amount of alcohol consumed is less than 10% [78, 89]. This inhibitory effect is attributed to discomfort due to the accumulation of acetaldehyde, which causes palpitations and nausea after consuming ethanol [80]. The modification of drinking behavior is reproducible in animal experiments; when *Aldh2* knockout mice and wild-type mice were given free access to either water or a 3% ethanol aqueous solution, the ethanol intake of the knockout mice was relatively small. Interestingly, the acetaldehyde concentration in the blood, liver, and brain tissues of these mice was equivalent to that of wild-type mice [42].

However, as previous animal experiments have suggested [17, 19, 93], acetaldehyde can promote dopamine release in the mesolimbic system and cause drinking addiction. It has also been reported that salsolinol, which is produced by the nonenzymatic condensation of acetaldehyde and dopamine, promotes ethanol intake through stimulation of the μ-opioid receptor [41]. It was reported that mood exaltation (positive feelings described as "great overall") and blood catecholamine concentrations after drinking were stronger and higher, respectively, in *ALDH2*1/*2* carriers than in *ALDH2*1/*1* carriers [2, 122] and that ALDH2 inhibitor increases ethanol-induced euphoria [7]. Thus, *ALDH2*2* does not always inhibit drinking habits. Indeed, 10%–20% of Japanese alcoholics carry the *ALDH2*1/*2* gene [36, 135], and these subjects generally consume an equivalent dose of ethanol to *ALDH2*1/*1*-carrying alcoholics [135].

3 Biological Remodeling for Homeostasis

In normally raised *Aldh2* knockout mice, obvious phenotypes, such as a shortened life span and congenital anomalies, are not observed [25, 77], despite the observable impact of ALDH2 in detoxifying endogenous aldehydes (see Sect. 2.1). This contradiction can be explained by the proposed mechanism, "defense mechanism induced by aldehyde exposure."

In a study on infants with Tetralogy of Fallot in China, aldehydes such as 4HNE and malondialdehyde (MDA) were detected at high concentrations in the myocardium of patients with the *ALDH2*2* allele. However, the induction of an anti-oxidative stress system due to glutathione production was detected, and clinical indicators, such as the cardiac troponin level in the blood and hospitalization period, were more promising in infants with the *ALDH2*2* allele than in infants without the *ALDH2*2* [149]. This study was designed to follow a discovery reported by Endo et al. [18]. Endo and colleagues conducted a transcriptome analysis using ALDH dysfunctional model animals and found an upregulated pathway through which the glucose metabolite 3-phosphoglycerate in the glycolysis system was consumed for glutathione production and reported it as "metabolic remodeling induced by aldehyde stress in the heart" [18].

Another kind of remodeling has also been observed in the liver: cytochrome P450 (CYP) 2E1, which has the ability to metabolize aldehydes [53, 119], is highly expressed in *Aldh2* knockout mice possibly by posttranslational modification [29, 73, 75]. Although CYP2E1 is a generator of oxidative stress due to a catalytic reaction [9], it also promotes stress responses, such as the activation of the nuclear factor-erythroid 2 p45-related factor (Nrf2) pathway [3, 68]. Indeed, antioxidant factors, including heme oxygenase 1 (HO-1), redox factor-1 (Ref-1), and manganese superoxide dismutase (MnSOD), were found to be induced in the liver of *Aldh2* knockout mice [54, 75]. It therefore seems to be reasonable for such an enzyme complex (i.e., CYP2E1, HO-1, Ref-1, and MnSOD) to fill the role of ALDH2 in the liver.

Such "remodeling" may be a commonly observed phenomenon in animals with difficulties managing antioxidant stress; in the liver of hepatocyte-specific glutathione-deficient mice, various detoxification and metabolic enzymes including HO-1, superoxide dismutase 1, and Nrf2 were upregulated, with accelerated phosphorylation of the upstream regulator AMP-activated protein kinase (a metabolic master switch) [11], which senses depletion of intermediate metabolites in glycolysis [148].

However, some studies have still reported unfavorable phenotypes in ALDH2-deficient models, in which remodeling for adaptation appears to have been unsuccessful. In untreated *Aldh2* knockout mice, a reduced hematopoietic stem cell ratio in the bone marrow [25] and an age-related cognitive decline and accompanying Alzheimer's disease-like pathology [16] have been observed. As such, individual organs/cells may exhibit different degrees of aldehyde tolerance, as the ALDH2 contribution rate for aldehyde detoxification differs among cell types [25]. Likewise, the capability of inducing alternative enzymes or stress-response systems is also expected to vary.

4 Various Phenotypes of *ALDH2*2*

The phenotypes of *ALDH2*2* are generally ambiguous, possibly due to the compensation system described above. However, phenotypes often become conspicuous due to lifestyle; aging; other environmental factors, such as occupational exposure to toxic chemicals; and other genetic vulnerabilities, increasing the significance of the *ALDH2* polymorphism in the field of preventive medicine [71, 72, 76]. Although investigations have largely been performed in the field of adult heath (Table 1), there are many other issues to be addressed. For example, environmental aldehyde exposure during the prenatal period is of concern for *ALDH2*2*-carrying fetuses, as ALDH2 in the fetal stage already plays a role in aldehyde detoxification [65, 99,

Table 1 Reported outcomes of *ALDH2*2*

Risk increased by *ALDH2*2*	(Possible mechanism)
Alcohol- and tobacco-related cancer	(Aldehyde-DNA formation, aldehyde interference in BRCA2)
Age-related cognitive dysfunction	(Glycation stress toward amyloid-β protein)
Vasospastic angina	(Aldehyde-induced spasm)
Accelerated progression of Fanconi anemia	(Exacerbated dysfunction of aldehyde detoxification)
Bone metabolism disorder in elderly women	
Risk reduced by *ALDH2*2*	
Energy metabolism-related traits	(Metabolic remodeling, inhibited gastric motility after the intake of ethanol)
Elevated blood pressure	(Increased vasodilation due to acetaldehyde exposure, metabolic remodeling)
Elevated serum uric acid	(Slower ATP production after the intake of ethanol)
Alcohol-induced AST, ALT, and GGT	(Upregulation of the anti-stress system and downregulation of fatty acid synthesis)
Mood and anxiety disorders	
Diabetic vascular disease and cardiac hypertrophy	(Blunted PI3K/Akt pathway, alleviation of age-related autophagy dysfunction)

139]. In addition, personalized prescriptions for *ALDH2*2* carriers will be required in the clinical setting, as ALDH2 is involved in the development of drug efficacy and side effects [30, 50, 60, 65, 67, 151], and some drugs interfere with the ALDH2 activity, such as acetaminophen [56, 63], salicylate [59], cyanamide [43, 112], and doxorubicin [24]. Furthermore, a lack of consideration about *ALDH2* polymorphisms in workers exposed to aldehydes or precursors is a problem, especially in countries with a high prevalence of this polymorphism.

4.1 Unfavorable Outcomes of ALDH2*2

The outcomes of *ALDH2*2* and the possible mechanisms are summarized in Table 1. The risk of alcohol-/smoking-related cancer is high in *ALDH2*2* carriers [1, 12, 15, 31, 48, 52, 86, 96, 100, 133, 134]. Several carcinogenic roles of aldehydes have been hypothesized [5, 85, 105, 144], including DNA adduct formation [70, 88] and the induction of genome instability via the interference of BRCA2, resulting in proteasome degradation [118].

The risk of Alzheimer's disease and cognitive dysfunction due to Parkinson's disease is also suggested to be relatively high in *ALHD2*2* carriers [46, 92, 123, 146], and the overlapping genetic variant apolipoprotein E gene ε4 allele (APOE4) reportedly further increases the risk of Alzheimer's disease [46, 92]. It is suggested that Alzheimer's disease is associated with glycation stress toward amyloid-β protein [107, 121] and that APOE4 exacerbates such stress due to decreased glucose metabolism [49, 126]. Since methylglyoxal produced in the early stage of glycation is the substance of ALDH2 (see Sect. 2.1), it is reasonable that *ALDH2*2* and *APOE4* possession results in an unfavorable phenotype. Age-related cognitive impairment due to a lack of ALDH2 activity was reproduced in *Aldh2* knockout mice, which exhibited age-dependent cognitive dysfunction accompanied by Alzheimer's disease-like pathology, such as increased amyloid-beta, phosphorylation of protein tau, and neurodegeneration [16].

The deterioration of many other diseases due to *ALHD2*2* has also been reported. For example, the risk of vasospastic angina is increased in *ALHD2*2* carriers [81–84], and this risk increases with drinking and smoking habits [82, 104]. Another study showed that *ALDH2*2* accelerates the progression of Fanconi anemia, a genomic instability disorder attributed to a series of genetic mutations [37], which seems to be a result of the exacerbation of aldehyde removal disorder [57]. There are also reports showing that the risk of osteoporosis in elderly women carrying *ALHD2*2/*2* is relatively high [114, 129].

4.2 *Favorable Outcomes of* ALDH2*2

Confounding the expectations of medical professionals, many reports have found that various physical and biochemical indicators are preferable in *ALDH2*2* carriers, such as the body mass index [47, 124], waist-to-hip ratio [14, 124], visceral fat volume, blood pressure [4, 39, 47, 113], blood sugar level [47, 116], blood triglyceride level [103, 117, 125], blood low-density lipoprotein cholesterol level [103], and serum uric acid level [66, 101, 102, 136]. These effects may reflect a smaller amount of alcohol consumption in *ALDH2*2* carriers even with an adjustment for a drinking habit (residual confounding), while, experimental approaches suggest the involvement of interactions between *ALDH2*2* and drinking habit or the effects of *ALDH2*2* itself [18, 23, 54, 106, 120, 130] (Table 1).

The serum levels of aspartate aminotransferase (AST), alanine aminotransferase (ALT), and γ-glutamyl-transpeptidase (GGT), which are indicators of misuse of alcohol, tend to remain low in *ALDH2*2* carriers [87, 115, 131]. This has been confirmed repeatedly in animal experiments with various methods of ethanol loading, such as binge and chronic administration, high-fat diet models, and additional administration of carbon tetrachloride. In every case, the serum ALT level in *Aldh2* knockout mice has been relatively low [54, 74], and this low ALT value lasts for as long as the ethanol treatment continues [77]. A common finding in animals and humans is that the ALT value even decreases below baseline levels when the ethanol intake is only a loading factor [74, 77, 115]. Because *ALDH2*2* carriers have a high risk of drinking-related cancer, it is important not to underestimate the amount of alcohol consumption by focusing solely on the serum AST, ALT, and GGT levels [76].

Various other diseases have been reported to show preferable outcomes in *ALHD2*2* subjects. As animal experiments suggest [94], the risk of depression/anxiety disorder is relatively low in Japanese subjects [141, 142]. Bipolar disorder shows a similar trend [61, 62], although the mechanism remains to be clarified. It has been reported that diabetic vascular disease is relatively mild in *ALDH2*2* carriers [40, 111], and consistent results in animal experiments have been obtained; delayed angiogenesis in *Aldh2* knockout mice was shown with an artificial ischemia model [152] (unfavorable aspect of the delayed angiogenesis). Furthermore, other experimental models [127, 150] and epidemiological investigations [149] regarding cardiac hypertrophy have obtained supportive findings. Blunted signaling for cell proliferation (e.g., PI3K/Akt signaling) [127] and the alleviation of age-related autophagy dysfunction [150] in *ALDH2*2* carriers have been suggested as possible mechanisms underlying these observations.

5 Conclusion and Perspective

The basic details of the prevalent gene polymorphism of *ALDH2* in East Asia, rs671, have been summarized. This polymorphism is important for disease prevention for the following reasons: (1) more than half a billion people carry the variant allele, *ALDH2**2*; (2) *ALDH2**2* shows various and bidirectional health effects; (3) the health effects appear due to overlapping environmental factors and each individual's vulnerability; and (4) ALDH2 substrates are commonly found in daily life and workplace environments as well as in prescription drugs. *ALDH2* genotype-based personalized measures enhance the efficiency of disease prevention and treatments; however, few attempts have been made so far, possibly because of the lack of supportive evidence. Another potential reason is that the biological mechanisms are not widely understood. One of the factors impeding study progress may be the strong modification of the drinking habit due to the *ALDH2* polymorphism; this is a serious obstacle to epidemiological surveys, as the target outcomes are often affected by subjects' drinking habits. To resolve these issues, large-scale studies with analyses stratified by drinking habit are required. An experimental approach that controls ethanol exposure is also an effective solution, but this has other associated limitations. In particular, experiments using the temporary knockdown of *ALDH2* or acetaldehyde loading may be inappropriate when the biological remodeling (see Sect. 3) has a primary effect on the outcomes. After overcoming these difficulties, further evidence should be accumulated in order to administer personalized and effective medicine based on *ALDH2* polymorphisms.

Acknowledgments This work was supported by JSPS Grant-in-Aid for Scientific Research (C) Grant Number 17K09115 and by the Uehara Memorial Foundation Research Grant.

Conflicts of Interest The authors declare no conflicts of interest.

References

1. Abe H, Aida Y, Seki N, Sugita T, Tomita Y, Nagano T, Itagaki M, Sutoh S, Nagatsuma K, Itoh K, Matsuura T, Aizawa Y (2015) Aldehyde dehydrogenase 2 polymorphism influences development to hepatocellular carcinoma in East Asian alcoholic liver cirrhosis. J Gastroenterol Hepatol 30(9):1376–1383. https://doi.org/10.1111/jgh.12948
2. Adachi J, Mizoi Y (1983) Acetaldehyde-mediated alcohol sensitivity and elevation of plasma catecholamine in man. Jpn J Pharmacol 33(3):531–539
3. Ahmad I, Shukla S, Singh D, Chauhan AK, Kumar V, Singh BK, Patel DK, Pandey HP, Singh C (2014) CYP2E1-mediated oxidative stress regulates HO-1 and GST expression in maneb- and paraquat-treated rat polymorphonuclear leukocytes. Mol Cell Biochem 393(1–2):209–222. https://doi.org/10.1007/s11010-014-2062-y
4. Amamoto K, Okamura T, Tamaki S, Kita Y, Tsujita Y, Kadowaki T, Nakamura Y, Ueshima H (2002) Epidemiologic study of the association of low-Km mitochondrial acetaldehyde dehydrogenase genotypes with blood pressure level and the prevalence of hypertension in a general population. Hypertens Res 25(6):857–864

5. Amanuma Y, Ohashi S, Itatani Y, Tsurumaki M, Matsuda S, Kikuchi O, Nakai Y, Miyamoto S, Oyama T, Kawamoto T, Whelan KA, Nakagawa H, Chiba T, Matsuda T, Muto M (2015) Protective role of ALDH2 against acetaldehyde-derived DNA damage in oesophageal squamous epithelium. Sci Rep 5:14142. https://doi.org/10.1038/srep14142
6. Brichac J, Ho KK, Honzatko A, Wang R, Lu X, Weiner H, Picklo MJ Sr (2007) Enantioselective oxidation of trans-4-hydroxy-2-nonenal is aldehyde dehydrogenase isozyme and Mg2+ dependent. Chem Res Toxicol 20(6):887–895. https://doi.org/10.1021/tx7000509
7. Brown ZW, Amit Z, Smith BR, Sutherland EA, Selvaggi N (1983) Alcohol-induced euphoria enhanced by disulfiram and calcium carbimide. Alcohol Clin Exp Res 7(3):276–278
8. Burgos-Barragan G, Wit N, Meiser J, Dingler FA, Pietzke M, Mulderrig L, Pontel LB, Rosado IV, Brewer TF, Cordell RL, Monks PS, Chang CJ, Vazquez A, Patel KJ (2017) Mammals divert endogenous genotoxic formaldehyde into one-carbon metabolism. Nature 548(7669):549–554. https://doi.org/10.1038/nature23481
9. Caro AA, Cederbaum AI (2004) Oxidative stress, toxicology, and pharmacology of CYP2E1. Annu Rev Pharmacol Toxicol 44:27–42. https://doi.org/10.1146/annurev.pharmtox.44.101802.121704
10. Chen NH, Djoko KY, Veyrier FJ, McEwan AG (2016) Formaldehyde stress responses in bacterial pathogens. Front Microbiol 7:257. https://doi.org/10.3389/fmicb.2016.00257
11. Chen Y, Singh S, Matsumoto A, Manna SK, Abdelmegeed MA, Golla S, Murphy RC, Dong H, Song BJ, Gonzalez FJ, Thompson DC, Vasiliou V (2016) Chronic glutathione depletion confers protection against alcohol-induced steatosis: implication for redox activation of AMP-activated protein kinase pathway. Sci Rep 6:29743. https://doi.org/10.1038/srep29743
12. Chen YJ, Chen C, Wu DC, Lee CH, Wu CI, Lee JM, Goan YG, Huang SP, Lin CC, Li TC, Chou YP, Wu MT (2006) Interactive effects of lifetime alcohol consumption and alcohol and aldehyde dehydrogenase polymorphisms on esophageal cancer risks. Int J Cancer 119(12):2827–2831
13. Chiang CP, Jao SW, Lee SP, Chen PC, Chung CC, Lee SL, Nieh S, Yin SJ (2012) Expression pattern, ethanol-metabolizing activities, and cellular localization of alcohol and aldehyde dehydrogenases in human large bowel: association of the functional polymorphisms of ADH and ALDH genes with hemorrhoids and colorectal cancer. Alcohol (Fayetteville, NY) 46(1):37–49. https://doi.org/10.1016/j.alcohol.2011.08.004
14. Cho Y, Shin SY, Won S, Relton CL, Davey Smith G, Shin MJ (2015) Alcohol intake and cardiovascular risk factors: a Mendelian randomisation study. Sci Rep 5:18422. https://doi.org/10.1038/srep18422
15. Cui R, Kamatani Y, Takahashi A, Usami M, Hosono N, Kawaguchi T, Tsunoda T, Kamatani N, Kubo M, Nakamura Y, Matsuda K (2009) Functional variants in ADH1B and ALDH2 coupled with alcohol and smoking synergistically enhance esophageal cancer risk. Gastroenterology 137(5):1768–1775. doi:S0016-5085(09)01425-5 [pii] https://doi.org/10.1053/j.gastro.2009.07.070
16. D'Souza Y, Elharram A, Soon-Shiong R, Andrew RD, Bennett BM (2015) Characterization of Aldh2 mice as an age-related model of cognitive impairment and Alzheimer's disease. Mol Brain 8(1):27. https://doi.org/10.1186/s13041-015-0117-y
17. Diana M, Peana AT, Sirca D, Lintas A, Melis M, Enrico P (2008) Crucial role of acetaldehyde in alcohol activation of the mesolimbic dopamine system. Ann N Y Acad Sci 1139:307–317. https://doi.org/10.1196/annals.1432.009
18. Endo J, Sano M, Katayama T, Hishiki T, Shinmura K, Morizane S, Matsuhashi T, Katsumata Y, Zhang Y, Ito H, Nagahata Y, Marchitti S, Nishimaki K, Wolf AM, Nakanishi H, Hattori F, Vasiliou V, Adachi T, Ohsawa I, Taguchi R, Hirabayashi Y, Ohta S, Suematsu M, Ogawa S, Fukuda K (2009) Metabolic remodeling induced by mitochondrial aldehyde stress stimulates tolerance to oxidative stress in the heart. Circ Res 105(11):1118–1127. https://doi.org/10.1161/CIRCRESAHA.109.206607

19. Enrico P, Sirca D, Mereu M, Peana AT, Lintas A, Golosio A, Diana M (2009) Acetaldehyde sequestering prevents ethanol-induced stimulation of mesolimbic dopamine transmission. Drug Alcohol Depend 100(3):265–271. https://doi.org/10.1016/j.drugalcdep.2008.10.010
20. Esterbauer H, Schaur RJ, Zollner H (1991) Chemistry and biochemistry of 4-hydroxynonenal, malonaldehyde and related aldehydes. Free Radic Biol Med 11(1):81–128
21. Farres J, Wang X, Takahashi K, Cunningham SJ, Wang TT, Weiner H (1994) Effects of changing glutamate 487 to lysine in rat and human liver mitochondrial aldehyde dehydrogenase. A model to study human (oriental type) class 2 aldehyde dehydrogenase. J Biol Chem 269(19):13854–13860
22. Fitzmaurice AG, Rhodes SL, Cockburn M, Ritz B, Bronstein JM (2014) Aldehyde dehydrogenase variation enhances effect of pesticides associated with Parkinson disease. Neurology 82(5):419–426. https://doi.org/10.1212/WNL.0000000000000083
23. Fox IH, Palella TD, Kelley WN (1987) Hyperuricemia: a marker for cell energy crisis. N Engl J Med 317(2):111–112. https://doi.org/10.1056/NEJM198707093170209
24. Gao Y, Xu Y, Hua S, Zhou S, Wang K (2015) ALDH2 attenuates Dox-induced cardiotoxicity by inhibiting cardiac apoptosis and oxidative stress. Int J Clin Exp Med 8(5):6794–6803
25. Garaycoechea JI, Crossan GP, Langevin F, Daly M, Arends MJ, Patel KJ (2012) Genotoxic consequences of endogenous aldehydes on mouse haematopoietic stem cell function. Nature 489(7417):571–575. https://doi.org/10.1038/nature11368
26. Goedde HW, Agarwal DP (1990) Pharmacogenetics of aldehyde dehydrogenase (ALDH). Pharmacol Ther 45(3):345–371
27. Goedde HW, Agarwal DP, Fritze G, Meier-Tackmann D, Singh S, Beckmann G, Bhatia K, Chen LZ, Fang B, Lisker R et al (1992) Distribution of ADH2 and ALDH2 genotypes in different populations. Hum Genet 88(3):344–346
28. Goedde HW, Agarwal DP, Harada S (1979) Alcohol metabolizing enzymes: studies of isozymes in human biopsies and cultured fibroblasts. Clin Genet 16(1):29–33
29. Gonzalez FJ (2007) The 2006 Bernard B. Brodie Award Lecture. Cyp2e1. Drug Metab Dispos 35(1):1–8. doi:dmd.106.012492 [pii] https://doi.org/10.1124/dmd.106.012492
30. Grilo NM, Charneira C, Pereira SA, Monteiro EC, Marques MM, Antunes AM (2014) Bioactivation to an aldehyde metabolite—possible role in the onset of toxicity induced by the anti-HIV drug abacavir. Toxicol Lett 224(3):416–423. https://doi.org/10.1016/j.toxlet.2013.10.036
31. Guo YM, Wang Q, Liu YZ, Chen HM, Qi Z, Guo QH (2008) Genetic polymorphisms in cytochrome P4502E1, alcohol and aldehyde dehydrogenases and the risk of esophageal squamous cell carcinoma in Gansu Chinese males. World J Gastroenterol 14(9):1444–1449
32. Hashibe M, Boffetta P, Zaridze D, Shangina O, Szeszenia-Dabrowska N, Mates D, Janout V, Fabianova E, Bencko V, Moullan N, Chabrier A, Hung R, Hall J, Canzian F, Brennan P (2006) Evidence for an important role of alcohol- and aldehyde-metabolizing genes in cancers of the upper aerodigestive tract. Cancer Epidemiol Biomark Prev 15(4):696–703. doi:15/4/696 [pii] https://doi.org/10.1158/1055-9965.EPI-05-0710
33. Hashimoto M, Watanabe M, Uematsu Y, Hattori S, Miyai N, Utsumi M, Oka M, Hayashida M, Kinoshita K, Arita M, Takeshita T (2016) Relationships of alcohol dehydrogenase 1B (ADH1B) and aldehyde dehydrogenase 2 (ALDH2) genotypes with alcohol sensitivity, drinking behavior and problem drinking in Japanese older men. Environ Health Prev Med. https://doi.org/10.1007/s12199-016-0507-5
34. Haug S, Braun-Falco M (2006) Restoration of fatty aldehyde dehydrogenase deficiency in Sjogren-Larsson syndrome. Gene Ther 13(13):1021–1026. https://doi.org/10.1038/sj.gt.3302743
35. Hayashida M, Kamada Y, Ota T, Kojima S, Iwao-Koizumi K, Murata S, Kinoshita K (2015) Associations between ALDH2 and ADH1B genotypes and ethanol-induced cutaneous erythema in young Japanese women. Nihon Eiseigaku Zasshi Jpn J Hyg 70(2):134–138. https://doi.org/10.1265/jjh.70.134

36. Higuchi S, Matsushita S, Murayama M, Takagi S, Hayashida M (1995) Alcohol and aldehyde dehydrogenase polymorphisms and the risk for alcoholism. Am J Psychiatry 152(8):1219–1221
37. Hira A, Yabe H, Yoshida K, Okuno Y, Shiraishi Y, Chiba K, Tanaka H, Miyano S, Nakamura J, Kojima S, Ogawa S, Matsuo K, Takata M, Yabe M (2013) Variant ALDH2 is associated with accelerated progression of bone marrow failure in Japanese Fanconi anemia patients. Blood 122(18):3206–3209. https://doi.org/10.1182/blood-2013-06-507962
38. Hu N, Zhang Y, Nair S, Culver BW, Ren J (2014) Contribution of ALDH2 polymorphism to alcoholism-associated hypertension. Recent Pat Endocr Metab Immune Drug Discov 8(3):180–185
39. Hui P, Nakayama T, Morita A, Sato N, Hishiki M, Saito K, Yoshikawa Y, Tamura M, Sato I, Takahashi T, Soma M, Izumi Y, Ozawa Y, Cheng Z (2007) Common single nucleotide polymorphisms in Japanese patients with essential hypertension: aldehyde dehydrogenase 2 gene as a risk factor independent of alcohol consumption. Hypertens Res 30(7):585–592
40. Idewaki Y, Iwase M, Fujii H, Ohkuma T, Ide H, Kaizu S, Jodai T, Kikuchi Y, Hirano A, Nakamura U, Kubo M, Kitazono T (2015) Association of genetically determined aldehyde dehydrogenase 2 activity with diabetic complications in relation to alcohol consumption in Japanese patients with Type 2 diabetes mellitus: the Fukuoka diabetes registry. PLoS One 10(11):e0143288. https://doi.org/10.1371/journal.pone.0143288
41. Israel Y, Quintanilla ME, Karahanian E, Rivera-Meza M, Herrera-Marschitz M (2015) The "first hit" toward alcohol reinforcement: role of ethanol metabolites. Alcohol Clin Exp Res 39(5):776–786. https://doi.org/10.1111/acer.12709
42. Isse T, Oyama T, Kitagawa K, Matsuno K, Matsumoto A, Yoshida A, Nakayama K, Nakayama K, Kawamoto T (2002) Diminished alcohol preference in transgenic mice lacking aldehyde dehydrogenase activity. Pharmacogenetics 12(8):621–626
43. Izaguirre G, Kikonyogo A, Pietruszko R (1998) Methylglyoxal as substrate and inhibitor of human aldehyde dehydrogenase: comparison of kinetic properties among the three isozymes. Comp Biochem Physiol B Biochem Mol Biol 119(4):747–754
44. Jackson B, Brocker C, Thompson DC, Black W, Vasiliou K, Nebert DW, Vasiliou V (2011) Update on the aldehyde dehydrogenase gene (ALDH) superfamily. Hum Genomics 5(4):283–303
45. Kalapos MP (1994) Methylglyoxal toxicity in mammals. Toxicol Lett 73(1):3–24
46. Kamino K, Nagasaka K, Imagawa M, Yamamoto H, Yoneda H, Ueki A, Kitamura S, Namekata K, Miki T, Ohta S (2000) Deficiency in mitochondrial aldehyde dehydrogenase increases the risk for late-onset Alzheimer's disease in the Japanese population. Biochem Biophys Res Commun 273(1):192–196
47. Kato N, Takeuchi F, Tabara Y, Kelly TN, Go MJ, Sim X, Tay WT, Chen CH, Zhang Y, Yamamoto K, Katsuya T, Yokota M, Kim YJ, Ong RT, Nabika T, Gu D, Chang LC, Kokubo Y, Huang W, Ohnaka K, Yamori Y, Nakashima E, Jaquish CE, Lee JY, Seielstad M, Isono M, Hixson JE, Chen YT, Miki T, Zhou X, Sugiyama T, Jeon JP, Liu JJ, Takayanagi R, Kim SS, Aung T, Sung YJ, Zhang X, Wong TY, Han BG, Kobayashi S, Ogihara T, Zhu D, Iwai N, Wu JY, Teo YY, Tai ES, Cho YS, He J (2011) Meta-analysis of genome-wide association studies identifies common variants associated with blood pressure variation in east Asians. Nat Genet 43(6):531–538. https://doi.org/10.1038/ng.834
48. Kato S, Tajiri T, Matsukura N, Matsuda N, Taniai N, Mamada H, Yoshida H, Kiyam T, Naito Z (2003) Genetic polymorphisms of aldehyde dehydrogenase 2, cytochrome p450 2E1 for liver cancer risk in HCV antibody-positive Japanese patients and the variations of CYP2E1 mRNA expression levels in the liver due to its polymorphism. Scand J Gastroenterol 38(8):886–893
49. Keeney JT, Ibrahimi S, Zhao L (2015) Human ApoE isoforms differentially modulate glucose and amyloid metabolic pathways in female brain: evidence of the mechanism of neuroprotection by ApoE2 and implications for Alzheimer's disease prevention and early intervention. J Alzheimers Dis 48(2):411–424. https://doi.org/10.3233/JAD-150348

50. Kim J, Chen CH, Yang J, Mochly-Rosen D (2017) Aldehyde dehydrogenase 2*2 knock-in mice show increased reactive oxygen species production in response to cisplatin treatment. J Biomed Sci 24(1):33. https://doi.org/10.1186/s12929-017-0338-8
51. Kitagawa K, Kawamoto T, Kunugita N, Tsukiyama T, Okamoto K, Yoshida A, Nakayama K, Nakayama K (2000) Aldehyde dehydrogenase (ALDH) 2 associates with oxidation of methoxyacetaldehyde; in vitro analysis with liver subcellular fraction derived from human and Aldh2 gene targeting mouse. FEBS Lett 476(3):306–311
52. Koyanagi YN, Ito H, Oze I, Hosono S, Tanaka H, Abe T, Shimizu Y, Hasegawa Y, Matsuo K (2016) Development of a prediction model and estimation of cumulative risk for upper aerodigestive tract cancer on the basis of the aldehyde dehydrogenase 2 genotype and alcohol consumption in a Japanese population. Eur J Cancer Prev. https://doi.org/10.1097/CEJ.0000000000000222
53. Kunitoh S, Imaoka S, Hiroi T, Yabusaki Y, Monna T, Funae Y (1997) Acetaldehyde as well as ethanol is metabolized by human CYP2E1. J Pharmacol Exp Ther 280(2):527–532
54. Kwon HJ, Won YS, Park O, Chang B, Duryee MJ, Thiele GE, Matsumoto A, Singh S, Abdelmegeed MA, Song BJ, Kawamoto T, Vasiliou V, Thiele GM, Gao B (2014) Aldehyde dehydrogenase 2 deficiency ameliorates alcoholic fatty liver but worsens liver inflammation and fibrosis in mice. Hepatology (Baltimore, MD). https://doi.org/10.1002/hep.27036
55. Lai CL, Yao CT, Chau GY, Yang LF, Kuo TY, Chiang CP, Yin SJ (2014) Dominance of the inactive Asian variant over activity and protein contents of mitochondrial aldehyde dehydrogenase 2 in human liver. Alcohol Clin Exp Res 38(1):44–50. https://doi.org/10.1111/acer.12215
56. Landin JS, Cohen SD, Khairallah EA (1996) Identification of a 54-kDa mitochondrial acetaminophen-binding protein as aldehyde dehydrogenase. Toxicol Appl Pharmacol 141(1):299–307. doi:S0041-008X(96)90287-2 [pii] https://doi.org/10.1006/taap.1996.0287
57. Langevin F, Crossan GP, Rosado IV, Arends MJ, Patel KJ (2011) Fancd2 counteracts the toxic effects of naturally produced aldehydes in mice. Nature 475(7354):53–58. https://doi.org/10.1038/nature10192
58. Larson HN, Zhou J, Chen Z, Stamler JS, Weiner H, Hurley TD (2007) Structural and functional consequences of coenzyme binding to the inactive Asian variant of mitochondrial aldehyde dehydrogenase: roles of residues 475 and 487. J Biol Chem 282(17):12940–12950. https://doi.org/10.1074/jbc.M607959200
59. Lee SL, Lee YP, Wu ML, Chi YC, Liu CM, Lai CL, Yin SJ (2015) Inhibition of human alcohol and aldehyde dehydrogenases by aspirin and salicylate: assessment of the effects on first-pass metabolism of ethanol. Biochem Pharmacol 95(1):71–79. https://doi.org/10.1016/j.bcp.2015.03.003
60. Lee SY, Chen SL, Chang YH, Chen PS, Huang SY, Tzeng NS, Wang LJ, Lee IH, Wang TY, Chen KC, Yang YK, Hong JS, Lu RB (2015) ALDH2 polymorphism, associated with attenuating negative symptoms in patients with schizophrenia treated with add-on dextromethorphan. J Psychiatr Res 69:50–56. https://doi.org/10.1016/j.jpsychires.2015.07.027
61. Lee SY, Chen SL, Chang YH, Chen SH, Chu CH, Huang SY, Tzeng NS, Wang CL, Lee IH, Yeh TL, Yang YK, Lu RB (2010) The ALDH2 and DRD2/ANKK1 genes interacted in bipolar II but not bipolar I disorder. Pharmacogenet Genomics 20(8):500–506. https://doi.org/10.1097/FPC.0b013e32833caa2b
62. Lee SY, Chen SL, Chang YH, Chu CH, Huang SY, Tzeng NS, Wang CL, Lin SH, Lee IH, Yeh TL, Yang YK, Lu RB (2012) The ALDH2 and 5-HT2A genes interacted in bipolar-I but not bipolar-II disorder. Prog Neuro-Psychopharmacol Biol Psychiatry 38(2):247–251. https://doi.org/10.1016/j.pnpbp.2012.04.005
63. Lee YP, Liao JT, Cheng YW, Wu TL, Lee SL, Liu JK, Yin SJ (2013) Inhibition of human alcohol and aldehyde dehydrogenases by acetaminophen: assessment of the effects on first-pass metabolism of ethanol. Alcohol (Fayetteville, NY) 47(7):559–565. https://doi.org/10.1016/j.alcohol.2013.09.001

64. Li H, Borinskaya S, Yoshimura K, Kal'ina N, Marusin A, Stepanov VA, Qin Z, Khaliq S, Lee MY, Yang Y, Mohyuddin A, Gurwitz D, Mehdi SQ, Rogaev E, Jin L, Yankovsky NK, Kidd JR, Kidd KK (2009) Refined geographic distribution of the oriental ALDH2*504Lys (nee 487Lys) variant. Ann Hum Genet 73(Pt 3):335–345
65. Li Y, Zhang D, Jin W, Shao C, Yan P, Xu C, Sheng H, Liu Y, Yu J, Xie Y, Zhao Y, Lu D, Nebert DW, Harrison DC, Huang W, Jin L (2006) Mitochondrial aldehyde dehydrogenase-2 (ALDH2) Glu504Lys polymorphism contributes to the variation in efficacy of sublingual nitroglycerin. J Clin Invest 116(2):506–511
66. Li Z, Zhou Z, Hou X, Lu D, Yuan X, Lu J, Wang C, Han L, Cui L, Liu Z, Chen J, Cheng X, Zhang K, Ji J, Jia Z, Ma L, Xin Y, Liu T, Yu Q, Ren W, Wang X, Li X, Mi QS, Shi Y, Li C (2017) Replication of gout/urate concentrations GWAS susceptibility loci associated with gout in a Han Chinese population. Sci Rep 7(1):4094. https://doi.org/10.1038/s41598-017-04127-4
67. Liu W, Zhai X, Wang W, Zheng B, Zhang Z, Fan X, Chen Y, Wang J (2018) Aldehyde dehydrogenase 2 activation ameliorates cyclophosphamide-induced acute cardiotoxicity via detoxification of toxic aldehydes and suppression of cardiac cell death. J Mol Cell Cardiol 121:134–144. https://doi.org/10.1016/j.yjmcc.2018.07.006
68. Lu Y, Zhang XH, Cederbaum AI (2012) Ethanol induction of CYP2A5: role of CYP2E1-ROS-Nrf2 pathway. Toxicol Sci 128(2):427–438. https://doi.org/10.1093/toxsci/kfs164
69. Marchitti SA, Brocker C, Stagos D, Vasiliou V (2008) Non-P450 aldehyde oxidizing enzymes: the aldehyde dehydrogenase superfamily. Expert Opin Drug Metab Toxicol 4(6):697–720
70. Matsuda T, Matsumoto A, Uchida M, Kanaly RA, Misaki K, Shibutani S, Kawamoto T, Kitagawa K, Nakayama KI, Tomokuni K, Ichiba M (2007) Increased formation of hepatic N2-ethylidene-2′-deoxyguanosine DNA adducts in aldehyde dehydrogenase 2-knockout mice treated with ethanol. Carcinogenesis 28(11):2363–2366. doi:bgm057 [pii] https://doi.org/10.1093/carcin/bgm057
71. Matsumoto A (2016) Fundamental properties of aldehyde dehydrogenase 2 (ALDH2) and the importance of the ALDH2 polymorphism. Nihon eiseigaku zasshi Jpn J Hyg 71(1):55–68. https://doi.org/10.1265/jjh.71.55
72. Matsumoto A (2018) Importance of an aldehyde dehydrogenase 2 polymorphism in preventive medicine. Nihon eiseigaku zasshi Jpn J Hyg 73(1):9–20. https://doi.org/10.1265/jjh.73.9
73. Matsumoto A, Kawamoto T, Horita M, Takahashi T, Isse T, Oyama T, Ichiba M (2007) Single-dose ethanol administration downregulates expression of cytochrome p450 2E1 mRNA in aldehyde dehydrogenase 2 knockout mice. Alcohol (Fayetteville, NY) 41(8):587–589. https://doi.org/10.1016/j.alcohol.2007.08.002
74. Matsumoto A, Kawamoto T, Mutoh F, Isse T, Oyama T, Kitagawa K, Nakayama KI, Ichiba M (2008) Effects of 5-week ethanol feeding on the liver of aldehyde dehydrogenase 2 knockout mice. Pharmacogenet Genomics 18(10):847–852
75. Matsumoto A, Thompson D, Chen Y, Vasiliou V, Kawamoto T, Ichiba M (2016) Heme oxygenase 1 protects ethanol-administered liver tissue in Aldh2 knockout mice. Alcohol (Fayetteville, NY) 52:49–54. https://doi.org/10.1016/j.alcohol.2016.02.004
76. Matsumoto A, Thompson DC, Chen Y, Kitagawa K, Vasiliou V (2016) Roles of defective ALDH2 polymorphism on liver protection and cancer development. Environ Health Prev Med 21(6):395–402. https://doi.org/10.1007/s12199-016-0579-2
77. Matsumoto A, Vasiliou V, Kawamoto T, Tanaka K, Ichiba M (2014) Ethanol reduces lifespan, body weight, and serum alanine aminotransferase level of aldehyde dehydrogenase 2 knockout mouse. Alcohol Clin Exp Res 38(7):1883–1893. https://doi.org/10.1111/acer.12462
78. Matsuo K, Wakai K, Hirose K, Ito H, Saito T, Tajima K (2006) Alcohol dehydrogenase 2 His47Arg polymorphism influences drinking habit independently of aldehyde dehydrogenase 2 Glu487Lys polymorphism: analysis of 2,299 Japanese subjects. Cancer Epidemiol Biomark Prev 15(5):1009–1013. https://doi.org/10.1158/1055-9965.EPI-05-0911
79. Mills PB, Struys E, Jakobs C, Plecko B, Baxter P, Baumgartner M, Willemsen MA, Omran H, Tacke U, Uhlenberg B, Weschke B, Clayton PT (2006) Mutations in antiquitin in individuals

with pyridoxine-dependent seizures. Nat Med 12(3):307–309. https://doi.org/10.1038/nm1366

80. Mizoi Y, Tatsuno Y, Adachi J, Kogame M, Fukunaga T, Fujiwara S, Hishida S, Ijiri I (1983) Alcohol sensitivity related to polymorphism of alcohol-metabolizing enzymes in Japanese. Pharmacol Biochem Behav 18(Suppl 1):127–133
81. Mizuno Y, Harada E, Morita S, Kinoshita K, Hayashida M, Shono M, Morikawa Y, Murohara T, Nakayama M, Yoshimura M, Yasue H (2015) East asian variant of aldehyde dehydrogenase 2 is associated with coronary spastic angina: possible roles of reactive aldehydes and implications of alcohol flushing syndrome. Circulation 131(19):1665–1673. https://doi.org/10.1161/CIRCULATIONAHA.114.013120
82. Mizuno Y, Hokimoto S, Harada E, Kinoshita K, Nakagawa K, Yoshimura M, Ogawa H, Yasue H (2016) Variant aldehyde dehydrogenase 2 (ALDH2*2) is a risk factor for coronary spasm and ST-segment elevation myocardial infarction. J Am Heart Assoc 5(5). https://doi.org/10.1161/JAHA.116.003247
83. Mizuno Y, Hokimoto S, Harada E, Kinoshita K, Yoshimura M, Yasue H (2016) Variant aldehyde dehydrogenase 2 (ALDH2*2) in east Asians interactively exacerbates tobacco smoking risk for coronary spasm—possible role of reactive aldehydes. Circ J 81(1):96–102. https://doi.org/10.1253/circj.CJ-16-0969
84. Mizuno Y, Morita S, Harada E, Shono M, Morikawa Y, Murohara T, Yasue H (2013) Alcohol flushing and positive ethanol patch test in patients with coronary spastic angina: possible role of aldehyde dehydrogenase 2 polymorphisms. Intern Med 52(23):2593–2598
85. Morita M, Oyama T, Kagawa N, Nakata S, Ono K, Sugaya M, Uramoto H, Yoshimatsu T, Hanagiri T, Sugio K, Kakeji Y, Yasumoto K (2005) Expression of aldehyde dehydrogenase 2 in the normal esophageal epithelium and alcohol consumption in patients with esophageal cancer. Front Biosci 10:2319–2324
86. Munaka M, Kohshi K, Kawamoto T, Takasawa S, Nagata N, Itoh H, Oda S, Katoh T (2003) Genetic polymorphisms of tobacco- and alcohol-related metabolizing enzymes and the risk of hepatocellular carcinoma. J Cancer Res Clin Oncol 129(6):355–360. https://doi.org/10.1007/s00432-003-0439-5
87. Murata C, Watanabe T, Furuya H, Sugioka Y, Mikurube H, Yokoyama A, Atsumi Y, Matsuoka K, Okazaki I (2003) Aldehyde dehydrogenase 2 and beta3-adrenergic receptor gene polymorphisms: their association with elevated liver enzymes and metabolic syndrome. Metab Clin Exp 52(9):1096–1101
88. Nagayoshi H, Matsumoto A, Nishi R, Kawamoto T, Ichiba M, Matsuda T (2009) Increased formation of gastric N(2)-ethylidene-2′-deoxyguanosine DNA adducts in aldehyde dehydrogenase-2 knockout mice treated with ethanol. Mutat Res 673 (1):74–77. doi:S1383-5718(08)00359-8 [pii] doi:https://doi.org/10.1016/j.mrgentox.2008.11.009
89. Nakamura Y, Amamoto K, Tamaki S, Okamura T, Tsujita Y, Ueno Y, Kita Y, Kinoshita M, Ueshima H (2002) Genetic variation in aldehyde dehydrogenase 2 and the effect of alcohol consumption on cholesterol levels. Atherosclerosis 164(1):171–177
90. Network NSG, International Stroke Genetics C (2016) Loci associated with ischaemic stroke and its subtypes (SiGN): a genome-wide association study. Lancet Neurol 15(2):174–184. https://doi.org/10.1016/S1474-4422(15)00338-5
91. O'Brien PJ, Siraki AG, Shangari N (2005) Aldehyde sources, metabolism, molecular toxicity mechanisms, and possible effects on human health. Crit Rev Toxicol 35(7):609–662
92. Ohta S, Ohsawa I (2006) Dysfunction of mitochondria and oxidative stress in the pathogenesis of Alzheimer's disease: on defects in the cytochrome c oxidase complex and aldehyde detoxification. J Alzheimers Dis 9(2):155–166
93. Orrico A, Hipolito L, Sanchez-Catalan MJ, Marti-Prats L, Zornoza T, Granero L, Polache A (2013) Efficacy of D-penicillamine, a sequestering acetaldehyde agent, in the prevention of alcohol relapse-like drinking in rats. Psychopharmacology 228(4):563–575. https://doi.org/10.1007/s00213-013-3065-1

94. Overstreet DH, Knapp DJ, Breese GR, Diamond I (2009) A selective ALDH-2 inhibitor reduces anxiety in rats. Pharmacol Biochem Behav 94(2):255–261. https://doi.org/10.1016/j.pbb.2009.09.004
95. Oze I, Matsuo K, Hosono S, Ito H, Kawase T, Watanabe M, Suzuki T, Hatooka S, Yatabe Y, Hasegawa Y, Shinoda M, Tajima K, Tanaka H (2010) Comparison between self-reported facial flushing after alcohol consumption and ALDH2 Glu504Lys polymorphism for risk of upper aerodigestive tract cancer in a Japanese population. Cancer Sci 101(8):1875–1880. https://doi.org/10.1111/j.1349-7006.2010.01599.x
96. Park JY, Matsuo K, Suzuki T, Ito H, Hosono S, Kawase T, Watanabe M, Oze I, Hida T, Yatabe Y, Mitsudomi T, Takezaki T, Tajima K, Tanaka H (2010) Impact of smoking on lung cancer risk is stronger in those with the homozygous aldehyde dehydrogenase 2 null allele in a Japanese population. Carcinogenesis 31(4):660–665. doi:bgq021 [pii] https://doi.org/10.1093/carcin/bgq021
97. Park MH, Igarashi K (2013) Polyamines and their metabolites as diagnostic markers of human diseases. Biomol Ther (Seoul) 21(1):1–9. https://doi.org/10.4062/biomolther.2012.097
98. Peng GS, Yin SJ (2009) Effect of the allelic variants of aldehyde dehydrogenase ALDH2*2 and alcohol dehydrogenase ADH1B*2 on blood acetaldehyde concentrations. Hum Genomics 3(2):121–127
99. Pikkarainen PH (1971) Aldehyde-oxidizing capacity during development in human and rat liver. Ann Med Exp Biol Fenn 49(3):151–156
100. Sakamoto T, Hara M, Higaki Y, Ichiba M, Horita M, Mizuta T, Eguchi Y, Yasutake T, Ozaki I, Yamamoto K, Onohara S, Kawazoe S, Shigematsu H, Koizumi S, Tanaka K (2006) Influence of alcohol consumption and gene polymorphisms of ADH2 and ALDH2 on hepatocellular carcinoma in a Japanese population. Int J Cancer 118(6):1501–1507
101. Sakiyama M, Matsuo H, Akashi A, Shimizu S, Higashino T, Kawaguchi M, Nakayama A, Naito M, Kawai S, Nakashima H, Sakurai Y, Ichida K, Shimizu T, Ooyama H, Shinomiya N (2017) Independent effects of ADH1B and ALDH2 common dysfunctional variants on gout risk. Sci Rep 7(1):2500. https://doi.org/10.1038/s41598-017-02528-z
102. Sakiyama M, Matsuo H, Nakaoka H, Yamamoto K, Nakayama A, Nakamura T, Kawai S, Okada R, Ooyama H, Shimizu T, Shinomiya N (2016) Identification of rs671, a common variant of ALDH2, as a gout susceptibility locus. Sci Rep 6:25360. https://doi.org/10.1038/srep25360
103. Sasakabe T, Wakai K, Kawai S, Hishida A, Naito M, Suzuki S, Nindita Y, Arisawa K, Kita Y, Hara M, Kuriyama N, Hirata A, Mikami H, Oze I, Kubo M, Tanaka H, Hamajima N (2017) Modification of the associations of alcohol intake with serum low-density lipoprotein cholesterol and triglycerides by ALDH2 and ADH1B polymorphisms in Japanese men. J Epidemiol. https://doi.org/10.2188/jea.JE20160189
104. Seki T, Okayama H, Isoyama S, Kagaya Y, Shirato K, Munakata K, Kanazawa M, Tamaki K, Hiramoto T, Okayama M, Kasahara S (1999) The role of alcohol dehydrogenase 2 and aldehyde dehydrogenase 2 genotypes in alcohol-induced vasospastic angina. Tohoku J Exp Med 187(4):311–322
105. Setshedi M, Wands JR, Monte SM (2010) Acetaldehyde adducts in alcoholic liver disease. Oxidative Med Cell Longev 3(3):178–185. https://doi.org/10.4161/oxim.3.3.12288
106. Shibata T, Yamashita H, Kawamura T, Jodai Y, Omori T, Sumi K, Ichikawa Y, Okubo M, Ishizuka T, Tahara T, Nagasaka M, Nakagawa Y, Hirata I, Ohmiya N, Nakao M (2016) The effect of consuming small volumes of beer on gastric motility and the involvement of gene polymorphisms. Biomed Res 37(5):305–310. https://doi.org/10.2220/biomedres.37.305
107. Smith MA, Taneda S, Richey PL, Miyata S, Yan SD, Stern D, Sayre LM, Monnier VM, Perry G (1994) Advanced Maillard reaction end products are associated with Alzheimer disease pathology. Proc Natl Acad Sci U S A 91(12):5710–5714
108. Stagos D, Chen Y, Brocker C, Donald E, Jackson BC, Orlicky DJ, Thompson DC, Vasiliou V (2010) Aldehyde dehydrogenase 1B1: molecular cloning and characterization of a novel

mitochondrial acetaldehyde-metabolizing enzyme. Drug Metab Dispos 38(10):1679–1687. https://doi.org/10.1124/dmd.110.034678
109. Steinmetz CG, Xie P, Weiner H, Hurley TD (1997) Structure of mitochondrial aldehyde dehydrogenase: the genetic component of ethanol aversion. Structure 5(5):701–711
110. Stewart MJ, Malek K, Crabb DW (1996) Distribution of messenger RNAs for aldehyde dehydrogenase 1, aldehyde dehydrogenase 2, and aldehyde dehydrogenase 5 in human tissues. J Investig Med 44(2):42–46
111. Suzuki Y, Taniyama M, Muramatsu T, Higuchi S, Ohta S, Atsumi Y, Matsuoka K (2004) ALDH2/ADH2 polymorphism associated with vasculopathy and neuropathy in type 2 diabetes. Alcohol Clin Exp Res 28(8 Suppl Proceedings):111S–116S
112. Svanas GW, Weiner H (1985) Enzymatic requirement for cyanamide inactivation of rat liver aldehyde dehydrogenase. Biochem Pharmacol 34(8):1197–1204
113. Takagi S, Baba S, Iwai N, Fukuda M, Katsuya T, Higaki J, Mannami T, Ogata J, Goto Y, Ogihara T (2001) The aldehyde dehydrogenase 2 gene is a risk factor for hypertension in Japanese but does not alter the sensitivity to pressor effects of alcohol: the Suita study. Hypertens Res 24(4):365–370
114. Takeshima K, Nishiwaki Y, Suda Y, Niki Y, Sato Y, Kobayashi T, Miyamoto K, Uchida H, Inokuchi W, Tsuji T, Funayama A, Nakamura M, Matsumoto M, Toyama Y, Miyamoto T (2017) A missense single nucleotide polymorphism in the ALDH2 gene, rs671, is associated with hip fracture. Sci Rep 7(1):428. https://doi.org/10.1038/s41598-017-00503-2
115. Takeshita T, Yang X, Morimoto K (2000) The ALDH2 genotype, alcohol intake, and liver-function biomarkers among Japanese male workers. Hum Genet 106(6):589–593
116. Takeuchi F, Isono M, Nabika T, Katsuya T, Sugiyama T, Yamaguchi S, Kobayashi S, Ogihara T, Yamori Y, Fujioka A, Kato N (2011) Confirmation of ALDH2 as a Major locus of drinking behavior and of its variants regulating multiple metabolic phenotypes in a Japanese population. Circ J 75(4):911–918
117. Tan A, Sun J, Xia N, Qin X, Hu Y, Zhang S, Tao S, Gao Y, Yang X, Zhang H, Kim ST, Peng T, Lin X, Li L, Mo L, Liang Z, Shi D, Huang Z, Huang X, Liu M, Ding Q, Trent JM, Zheng SL, Mo Z, Xu J (2012) A genome-wide association and gene-environment interaction study for serum triglycerides levels in a healthy Chinese male population. Hum Mol Genet 21(7):1658–1664. https://doi.org/10.1093/hmg/ddr587
118. Tan SLW, Chadha S, Liu Y, Gabasova E, Perera D, Ahmed K, Constantinou S, Renaudin X, Lee M, Aebersold R, Venkitaraman AR (2017) A class of environmental and endogenous toxins induces BRCA2 haploinsufficiency and genome instability. Cell 169(6):1105–1118, e1115. https://doi.org/10.1016/j.cell.2017.05.010
119. Terelius Y, Norsten-Hoog C, Cronholm T, Ingelman-Sundberg M (1991) Acetaldehyde as a substrate for ethanol-inducible cytochrome P450 (CYP2E1). Biochem Biophys Res Commun 179(1):689–694
120. Toda N, Konishi M, Miyazaki M, Komura S (1983) The effects of ethanol and acetaldehyde on dog arterial smooth muscle. J Stud Alcohol 44(1):1–16
121. Vitek MP, Bhattacharya K, Glendening JM, Stopa E, Vlassara H, Bucala R, Manogue K, Cerami A (1994) Advanced glycation end products contribute to amyloidosis in Alzheimer disease. Proc Natl Acad Sci U S A 91(11):4766–4770
122. Wall TL, Thomasson HR, Schuckit MA, Ehlers CL (1992) Subjective feelings of alcohol intoxication in Asians with genetic variations of ALDH2 alleles. Alcohol Clin Exp Res 16(5):991–995
123. Wang B, Wang J, Zhou S, Tan S, He X, Yang Z, Xie YC, Li S, Zheng C, Ma X (2008) The association of mitochondrial aldehyde dehydrogenase gene (ALDH2) polymorphism with susceptibility to late-onset Alzheimer's disease in Chinese. J Neurol Sci 268(1–2):172–175
124. Wang T, Ma X, Peng D, Zhang R, Sun X, Chen M, Yan J, Wang S, Yan D, He Z, Jiang F, Bao Y, Hu C, Jia W (2016) Effects of obesity related genetic variations on visceral and subcutaneous fat distribution in a Chinese population. Sci Rep 6:20691. https://doi.org/10.1038/srep20691

125. Wang Y, Zhang Y, Zhang J, Tang X, Qian Y, Gao P, Zhu D (2013) Association of a functional single-nucleotide polymorphism in the ALDH2 gene with essential hypertension depends on drinking behavior in a Chinese Han population. J Hum Hypertens 27(3):181–186. https://doi.org/10.1038/jhh.2012.15
126. Wu L, Zhang X, Zhao L (2018) Human ApoE isoforms differentially modulate brain glucose and ketone body metabolism: implications for Alzheimer's disease risk reduction and early intervention. J Neurosci 38(30):6665–6681. https://doi.org/10.1523/JNEUROSCI.2262-17.2018
127. Xia G, Fan F, Liu M, Wang S, Wu J, Shen C, Han S, Wang C, Jia J, Zou Y, Hu K, Ge J, Sun A (2016) Aldehyde dehydrogenase 2 deficiency blunts compensatory cardiac hypertrophy through modulating Akt phosphorylation early after transverse aorta constriction in mice. Biochim Biophys Acta 1862(9):1587–1593. https://doi.org/10.1016/j.bbadis.2016.05.019
128. Xiao Q, Weiner H, Crabb DW (1996) The mutation in the mitochondrial aldehyde dehydrogenase (ALDH2) gene responsible for alcohol-induced flushing increases turnover of the enzyme tetramers in a dominant fashion. J Clin Invest 98(9):2027–2032. https://doi.org/10.1172/JCI119007
129. Yamaguchi J, Hasegawa Y, Kawasaki M, Masui T, Kanoh T, Ishiguro N, Hamajima N (2006) ALDH2 polymorphisms and bone mineral density in an elderly Japanese population. Osteoporos Int 17(6):908–913. https://doi.org/10.1007/s00198-006-0077-2
130. Yamanaka H, Kamatani N, Hakoda M, Terai C, Kawaguchi R, Kashiwazaki S (1994) Analysis of the genotypes for aldehyde dehydrogenase 2 in Japanese patients with primary gout. Adv Exp Med Biol 370:53–56
131. Yang X, Takeshita T, Hirata M, Sato S, Morimoto K (1999) Effects of alcohol intake and low Km aldehyde dehydrogenase on hepatic function in a high hepatitis C virus-prevalent Japanese island population. Alcohol Clin Exp Res 23(4):751–756
132. Yokoyama A, Muramatsu T, Ohmori T, Kumagai Y, Higuchi S, Ishii H (1997) Reliability of a flushing questionnaire and the ethanol patch test in screening for inactive aldehyde dehydrogenase-2 and alcohol-related cancer risk. Cancer Epidemiol Biomark Prev 6(12):1105–1107
133. Yokoyama A, Muramatsu T, Ohmori T, Yokoyama T, Okuyama K, Takahashi H, Hasegawa Y, Higuchi S, Maruyama K, Shirakura K, Ishii H (1998) Alcohol-related cancers and aldehyde dehydrogenase-2 in Japanese alcoholics. Carcinogenesis 19(8):1383–1387
134. Yokoyama A, Omori T, Yokoyama T, Tanaka Y, Mizukami T, Matsushita S, Higuchi S, Takahashi H, Maruyama K, Ishii H, Hibi T (2005) Esophageal melanosis, an endoscopic finding associated with squamous cell neoplasms of the upper aerodigestive tract, and inactive aldehyde dehydrogenase-2 in alcoholic Japanese men. J Gastroenterol 40(7):676–684. https://doi.org/10.1007/s00535-005-1610-3
135. Yokoyama A, Yokoyama T, Brooks PJ, Mizukami T, Matsui T, Kimura M, Matsushita S, Higuchi S, Maruyama K (2014) Macrocytosis, macrocytic anemia, and genetic polymorphisms of alcohol dehydrogenase-1B and aldehyde dehydrogenase-2 in Japanese alcoholic men. Alcohol Clin Exp Res 38(5):1237–1246. https://doi.org/10.1111/acer.12372
136. Yokoyama A, Yokoyama T, Mizukami T, Matsui T, Kimura M, Matsushita S, Higuchi S, Maruyama K (2016) Alcohol dehydrogenase-1B (rs1229984) and aldehyde dehydrogenase-2 (rs671) genotypes and alcoholic ketosis are associated with the serum uric acid level in Japanese alcoholic men. Alcohol Alcohol (Oxford, Oxfordshire) 51(3):268–274. https://doi.org/10.1093/alcalc/agv123
137. Yokoyama T, Yokoyama A, Kato H, Tsujinaka T, Muto M, Omori T, Haneda T, Kumagai Y, Igaki H, Yokoyama M, Watanabe H, Yoshimizu H (2003) Alcohol flushing, alcohol and aldehyde dehydrogenase genotypes, and risk for esophageal squamous cell carcinoma in Japanese men. Cancer Epidemiol Biomark Prev 12(11 Pt 1):1227–1233
138. Yoshida A, Hsu LC, Yasunami M (1991) Genetics of human alcohol-metabolizing enzymes. Prog Nucleic Acid Res Mol Biol 40:255–287

139. Yoshida A, Shibuya A, Dave V, Nakayama M, Hayashi A (1990) Developmental changes of aldehyde dehydrogenase isozymes in human livers: mitochondrial ALDH2 isozyme is expressed in fetal livers. Experientia 46(7):747–750
140. Yoshida M, Tomitori H, Machi Y, Hagihara M, Higashi K, Goda H, Ohya T, Niitsu M, Kashiwagi K, Igarashi K (2009) Acrolein toxicity: comparison with reactive oxygen species. Biochem Biophys Res Commun 378(2):313–318. https://doi.org/10.1016/j.bbrc.2008.11.054
141. Yoshimasu K, Mure K, Hashimoto M, Takemura S, Tsuno K, Hayashida M, Kinoshita K, Takeshita T, Miyashita K (2015) Genetic alcohol sensitivity regulated by ALDH2 and ADH1B polymorphisms as indicator of mental disorders in Japanese employees. Alcohol Alcohol (Oxford, Oxfordshire) 50(1):39–45. https://doi.org/10.1093/alcalc/agu080
142. Yoshimasu K, Mure K, Hashimoto M, Takemura S, Tsuno K, Hayashida M, Kinoshita K, Takeshita T, Miyashita K (2015) Genetic alcohol sensitivity regulated by ALDH2 and ADH1B polymorphisms is strongly associated with depression and anxiety in Japanese employees. Drug Alcohol Depend 147:130–136. https://doi.org/10.1016/j.drugalcdep.2014.11.034
143. Yoval-Sanchez B, Rodriguez-Zavala JS (2012) Differences in susceptibility to inactivation of human aldehyde dehydrogenases by lipid peroxidation byproducts. Chem Res Toxicol 25(3):722–729. https://doi.org/10.1021/tx2005184
144. Yu HS, Oyama T, Isse T, Kitagawa K, Pham TT, Tanaka M, Kawamoto T (2010) Formation of acetaldehyde-derived DNA adducts due to alcohol exposure. Chem Biol Interact 188 (3):367–375. doi:S0009-2797(10)00525-9 [pii] https://doi.org/10.1016/j.cbi.2010.08.005
145. Yu R, Lai Y, Hartwell HJ, Moeller BC, Doyle-Eisele M, Kracko D, Bodnar WM, Starr TB, Swenberg JA (2015) Formation, accumulation, and hydrolysis of endogenous and exogenous formaldehyde-induced DNA damage. Toxicol Sci 146(1):170–182. https://doi.org/10.1093/toxsci/kfv079
146. Yu RL, Tan CH, Lu YC, Wu RM (2016) Aldehyde dehydrogenase 2 is associated with cognitive functions in patients with Parkinson's disease. Sci Rep 6:30424. https://doi.org/10.1038/srep30424
147. Yu YH, Liao PR, Guo CJ, Chen CH, Mochly-Rosen D, Chuang LM (2016) PKC-ALDH2 pathway plays a novel role in adipocyte differentiation. PLoS One 11(8):e0161993. https://doi.org/10.1371/journal.pone.0161993
148. Zhang CS, Hawley SA, Zong Y, Li M, Wang Z, Gray A, Ma T, Cui J, Feng JW, Zhu M, Wu YQ, Li TY, Ye Z, Lin SY, Yin H, Piao HL, Hardie DG, Lin SC (2017) Fructose-1,6-bisphosphate and aldolase mediate glucose sensing by AMPK. Nature 548(7665):112–116. https://doi.org/10.1038/nature23275
149. Zhang H, Gong DX, Zhang YJ, Li SJ, Hu S (2012) Effect of mitochondrial aldehyde dehydrogenase-2 genotype on cardioprotection in patients with congenital heart disease. Eur Heart J 33(13):1606–1614. https://doi.org/10.1093/eurheartj/ehs061
150. Zhang Y, Wang C, Zhou J, Sun A, Hueckstaedt LK, Ge J, Ren J (2017) Complex inhibition of autophagy by mitochondrial aldehyde dehydrogenase shortens lifespan and exacerbates cardiac aging. Biochim Biophys Acta 1863(8):1919–1932. https://doi.org/10.1016/j.bbadis.2017.03.016
151. Zhou L, Ishizaki H, Spitzer M, Taylor KL, Temperley ND, Johnson SL, Brear P, Gautier P, Zeng Z, Mitchell A, Narayan V, McNeil EM, Melton DW, Smith TK, Tyers M, Westwood NJ, Patton EE (2012) ALDH2 mediates 5-nitrofuran activity in multiple species. Chem Biol 19(7):883–892. https://doi.org/10.1016/j.chembiol.2012.05.017
152. Zhu H, Sun A, Zhu H, Li Z, Huang Z, Zhang S, Ma X, Zou Y, Hu K, Ge J (2014) Aldehyde dehydrogenase-2 is a host factor required for effective bone marrow mesenchymal stem cell therapy. Arterioscler Thromb Vasc Biol 34(4):894–901. https://doi.org/10.1161/ATVBAHA.114.303241

Aldehyde Dehydrogenase 2 and Heart Failure

Wenjia Li, Cheng Shen, Lei Yin, Junbo Ge, and Aijun Sun

1 Introduction

Heart failure (HF) is a structural or functional cardiac abnormal syndrome characterized with series of symptoms and signs such as breathlessness, fatigue, pulmonary crackles, and peripheral edema. Being a terminal phase of most myocardial lesions, HF has become a leading cause of mobility and mortality worldwide, associated with heavy clinical burden and economic costs affecting over 23 million people [14]. There is an increase to 5.5% with systolic dysfunction and an increase to 36.0% with diastolic dysfunction in people 60 years or older [85]. The costs accompanied with heart failure stand 2–3% of the total healthcare system expenditure in high-income countries and are expected to increase >2-fold in the next 2 decades [34].

W. Li
Shanghai Institute of Cardiovascular Diseases, Zhongshan Hospital, Fudan University, Shanghai, China

C. Shen
Department of Cardiology, The Affiliated Hospital of Jining Medical University, Jining, Shandong, China

L. Yin
Department of Urology, Shanghai Tenth People's Hospital, Tongji University, Shanghai, China

J. Ge
Department of Cardiology, Shanghai Institute of Cardiovascular Diseases, Zhongshan Hospital, Fudan University, Shanghai, China

Institute of Biomedical Science, Fudan University, Shanghai, China

A. Sun (✉)
Shanghai Institute of Cardiovascular Diseases, Zhongshan Hospital, Fudan University, Shanghai, China

Institute of Biomedical Science, Fudan University, Shanghai, China
e-mail: sun.aijun@zs-hospital.sh.cn

J. Ren et al. (eds.), *Aldehyde Dehydrogenases*, Advances in Experimental Medicine and Biology 1193, https://doi.org/10.1007/978-981-13-6260-6_5

The pathological mechanisms of HF include hemodynamic abnormalities, neuroendocrine cytokine system over-activation, bioenergetics defects, signal transduction pathway dysfunction, and abnormal calcium homeostasis [54]. In addition, oxidative stress and inflammatory disorders also contribute to the pathogenic process [58, 87]. HF is caused by multiple different etiologies; however, all causes have final common pathways, at least in part, independent of the original cause. Based on previous clinical trials of neurohormonal therapies, neurohormonal activation plays a pivotal role in its pathophysiology [56, 57]. Any cardiovascular diseases can cause cardiac injury; as a consequence, myocyte cells loss and the remaining myocytes become eccentric hypertrophy. Followed by neurohormonal activation, the left ventricle changes from elliptical to spherical and is characterized by functional mitral regurgitation. Afterward, left ventricular remodeling occurs with fibrosis and ventricular dilatation in which process myocardial oxygen consumption increases and myocardial contraction efficiency reduces [10, 69]. As the course progresses, concomitant renal dysfunction and gut congestion cause reduced response to diuretics and inflammatory activation, leading to worse outcomes [58, 84]. Existing therapies including angiotensin-converting enzyme (ACE) inhibitors, angiotensin receptor blockers (ARBs), β-blockers, mineralocorticoid receptor antagonists, and advanced device therapies provide symptomatic and clinical benefits; however, they do not completely solve the molecular abnormalities. Thus, it is urgent to find other effective therapeutic targets.

Recently, energetic impairment in the pathophysiology and progression of HF has caused lots of interest. Mitochondria accounts for 30% of the volume of cardiomyocytes whose dysfunction has been recognized as the key link in the development of heart failure [69]. Mitochondrial dysfunctions cover altered utilization of metabolic substrates, increased formation of reactive oxygen species, impaired mitochondrial electron transport chain activity, abnormal mitochondrial dynamics, and altered ion homeostasis [4]. In view of the above evidences, mitochondria appear to be the main target for direct improvement of cardiac function. Acetaldehyde dehydrogenase 2 (ALDH2) is firstly regarded as an alcohol metabolism enzyme, which mainly distributed in the mitochondria. Increasing evidences have supported its important cardioprotective role [44, 53, 61]. In the following review, we will summarize its beneficial role and novel insights in heart failure due to different etiologies, as well as the potential therapeutic measures.

2 Characteristics of ALDH2

Acetaldehyde dehydrogenase (ALDH) is an enzyme superfamily responsible for the catalytic oxidation of acetaldehyde to acetic acid both physiologically and pathologically [38]. It has been identified 19 isozymes till now. ALDH2 is the most active isozyme and its molecule form is a 56 kDa tetramer. The precursor protein of ALDH2 encodes from chromosome 12, enters the mitochondria under the guidance of the signal peptide, and cleaves off the signal peptide to locate in mitochondria [28]. ALDH2 is widely distributed in human liver, kidney, heart, lung, brain, and other tissues. It contains three domains including coenzyme-binding or NAD^+-binding domain, catalytic domain, and oligomerization domain [61]. ALDH2 also processes three kinds of enzyme activities including dehydrogenase, esterase, and reductase activities detoxifying aldehyde, 4-nitrophenyl acetate, and nitroglycerin, respectively [55].

A total of 84 single nucleotide polymorphism (SNP) loci have been found on human ALDH2 gene [46]. The foremost is rs671 G > A mutation which presents in exon 12 and results in amino acid substitutions of p.Glu504Lys [74]. The genotypes of ALDH2 comprise wild homozygous (GG also known as ALDH2 *1/*1), owning normal catalytic activity of the enzyme; mutant heterozygous (GA also known as ALDH2 *1/*2), processing a decrease in enzyme catalytic activity with only 10%–45% of the wild homozygous; and mutant homozygotes (AA type also known as ALDH2 *2/*2) having only 1%–5% enzyme catalytic activity of the wild homozygous [15]. In general, the prevalence of the genetic polymorphism (ALDH2 *2/*2) is found in nearly 8% of world populations and 40% of East Asian population [103].

3 Beneficial Role and Mechanisms of ALDH2 in Heart Failure

Experimental and clinical studies have proved that ALDH2 plays a pivotal role in heart failure via maintaining cellular homeostasis [13, 17]. Evidences from our group and others have verified ALDH2 is closely related to various etiologies of heart failure, such as coronary artery disease (CAD) [80], hypertension [71], alcoholism [73], and other susceptibilities [102, 106]. Next we will review the pathological role and underlying mechanisms of ALDH2 involved in heart failure.

3.1 ALDH2 and Ischemic Heart Failure

Numerous studies have shown that ALDH2 has cardioprotective effects in ischemic with or without reperfusion myocardial injury. The expression of ALDH2 was found decreased in infarction border zone [90]. Cardiac function deteriorated in ALDH2 knockout posts myocardial infarction (MI) mice as evidenced by increased left ventricle (LV) cavity, LV end-diastolic pressure, and infarct size [80]. Meanwhile, there were more apoptotic cells in the non-infarcted LV region as compared with wild-type (WT) MI mice. In contrast, overexpression of ALDH2 in the heart could alleviate these injures. Ma, et al. found that ALDH2 induced autophagy during ischemia and inhibited autophagy during reperfusion which reduced hypoxic and reoxygenation cell death [53]. These data demonstrated that ALDH2 deficiency aggravated mitochondrial dysfunction and increased cardiomyocyte apoptotic cell death.

DNA methylation is a process adding methyl groups to cytosine residues into DNA sequences, which can prevent transcription factors from entering the gene regulatory region, thereby inhibiting gene transcription [65]. Increasing studies indicated that DNA methylation provided a potential molecular basis on energy metabolism between environmental and genetic factors interaction and might contribute to myocardial injury [40]. We also found DNA methylation at CpG sites (CpG1, CpG2, and CpG7) in the upstream sequence of ALDH2 promoter was upregulated post MI. These abnormal hypermethylations at the CpG sites downregulated ALDH2 enzyme activity and aggravated ischemic damages [90]. DNA methylation reflected the upstream mechanism of ALDH2 regulating cardiac function after ischemia.

We also demonstrated the downstream mechanisms of ALDH2 in protecting cardiac from ischemic injures. ALDH2 could increase the intracellular levels of 4-HNE, which could exacerbate apoptosis by inhibition of HSP70, phosphorylation of JNK, and activation of p53 [80]. 4-HNE is also a diffusible product of membrane lipid peroxidation and relates to oxidative stress-induced cell death [70]. Mitochondrial-derived ROS also attacked polyunsaturated fatty acids, leading to membrane lipid peroxidation, thereby increasing reactive aldehydes [7, 8]. We demonstrated both experimentally and clinically that ALDH2 was vital in regulating microenvironment homeostasis. ALDH2 could promote angiogenesis post chronic ischemia. ALDH2 deficiency inhibited tubelike construction formation of hypoxia endothelial cell through HIF-1α/VEGF pathway, deteriorating perfusion recovery in ischemia tissue, while overexpression of ALDH2 promoted angiogenesis. Furthermore, clinical data suggested that the dysfunction of ALDH2 due to gene variant was an unfavorable factor for revascularization in patients with chronic total occlusion (CTO). Therefore, targeting ALDH2 activity may be a potential therapeutic strategy for chronic ischemic heart failure, and we have used this achievement to guide clinical decision-making [49].

3.2 *ALDH2 and Stress-Induced Heart Failure*

Endoplasmic reticulum (ER) stress refers to an increase in unfolded and misfolded proteins in the ER that disrupts the homeostasis in response to cellular stressors, such as heat, hypoxia, metabolic starvation, angiotensin II, and tumor necrosis factor-α [29, 42]. Substantial evidences indicated ES stress as important target for the treatment of cardiovascular disease, including ischemia/reperfusion injury, atherosclerosis, cardiac hypertrophy, and heart failure [29, 82]. Our results suggested that ALDH2 deficiency aggravated cardiac contractile dysfunction following activation of ES stress, manifested as descend of ejection fraction and fractional shortening. NADPH oxidase (p47phox subunit) increased in ALDH2 knockout mice, suggesting that ALDH2 might regulate Akt signaling pathway through p47phox NADPH oxidase-dependent manner against ER stress and ER stress-induced apoptosis [47].

Persistent pressure overload such as refractory hypertension is a significant risk factor for heart failure and sudden death. Series of pathological cardiac remodeling were characterized by increased myocardial cells, "fetal gene program" activation, cytoskeletal reorganization, and irreversible systolic dysfunction [35, 51]. Autophagy is an important homeostatic pathway in degrading damaged proteins and intracellular organelles [24]. It has been found to be involved in pressure-induced heart failure [22]. Autophagy-related signaling pathways involved in pressure overload included the classic AMPK-mTOR-autophagy, Beclin-1-dependent pathway, Akt/mTOR/FoxO3a signal pathway, and PI3K/Akt signaling [48, 89, 96, 98]. In early compensated cardiac hypertrophy after transverse aortic constriction (TAC), autophagy played an adverse role with worse cardiac function and severer mitochondria damage. In this process, ALDH2 acted through the regulation of PI3K/PTEN/Akt signaling [95]. Besides, ALDH2 deficiency further inhibited autophagy during decompensated cardiac hypertrophy accompanied with inactive Beclin-1-dependent autophagy signaling [71]. Therefore, autophagy works in both early stage and late stage of ALDH2 regulation in pressure-overload adaptive response.

3.3 ALDH2 and Alcoholic Heart Failure

Alcoholic cardiomyopathy is characterized by a dilated left ventricle and reduced myocardial contractility due to a long-term history of heavy alcohol consumption. It was estimated that about one-third of alcoholics suffered from varying degrees of alcoholic cardiomyopathy and approximately half of them resulted in death within 4 years [43, 101]. Mitochondrial defects, cell death, heart rate variability, and cardiac remodeling would eventually result in heart failure [31].

As early as the twentieth century, it was found about 50% of Asians had facial flushing after drinking alcohol, and these individuals were tested having an inactive form of mitochondrial ALDH2 [88]. rs671 in ALDH2 gene was the most influential genetic variant linked to alcohol consumption [83]. ALDH2 deficiency aggravated alcoholic myocardiopathy by weakening acetaldehyde-biogenic amine condensation products detoxification, breaking intracellular Ca^{2+} homeostasis, increasing apoptosis, upregulating autophagy, as well as impairing mitochondrial function [52, 66, 72, 73, 101]. Besides, our study explored the role of ALDH2 in low-to-moderate alcohol consumption. On one hand, we confirmed the cardioprotective effects of low-to-moderate alcohol consumption which manifested by elevated HDL-c levels and upregulated HO-1 expression in the myocardium. However, the benefits were disrupted when ALDH2 was deficient, possibly by activating ROS-dependent apoptosis and RIP1/RIP3/MLKL-mediated necrosis [73]. A Guangzhou biobank cohort study genotyped rs671 of ALDH2 in 4867 men. Diastolic blood pressure and HDL cholesterol which associated with ALDH2 variants were attenuated after adjusting for alcohol use. The result suggested the apparent associations between physical activity and alcohol use in ALDH2 variants population [3]. Even moderate alcohol use was found associated with subclinical adverse effects with greater left ventricle mass and more impaired diastolic functions in subjects carrying ALDH2 variants, especially among East Asians [36]. Individuals with inactive isoforms of ALDH2 should be warned to avoid drinking alcohol, even for social or occupational promotion.

3.4 ALDH2 and Diabetes Mellitus-Related Heart Failure

The increasing morbidity and mortality of heart failure are related to the increase in aging, obesity, and diabetes mellitus in a large part. The prognosis of heart failure in patients with diabetes is much worse than that in patients without diabetes. The early symptom of diabetic cardiomyopathy is diastolic dysfunction and can gradually develop into systolic dysfunction. Metabolic disorder is an important feature of diabetic cardiomyopathy, manifested as reduced glucose uptake and increased fatty acid utilization accompanied by oxidative stress, inflammation, cardiomyocyte apoptosis, and myocardial fibrosis [37]. Beyond these, mitochondrial dysfunction, impaired mitochondrial and cardiomyocyte calcium handling, endoplasmic reticulum stress, and reduced nitric oxide bioavailability were also implicated in the development and progression of diabetic cardiomyopathy [39].

A genome-wide association study contained 12,720 participants found rs671 (ALDH2) was associated with metabolic syndrome (MetS) in Han Chinese. What's more, the effects of rs671 on metabolic components were significantly correlated with drinking [108]. Previous studies showed that ALDH2 improved the contractile function of advanced diabetic cardiomyopathy by regulating Ca^{2+} homeostasis and autophagy [32, 104]. Our data found that ALDH2 deficiency impaired diastolic function in early stage of diabetic cardiomyopathy, while cardiac contractile function remained normal. In this stage, ALDH2 deficiency disrupted energy metabolism with increased AMP/ATP and ADP/ATP and decreased PCr/ATP ratio, which in turn induced activation of energy regulatory LKB1/AMPK pathway. The progressive accumulation of phosphatidylcholine and phosphatidylinositol in heart tissue induced metabolic homeostasis disequilibrium and led to deterioration of diastolic function [20, 91].

3.5 ALDH2 and Aging-Related Heart Failure

Aging is an irreversible biological process. In the cardiovascular area, age-dependent increases include left ventricular hypertrophy, diastolic dysfunction, atrial fibrillation, as well as vascular intimal thickening and vessel stiffness. Aging-related cardiac dysfunction is characterized with loss of cardiac contractile reserve, increased fibrosis and remodeling, impaired cardiomyocyte proteostasis, and loss of autophagy [21, 50]. Heart failure can be regarded as an aging-related phenotype.

Aging-associated cardiac pathological changes involve oxidative stress, short telomere defect, mitochondrial damage, intracellular Ca^{2+} mishandling, etc. [2, 99].

Numerous studies suggested ALDH2 participated in the process of aging and age-related cardiovascular diseases [18, 106]; however, whether ALDH2 is beneficial or detrimental is still controversial. Wu found that ALDH2 ablation led to cardiac aging and sustained usage of Alda-1 (a specific activator of ALDH2) abrogated the aging effect [94]. ALDH2 activity was discovered to be significantly decreased in aged hearts which also demonstrated the benefits of activation of ALDH2 on retarding the aging process. In the meantime, ALDH2 was also discovered to exert age-dependent vasoprotective effects with decreased mitochondrial ROS formation and oxidative mtDNA damage [92]. On the contrary, our lab found ALDH2 overexpression such as using Alda-1 accentuated aging-related cardiomyocyte. The dysfunction was characterized by increased contractile dysfunction, oxidative stress, intracellular Ca^{2+} mishandling, and mitochondrial injury [105]. Moreover, AMPK/Sirt1 signaling cascades were found taken part in ALDH2-accenuated cardiac aging [106]. In the light of the debates, further epidemiological of different races or experimental studies will be needed to provide more evidence about the effect of ALDH2 in aging.

3.6 ALDH2 and Drug-Induced Heart Failure

Several chemical agents and drugs could impair cardiac mitochondrial function via destroying mitochondrial respiratory chain (e.g., uncoupling) or inhibiting mitochondrial enzymes. The most common agents are anticancer drugs such as anthracycline doxorubicin (DOX), cisplatin, and Trisenox; antiviral compound azidothymidine (AZT zidovudine); and several oral antidiabetic drugs such as Avandia [86].

The chemotherapy drug DOX is frequently found in inducing cardiotoxic. Left ventricular systolic pressure would significantly reduce, and left ventricular end-diastolic pressure would overtly increase after DOX treatment. ALDH2 attenuated this cardiotoxicity by inhibiting oxidative stress, decreasing the expression and activity of NADPH oxidase 2, and reducing myocardial apoptosis. In addition, DOX-induced myocardial dysfunction was severer with increased levels of 4-HNE and autophagy in ALDH2 knockout mice. Besides, these symptoms could be improved when ALDH2 activity was restored, suggesting that inhibition of 4-HNE and autophagy may be the possible mechanisms of ALDH2 against DOX-induced cardiac dysfunction [79].

Oxidative stress-induced cardiomyocyte apoptosis is also a main part in the pathogenesis of heart failure [59]. ROS production and accumulation caused intracellular redox imbalance, leading to mitochondrial dysfunction and decreased production of ATP [26]. We elucidated the relationship between ALDH2 deficiency and oxidative stress-induced apoptosis in an antimycin-induced heart failure model. Inhibiting ALDH2 activity by daidzin increased intracellular ROS levels and apoptosis in which associated with the upregulated phosphorylation of ERK1/2, JNK, and p38-MAPK [102]. Thus, these data suggested the beneficial role of ALDH2 in drug-induced heart failure.

4 ALDH2-Related Therapy

The following mediators have been reported mediating the cardiac dysfunction in ALDH2 deficiency: (1) ROS and toxic aldehydes [53]; (2) apoptosis pathways [80], involving caspase 3, Bcl- 2; (3) oxidative stress signaling cascade [1], involving MAP kinase cascades ERK1/2, SAPK/JNK, and p38 MAP kinase; and (4) autophagy [1], involving Beclin-1 and AMPK-mTOR. The indispensable role of ALDH2 in the pathogenesis of heart failure sheds light on the development of potential therapeutic target of it (Fig. 1). Here we summarized activators of ALDH2 and several other aspects which have already been proved to have therapeutic effects or just have a therapeutic potential on heart failure.

4.1 ALDH2 Activator

Alda-1, a small molecule activator of ALDH2, is a potential new therapeutic candidate. Alda-1 was reported to exert its cardioprotective effect through reducing oxidative stress, restoring calcium and CaMKII homeostasis, and detoxifying $O_2^{\bullet -}$ induced reactive aldehydes to less reactive acids [16, 93]. It was also been reported

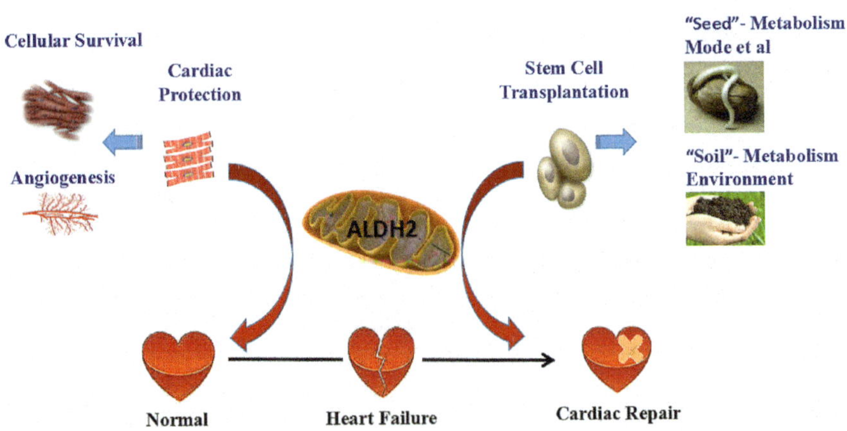

Fig. 1 A scheme depicting how ALDH2 plays the protective role in heart failure. ALDH2 protects myocardium by maintaining myocardial cell survival and angiogenesis. At the same time, ALDH2 can determine the efficacy of stem cell transplantation by increasing the quality of seed stem cells and metabolic microenvironment

to ameliorate pulmonary vascular remodeling in pulmonary arterial hypertension and inhibit atherosclerosis in apolipoprotein E-knockout mice [76, 97]. In addition, it played a protective role in cardiac dysfunction induced by abnormal glucose and lipid metabolism as well as DOX or 4-HNE-induced autophagic reduction and cell death [30, 79]. However, the efficiency of the activation by Alda-1 is variant, for example, Alda-1 increases acetaldehyde oxidation in wild-type ALDH2 (ALDH2*1) and East Asian variant of ALDH2 (ALDH2*2) approximately 1.5- and 6-fold, respectively [6]. Thus, the patients undergoing pathological processes such as cardiac ischemia with wild type or mutant ALDH2 might gain more benefits with the treatment of pharmacologic enhancement of ALDH2 activity [16]. Belmont conducted kinetic experiments to characterize Alda-1 on the properties of ALDH2 and found it was a complex behavior, where Alda-1 acted as inhibitor in low concentrations of aldehyde and as an activator in high concentrations [5]. Moreover, Alda-1 has exogenous and toxic characters. All factors should be taken into consideration when considering Alda-1 as an exogenous stimulator of ALDH2, including the working concentration, half-life period, continuity of stimulation, toxic, and side effects.

Protein kinase C ε (PKCε) has been found as another ALDH2 activator via direct phosphorylation. In vitro, phosphorylation of wild-type ALDH2 recombinant protein was reported increased its enzymatic activity [16]. The enzymatic activity of the phosphorylated ALDH2*2 is 270% of the non-phosphorylated ALDH2*2 [60]. Following phosphorylation by recombinant εPKC, there was an increase of 70% of the ALDH2 activity. Similar to Alda-1, the effect of εPKC phosphorylation was more pronounced on ALDH2*2 mutant enzyme. T185E, S279E, and T412E were three common phosphomimetic mutations sites by εPKC in the protection of ALDH2 against reactive aldehydes. Treatment with PKC activator upregulated ALDH2 activity, while applying PKC inhibitor had the opposite effect. Some study demonstrated that PKCε-ALDH2 interaction had disincentive effects in 4-HNE-induced aberrant PPARγ regulation, which suggest that PKCε-ALDH2 regulatory axis may be a therapeutic target for treating metabolic syndrome [100].

4.2 *MicroRNAs*

MicroRNAs are a class of endogenous interfering RNAs whose primary function are regulating the expression of genes. Evidences indicated that microRNAs could participate in diverse pathophysiological processes of cardiovascular disease, including hypertrophy, apoptosis, cardiac conduction, fibrosis, and angiogenesis [25, 75]. MiR-34a has been elucidated in many cell lines and was found related to apoptosis, energy metabolism, lipid metabolism, aging, and stem cell division [12, 33]. Bioinformatics analysis produced a protein-protein interaction network in HepG2 cells and revealed that ALDH2 was a potential target of miR-34a [19]. Our study found that increased circulating miR-34a could decrease ALDH2 activity and increase cardiomyocyte apoptosis post-MI injury [23]. Although miR-34a is not a cardiac-specific miRNA, the expression abundance of miR-34a in myocardium ranks third in all tissues [9], and it is reasonable to regard miR-34a as a diagnostic marker for MI. MiR-28 could also promote ischemia via inhibition of ALDH2 expression in myocardium [45]. In addition, Shen predicted ALDH2 as target genes of miR-224 [72]. The target sites existed in 3'UTR of ALDH2 suggesting that miR-224 downregulated the expressions of ALDH2 and finally regulated target genes in lipid metabolism. MicroRNAs might be new diagnostic indicators and therapeutic targets for heart failure patients. But it requires further studies in large cohort to assess the specificity and sensitivity of them.

4.3 Mesenchymal Stem Cell (MSC) Therapy

Mesenchymal stem cell (MSC) therapy is a promising approach in alleviating ischemic injury and promoting tissue regeneration [64]. A present meta-analysis including 64 studies strongly supported the potential of MSCs therapy for ischemic stroke [68]. Up to now, various strategies have been used to increase transplant effects in ischemic diseases, including tissue engineering scaffolds, genetic modification, and hypoxia-based pretreatment [27]; however, implanted cell dysplasia is still a problem [78]. Many factors could affect the efficiency including regenerative cell source, injectable delivery vehicles, and microenvironmental signals [67, 77]. Our results showed that host ALDH2 affected the survival of transplanted MSCs. Protein array analysis also revealed that ALDH2$^{-/-}$ tissues expressed low levels of angiogenic factors, including cysteine-rich angiogenic inducer 61, endoglin, epidermal growth factor, fibroblast growth factor-1, angiopoietin-1, matrix metallopeptidase-3/-9, and insulin-like growth factor binding protein, all of which could enhance the tolerance of engrafted MSCs during vasculogenesis in hypoxia injury [107]. Thus, ALDH2 may be regarded as a homeostatic mediator of microenvironment by increasing local capillary density and energy supply and decreasing oxidative stress after ischemia.

4.4 ALDH Bright Cells

Autologous bone marrow-derived aldehyde dehydrogenase bright (ALDHbr) cells isolated by flow sorting express high activity of ALDH [41]. They have been applied in clinical practice to repair tissue damage and have been proven safe and efficient in patients with chronic myocardial ischemia [62]. Our observations supported the effective therapeutic effect of ALDHbr cells on ischemic myocardium; in addition, we demonstrated ALDH2 as a key mediator in the process [81]. Weakened glycolysis, mitochondrial respiratory abnormalities, and increased mitochondrial ROS gave rise to the diminished therapeutic efficacy of ALDHbr cells in ALDH2 deficiency mice rather than oxidative phosphorylation impairment. The results gave us a hint that ALDH2 activity was a pivotal precondition in the efficacy of ALDHbr cell therapy; therefore individuals with loss of ALDH2 function are unsuitable for ALDHbr cell therapy. However, recently a clinical trial didn't find significant positive outcomes of ALDHbr cells in patients with peripheral artery disease [63]. Future investigational trial tests about cell therapy should be carried out to find new anatomic and perfusion insights.

5 Conclusion

Just as Braunwald said, heart failure is the last battle of cardiovascular diseases [11]. It is urgent to find a novel prospective to illustrate the mechanism of heart failure and to improve the prognosis. This review summarizes the roles of ALDH2 gene polymorphisms and ALDH2 enzyme activity in heart failure induced by multiple causes, such as ischemic injury, hypertension, alcohol, diabetes, and aging. Except the controversial role of ALDH2 in aging-related cardiac dysfunction, studies have suggested a cardioprotective role of ALDH2 to counteract cardiac dysfunction due to different etiologies. Emerging evidences provided new insight in understanding the epigenetic and transcriptional regulation of ALDH2 as well as the effect of ALDH2 in stem cell transplantation. Strategies aim to enhancing ALDH2 activity or expression, as well as improving mitochondrial function, will bring novel prospects for the treatment of heart failure.

References

1. Adameova A, Goncalvesova E, Szobi A, Dhalla NS (2016) Necroptotic cell death in failing heart: relevance and proposed mechanisms. Heart Fail Rev 21(2):213–221. https://doi.org/10.1007/s10741-016-9537-8
2. Armanios M, Blackburn EH (2012) The telomere syndromes. Nat Rev Genet 13(10):693–704. https://doi.org/10.1038/nrg3246
3. Au YS, Jiang C, Cheng KK, Liu B, Zhang W, Lam TH, Leung GM, Schooling CM (2013) Is aldehyde dehydrogenase 2 a credible genetic instrument for alcohol use in Mendelian randomization analysis in southern Chinese men? Int J Epidemiol 42(1):318–328. https://doi.org/10.1093/ije/dys221
4. Bayeva M, Gheorghiade M, Ardehali H (2013) Mitochondria as a therapeutic target in heart failure. J Am Coll Cardiol 61(6):599–610. https://doi.org/10.1016/j.jacc.2012.08.1021
5. Belmont-Diaz JA, Yoval-Sanchez B, Calleja-Castaneda LF, Pardo VJ, Rodriguez-Zavala JS (2016) Alda-1 modulates the kinetic properties of mitochondrial aldehyde dehydrogenase (ALDH2). FEBS J 283(19):3637–3650. https://doi.org/10.1111/febs.13833
6. Beretta M, Gorren AC, Wenzl MV, Weis R, Russwurm M, Koesling D, Schmidt K, Mayer B (2010) Characterization of the east Asian variant of aldehyde dehydrogenase-2: bioactivation of nitroglycerin and effects of Alda-1. J Biol Chem 285(2):943–952. https://doi.org/10.1074/jbc.M109.014548
7. Bhatnagar A (1995) Electrophysiological effects of 4-hydroxynonenal, an aldehydic product of lipid peroxidation, on isolated rat ventricular myocytes. Circ Res 76(2):293–304

8. Boldogh I, Bacsi A, Choudhury BK, Dharajiya N, Alam R, Hazra TK, Mitra S, Goldblum RM, Sur S (2005) ROS generated by pollen NADPH oxidase provide a signal that augments antigen-induced allergic airway inflammation. J Clin Invest 115(8):2169–2179. https://doi.org/10.1172/JCI24422
9. Bommer GT, Gerin I, Feng Y, Kaczorowski AJ, Kuick R, Love RE, Zhai Y, Giordano TJ, Qin ZS, Moore BB, MacDougald OA, Cho KR, Fearon ER (2007) 53-mediated activation of miRNA34 candidate tumor-suppressor genes. Curr Biol 17(15):1298–1307. https://doi.org/10.1016/j.cub.2007.06.068
10. Braunwald E (2013) Heart failure. JACC Heart Fail 1(1):1–20. https://doi.org/10.1016/j.jchf.2012.10.002
11. Braunwald E (2015) The war against heart failure: the lancet lecture. Lancet 385(9970):812–824. https://doi.org/10.1016/S0140-6736(14)61889-4
12. Bu P, Wang L, Chen KY, Srinivasan T, Murthy PK, Tung KL, Varanko AK, Chen HJ, Ai Y, King S, Lipkin SM, Shen X (2016) A miR-34a-numb feedforward loop triggered by inflammation regulates asymmetric stem cell division in intestine and Colon cancer. Cell Stem Cell 18(2):189–202. https://doi.org/10.1016/j.stem.2016.01.006
13. Budas GR, Disatnik MH, Mochly-Rosen D (2009) Aldehyde dehydrogenase 2 in cardiac protection: a new therapeutic target? Trends Cardiovasc Med 19(5):158–164. https://doi.org/10.1016/j.tcm.2009.09.003
14. Bui AL, Horwich TB, Fonarow GC (2011) Epidemiology and risk profile of heart failure. Nat Rev Cardiol 8(1):30–41. https://doi.org/10.1038/nrcardio.2010.165
15. Chen Z, Foster MW, Zhang J, Mao L, Rockman HA, Kawamoto T, Kitagawa K, Nakayama KI, Hess DT, Stamler JS (2005) An essential role for mitochondrial aldehyde dehydrogenase in nitroglycerin bioactivation. Proc Natl Acad Sci U S A 102(34):12159–12164. https://doi.org/10.1073/pnas.0503723102
16. Chen CH, Budas GR, Churchill EN, Disatnik MH, Hurley TD, Mochly-Rosen D (2008) Activation of aldehyde dehydrogenase-2 reduces ischemic damage to the heart. Science 321(5895):1493–1495. https://doi.org/10.1126/science.1158554
17. Chen CH, Sun L, Mochly-Rosen D (2010) Mitochondrial aldehyde dehydrogenase and cardiac diseases. Cardiovasc Res 88(1):51–57. https://doi.org/10.1093/cvr/cvq192
18. Chen CH, Ferreira JC, Gross ER, Mochly-Rosen D (2014) Targeting aldehyde dehydrogenase 2: new therapeutic opportunities. Physiol Rev 94(1):1–34. https://doi.org/10.1152/physrev.00017.2013
19. Cheng J, Zhou L, Xie QF, Xie HY, Wei XY, Gao F, Xing CY, Xu X, Li LJ, Zheng SS (2010) The impact of miR-34a on protein output in hepatocellular carcinoma HepG2 cells. Proteomics 10(8):1557–1572. https://doi.org/10.1002/pmic.200900646
20. Christoffersen C, Bollano E, Lindegaard ML, Bartels ED, Goetze JP, Andersen CB, Nielsen LB (2003) Cardiac lipid accumulation associated with diastolic dysfunction in obese mice. Endocrinology 144(8):3483–3490. https://doi.org/10.1210/en.2003-0242
21. Cuervo AM, Bergamini E, Brunk UT, Droge W, Ffrench M, Terman A (2005) Autophagy and aging: the importance of maintaining "clean" cells. Autophagy 1(3):131–140
22. Delbridge L, Mellor KM, Taylor DJ, Gottlieb RA (2017) Myocardial stress and autophagy: mechanisms and potential therapies. Nat Rev Cardiol 14(7):412–425. https://doi.org/10.1038/nrcardio.2017.35
23. Fan F, Sun A, Zhao H, Liu X, Zhang W, Jin X, Wang C, Ma X, Shen C, Zou Y, Hu K, Ge J (2013) MicroRNA-34a promotes cardiomyocyte apoptosis post myocardial infarction through down-regulating aldehyde dehydrogenase 2. Curr Pharm Des 19(27):4865–4873
24. Galluzzi L, Baehrecke EH, Ballabio A, Boya P, Bravo-San PJ, Cecconi F, Choi AM, Chu CT, Codogno P, Colombo MI, Cuervo AM, Debnath J, Deretic V, Dikic I, Eskelinen EL, Fimia GM, Fulda S, Gewirtz DA, Green DR, Hansen M, Harper JW, Jaattela M, Johansen T, Juhasz G, Kimmelman AC, Kraft C, Ktistakis NT, Kumar S, Levine B, Lopez-Otin C, Madeo F, Martens S, Martinez J, Melendez A, Mizushima N, Munz C, Murphy LO, Penninger JM, Piacentini M, Reggiori F, Rubinsztein DC, Ryan KM, Santambrogio L, Scorrano L, Simon

AK, Simon HU, Simonsen A, Tavernarakis N, Tooze SA, Yoshimori T, Yuan J, Yue Z, Zhong Q, Kroemer G (2017) Molecular definitions of autophagy and related processes. EMBO J 36(13):1811–1836. https://doi.org/10.15252/embj.201796697
25. Ghildiyal M, Zamore PD (2009) Small silencing RNAs: an expanding universe. Nat Rev Genet 10(2):94–108. https://doi.org/10.1038/nrg2504
26. Giordano FJ (2005) Oxygen, oxidative stress, hypoxia, and heart failure. J Clin Invest 115(3):500–508. https://doi.org/10.1172/JCI24408
27. Gnecchi M, He H, Liang OD, Melo LG, Morello F, Mu H, Noiseux N, Zhang L, Pratt RE, Ingwall JS, Dzau VJ (2005) Paracrine action accounts for marked protection of ischemic heart by Akt-modified mesenchymal stem cells. Nat Med 11(4):367–368. https://doi.org/10.1038/nm0405-367
28. Goedde HW, Agarwal DP (1990) Pharmacogenetics of aldehyde dehydrogenase (ALDH). Pharmacol Ther 45(3):345–371
29. Groenendyk J, Sreenivasaiah PK, Kim DH, Agellon LB, Michalak M (2010) Biology of endoplasmic reticulum stress in the heart. Circ Res 107(10):1185–1197. https://doi.org/10.1161/CIRCRESAHA.110.227033
30. Gu X, Fang T, Kang P, Hu J, Yu Y, Li Z, Cheng X, Gao Q (2017) Effect of ALDH2 on high glucose-induced cardiac fibroblast oxidative stress, apoptosis, and fibrosis. Oxid Med Cell Longev 2017:9257967. https://doi.org/10.1155/2017/9257967
31. Guo R, Ren J (2010) Alcohol and acetaldehyde in public health: from marvel to menace. Int J Environ Res Public Health 7(4):1285–1301. https://doi.org/10.3390/ijerph7041285
32. Guo Y, Yu W, Sun D, Wang J, Li C, Zhang R, Babcock SA, Li Y, Liu M, Ma M, Shen M, Zeng C, Li N, He W, Zou Q, Zhang Y, Wang H (2015) A novel protective mechanism for mitochondrial aldehyde dehydrogenase (ALDH2) in type i diabetes-induced cardiac dysfunction: role of AMPK-regulated autophagy. Biochim Biophys Acta 1852(2):319–331. https://doi.org/10.1016/j.bbadis.2014.05.017
33. Han HS, Choi BH, Kim JS, Kang G, Koo SH (2017) Hepatic Crtc2 controls whole body energy metabolism via a miR-34a-Fgf21 axis. Nat Commun 8(1):1878. https://doi.org/10.1038/s41467-017-01878-6
34. Heidenreich PA, Albert NM, Allen LA, Bluemke DA, Butler J, Fonarow GC, Ikonomidis JS, Khavjou O, Konstam MA, Maddox TM, Nichol G, Pham M, Pina IL, Trogdon JG (2013) Forecasting the impact of heart failure in the United States: a policy statement from the American Heart Association. Circ Heart Fail 6(3):606–619. https://doi.org/10.1161/HHF.0b013e318291329a
35. Hein S, Arnon E, Kostin S, Schonburg M, Elsasser A, Polyakova V, Bauer EP, Klovekorn WP, Schaper J (2003) Progression from compensated hypertrophy to failure in the pressure-overloaded human heart: structural deterioration and compensatory mechanisms. Circulation 107(7):984–991
36. Hung CL, Chang SC, Chang SH, Chi PC, Lai YJ, Wang SW, Wu YJ, Yeh HI, Lin SJ, Chen CH, Mochly-Rosen D, Wang LY (2017) Genetic polymorphisms of alcohol metabolizing enzymes and alcohol consumption are associated with asymptomatic cardiac remodeling and subclinical systolic dysfunction in Large community-dwelling Asians. Alcohol Alcohol 52(6):638–646. https://doi.org/10.1093/alcalc/agx049
37. Isfort M, Stevens SC, Schaffer S, Jong CJ, Wold LE (2014) Metabolic dysfunction in diabetic cardiomyopathy. Heart Fail Rev 19(1):35–48. https://doi.org/10.1007/s10741-013-9377-8
38. Jackson B, Brocker C, Thompson DC, Black W, Vasiliou K, Nebert DW, Vasiliou V (2011) Update on the aldehyde dehydrogenase gene (ALDH) superfamily. Hum Genomics 5(4):283–303
39. Jia G, Hill MA, Sowers JR (2018) Diabetic cardiomyopathy: an update of mechanisms contributing to this clinical entity. Circ Res 122(4):624–638. https://doi.org/10.1161/CIRCRESAHA.117.311586
40. Jiang Y, Xia W, Yang J, Zhu Y, Chang H, Liu J, Huo W, Xu B, Chen X, Li Y, Xu S (2015) BPA-induced DNA hypermethylation of the master mitochondrial gene PGC-1alpha

contributes to cardiomyopathy in male rats. Toxicology 329:21–31. https://doi.org/10.1016/j.tox.2015.01.001

41. Jones RJ, Barber JP, Vala MS, Collector MI, Kaufmann SH, Ludeman SM, Colvin OM, Hilton J (1995) Assessment of aldehyde dehydrogenase in viable cells. Blood 85(10):2742–2746
42. Kim I, Xu W, Reed JC (2008) Cell death and endoplasmic reticulum stress: disease relevance and therapeutic opportunities. Nat Rev Drug Discov 7(12):1013–1030. https://doi.org/10.1038/nrd2755
43. Laonigro I, Correale M, Di Biase M, Altomare E (2009) Alcohol abuse and heart failure. Eur J Heart Fail 11(5):453–462. https://doi.org/10.1093/eurjhf/hfp037
44. Li Y, Zhang D, Jin W, Shao C, Yan P, Xu C, Sheng H, Liu Y, Yu J, Xie Y, Zhao Y, Lu D, Nebert DW, Harrison DC, Huang W, Jin L (2006) Mitochondrial aldehyde dehydrogenase-2 (ALDH2) Glu504Lys polymorphism contributes to the variation in efficacy of sublingual nitroglycerin. J Clin Invest 116(2):506–511. https://doi.org/10.1172/JCI26564
45. Li SP, Liu B, Song B, Wang CX, Zhou YC (2015) miR-28 promotes cardiac ischemia by targeting mitochondrial aldehyde dehydrogenase 2 (ALDH2) in mus musculus cardiac myocytes. Eur Rev Med Pharmacol Sci 19(5):752–758
46. Li R, Zhao Z, Sun M, Luo J, Xiao Y (2016) ALDH2 gene polymorphism in different types of cancers and its clinical significance. Life Sci 147:59–66. https://doi.org/10.1016/j.lfs.2016.01.028
47. Liao J, Sun A, Xie Y, Isse T, Kawamoto T, Zou Y, Ge J (2012) Aldehyde dehydrogenase-2 deficiency aggravates cardiac dysfunction elicited by endoplasmic reticulum stress induction. Mol Med 18:785–793. https://doi.org/10.2119/molmed.2011.00466
48. Lin H, Li HF, Chen HH, Lai PF, Juan SH, Chen JJ, Cheng CF (2014) Activating transcription factor 3 protects against pressure-overload heart failure via the autophagy molecule Beclin-1 pathway. Mol Pharmacol 85(5):682–691. https://doi.org/10.1124/mol.113.090092
49. Liu X, Sun X, Liao H, Dong Z, Zhao J, Zhu H, Wang P, Shen L, Xu L, Ma X, Shen C, Fan F, Wang C, Hu K, Zou Y, Ge J, Ren J, Sun A (2015) Mitochondrial aldehyde dehydrogenase 2 regulates revascularization in chronic ischemia: potential impact on the development of coronary collateral Circulation. Arterioscler Thromb Vasc Biol 35(10):2196–2206. https://doi.org/10.1161/ATVBAHA.115.306012
50. Lopez-Otin C, Blasco MA, Partridge L, Serrano M, Kroemer G (2013) The hallmarks of aging. Cell 153(6):1194–1217. https://doi.org/10.1016/j.cell.2013.05.039
51. Lorell BH, Carabello BA (2000) Left ventricular hypertrophy: pathogenesis, detection, and prognosis. Circulation 102(4):470–479
52. Ma H, Li J, Gao F, Ren J (2009) Aldehyde dehydrogenase 2 ameliorates acute cardiac toxicity of ethanol: role of protein phosphatase and forkhead transcription factor. J Am Coll Cardiol 54(23):2187–2196. https://doi.org/10.1016/j.jacc.2009.04.100
53. Ma H, Guo R, Yu L, Zhang Y, Ren J (2011) Aldehyde dehydrogenase 2 (ALDH2) rescues myocardial ischaemia/reperfusion injury: role of autophagy paradox and toxic aldehyde. Eur Heart J 32(8):1025–1038. https://doi.org/10.1093/eurheartj/ehq253
54. Mann DL, Bristow MR (2005) Mechanisms and models in heart failure: the biomechanical model and beyond. Circulation 111(21):2837–2849. https://doi.org/10.1161/CIRCULATIONAHA.104.500546
55. Matsumoto A (2016) Fundamental Properties of Aldehyde Dehydrogenase 2 (ALDH2) and the Importance of the ALDH2 Polymorphism. Nihon Eiseigaku Zasshi 71(1):55–68. https://doi.org/10.1265/jjh.71.55
56. McMurray JJ (2015) Neprilysin inhibition to treat heart failure: a tale of science, serendipity, and second chances. Eur J Heart Fail 17(3):242–247. https://doi.org/10.1002/ejhf.250
57. McMurray JJ, Packer M, Desai AS, Gong J, Lefkowitz MP, Rizkala AR, Rouleau JL, Shi VC, Solomon SD, Swedberg K, Zile MR (2014) Angiotensin-neprilysin inhibition versus enalapril in heart failure. N Engl J Med 371(11):993–1004. https://doi.org/10.1056/NEJMoa1409077
58. Metra M, Teerlink JR (2017) Heart failure. Lancet 390(10106):1981–1995. https://doi.org/10.1016/S0140-6736(17)31071-1

59. Munzel T, Gori T, Keaney JJ, Maack C, Daiber A (2015) Pathophysiological role of oxidative stress in systolic and diastolic heart failure and its therapeutic implications. Eur Heart J 36(38):2555–2564. https://doi.org/10.1093/eurheartj/ehv305
60. Nene A, Chen CH, Disatnik MH, Cruz L, Mochly-Rosen D (2017) Aldehyde dehydrogenase 2 activation and coevolution of its epsilonPKC-mediated phosphorylation sites. J Biomed Sci 24(1):3. https://doi.org/10.1186/s12929-016-0312-x
61. Pang J, Wang J, Zhang Y, Xu F, Chen Y (2017) Targeting acetaldehyde dehydrogenase 2 (ALDH2) in heart failure-recent insights and perspectives. Biochim Biophys Acta 1863(8):1933–1941. https://doi.org/10.1016/j.bbadis.2016.10.004
62. Perin EC, Silva GV, Zheng Y, Gahremanpour A, Canales J, Patel D, Fernandes MR, Keller LH, Quan X, Coulter SA, Moore WH, Herlihy JP, Willerson JT (2012) Randomized, double-blind pilot study of transendocardial injection of autologous aldehyde dehydrogenase-bright stem cells in patients with ischemic heart failure. Am Heart J 163(3):415–421. https://doi.org/10.1016/j.ahj.2011.11.020
63. Perin EC, Murphy MP, March KL, Bolli R, Loughran J, Yang PC, Leeper NJ, Dalman RL, Alexander J, Henry TD, Traverse JH, Pepine CJ, Anderson RD, Berceli S, Willerson JT, Muthupillai R, Gahremanpour A, Raveendran G, Velasquez O, Hare JM, Hernandez SI, Kasi VS, Hiatt WR, Ambale-Venkatesh B, Lima JA, Taylor DA, Resende M, Gee AP, Durett AG, Bloom J, Richman S, G'Sell P, Williams S, Khan F, Gyang RE, Santoso MR, Goldman J, Leach D, Handberg E, Cheong B, Piece N, DiFede D, Bruhn-Ding B, Caldwell E, Bettencourt J, Lai D, Piller L, Simpson L, Cohen M, Sayre SL, Vojvodic RW, Moye L, Ebert RF, Simari RD, Hirsch AT (2017) Evaluation of cell therapy on exercise performance and limb perfusion in peripheral artery disease: the CCTRN PACE trial (patients with intermittent claudication injected with ALDH bright cells). Circulation 135(15):1417–1428. https://doi.org/10.1161/CIRCULATIONAHA.116.025707
64. Ranganath SH, Levy O, Inamdar MS, Karp JM (2012) Harnessing the mesenchymal stem cell secretome for the treatment of cardiovascular disease. Cell Stem Cell 10(3):244–258. https://doi.org/10.1016/j.stem.2012.02.005
65. Rau CD, Vondriska TM (2017) DNA methylation and human heart failure: mechanisms or prognostics. Circulation 136(16):1545–1547. https://doi.org/10.1161/CIRCULATIONAHA.117.029840
66. Ren J (2007) Acetaldehyde and alcoholic cardiomyopathy: lessons from the ADH and ALDH2 transgenic models. Novartis Found Symp 285:69–76, 76–79, 198–199
67. Russo V, Young S, Hamilton A, Amsden BG, Flynn LE (2014) Mesenchymal stem cell delivery strategies to promote cardiac regeneration following ischemic injury. Biomaterials 35(13):3956–3974. https://doi.org/10.1016/j.biomaterials.2014.01.075
68. Sarmah D, Agrawal V, Rane P, Bhute S, Watanabe M, Kalia K, Ghosh Z, Dave KR, Yavagal DR, Bhattacharya P (2017) Mesenchymal stem cell therapy in ischemic stroke: a meta-analysis of preclinical studies. Clin Pharmacol Ther. https://doi.org/10.1002/cpt.927
69. Shah AM, Mann DL (2011) In search of new therapeutic targets and strategies for heart failure: recent advances in basic science. Lancet 378(9792):704–712. https://doi.org/10.1016/S0140-6736(11)60894-5
70. Sharma A, Sharma R, Chaudhary P, Vatsyayan R, Pearce V, Jeyabal PV, Zimniak P, Awasthi S, Awasthi YC (2008) 4-Hydroxynonenal induces p53-mediated apoptosis in retinal pigment epithelial cells. Arch Biochem Biophys 480(2):85–94. https://doi.org/10.1016/j.abb.2008.09.016
71. Shen C, Wang C, Fan F, Yang Z, Cao Q, Liu X, Sun X, Zhao X, Wang P, Ma X, Zhu H, Dong Z, Zou Y, Hu K, Sun A, Ge J (2015) Acetaldehyde dehydrogenase 2 (ALDH2) deficiency exacerbates pressure overload-induced cardiac dysfunction by inhibiting Beclin-1 dependent autophagy pathway. Biochim Biophys Acta 1852(2):310–318. https://doi.org/10.1016/j.bbadis.2014.07.014
72. Shen B, Pan Q, Yang Y, Gao Y, Liu X, Li W, Han Y, Yuan X, Qu Y, Zhao Z (2017a) miR-224 Affects Mammary Epithelial Cell Apoptosis and Triglyceride Production by Downregulating

ACADM and ALDH2 Genes. DNA Cell Biol 36(1):26–33. https://doi.org/10.1089/dna.2016.3540

73. Shen C, Wang C, Han S, Wang Z, Dong Z, Zhao X, Wang P, Zhu H, Sun X, Ma X, Zhu H, Zou Y, Hu K, Ge J, Sun A (2017b) Aldehyde dehydrogenase 2 deficiency negates chronic low-to-moderate alcohol consumption-induced cardioprotecion possibly via ROS-dependent apoptosis and RIP1/RIP3/MLKL-mediated necroptosis. Biochim Biophys Acta 1863(8):1912–1918. https://doi.org/10.1016/j.bbadis.2016.11.016
74. Singh S, Arcaroli J, Thompson DC, Messersmith W, Vasiliou V (2015) Acetaldehyde and retinaldehyde-metabolizing enzymes in colon and pancreatic cancers. Adv Exp Med Biol 815:281–294. https://doi.org/10.1007/978-3-319-09614-8_16
75. Small EM, Frost RJ, Olson EN (2010) MicroRNAs add a new dimension to cardiovascular disease. Circulation 121(8):1022–1032. https://doi.org/10.1161/CIRCULATIONAHA.109.889048
76. Stachowicz A, Olszanecki R, Suski M, Wisniewska A, Toton-Zuranska J, Madej J, Jawien J, Bialas M, Okon K, Gajda M, Glombik K, Basta-Kaim A, Korbut R (2014) Mitochondrial aldehyde dehydrogenase activation by Alda-1 inhibits atherosclerosis and attenuates hepatic steatosis in apolipoprotein E-knockout mice. J Am Heart Assoc 3(6):e1329. https://doi.org/10.1161/JAHA.114.001329
77. Sui BD, Hu CH, Zheng CX, Jin Y (2016) Microenvironmental views on mesenchymal stem cell differentiation in aging. J Dent Res 95(12):1333–1340. https://doi.org/10.1177/0022034516653589
78. Sui BD, Hu CH, Liu AQ, Zheng CX, Xuan K, Jin Y (2017) Stem cell-based bone regeneration in diseased microenvironments: challenges and solutions. Biomaterials. https://doi.org/10.1016/j.biomaterials.2017.10.046
79. Sun A, Cheng Y, Zhang Y, Zhang Q, Wang S, Tian S, Zou Y, Hu K, Ren J, Ge J (2014a) Aldehyde dehydrogenase 2 ameliorates doxorubicin-induced myocardial dysfunction through detoxification of 4-HNE and suppression of autophagy. J Mol Cell Cardiol 71:92–104. https://doi.org/10.1016/j.yjmcc.2014.01.002
80. Sun A, Zou Y, Wang P, Xu D, Gong H, Wang S, Qin Y, Zhang P, Chen Y, Harada M, Isse T, Kawamoto T, Fan H, Yang P, Akazawa H, Nagai T, Takano H, Ping P, Komuro I, Ge J (2014b) Mitochondrial aldehyde dehydrogenase 2 plays protective roles in heart failure after myocardial infarction via suppression of the cytosolic JNK/p53 pathway in mice. J Am Heart Assoc 3(5):e779. https://doi.org/10.1161/JAHA.113.000779
81. Sun X, Zhu H, Dong Z, Liu X, Ma X, Han S, Lu F, Wang P, Qian S, Wang C, Shen C, Zhao X, Zou Y, Ge J, Sun A (2017) Mitochondrial aldehyde dehydrogenase-2 deficiency compromises therapeutic effect of ALDH bright cell on peripheral ischemia. Redox Biol 13:196–206. https://doi.org/10.1016/j.redox.2017.05.018
82. Tabas I (2010) The role of endoplasmic reticulum stress in the progression of atherosclerosis. Circ Res 107(7):839–850. https://doi.org/10.1161/CIRCRESAHA.110.224766
83. Takeshita T, Morimoto K, Mao XQ, Hashimoto T, Furuyama J (1993) Phenotypic differences in low km aldehyde dehydrogenase in Japanese workers. Lancet 341(8848):837–838
84. Valentova M, von Haehling S, Bauditz J, Doehner W, Ebner N, Bekfani T, Elsner S, Sliziuk V, Scherbakov N, Murin J, Anker SD, Sandek A (2016) Intestinal congestion and right ventricular dysfunction: a link with appetite loss, inflammation, and cachexia in chronic heart failure. Eur Heart J 37(21):1684–1691. https://doi.org/10.1093/eurheartj/ehw008
85. van Riet EE, Hoes AW, Wagenaar KP, Limburg A, Landman MA, Rutten FH (2016) Epidemiology of heart failure: the prevalence of heart failure and ventricular dysfunction in older adults over time. A systematic review. Eur J Heart Fail 18(3):242–252. https://doi.org/10.1002/ejhf.483
86. Varga ZV, Ferdinandy P, Liaudet L, Pacher P (2015) Drug-induced mitochondrial dysfunction and cardiotoxicity. Am J Physiol Heart Circ Physiol 309(9):H1453–H1467. https://doi.org/10.1152/ajpheart.00554.2015

87. von Hardenberg A, Maack C (2017) Mitochondrial therapies in heart failure. Handb Exp Pharmacol 243:491–514. https://doi.org/10.1007/164_2016_123
88. Wall TL, Gallen CC, Ehlers CL (1993) Effects of alcohol on the EEG in Asian men with genetic variations of ALDH2. Biol Psychiatry 34(1–2):91–99
89. Wang F, Ye P (2014) Improving heart function by modulating myocardiocyte autophagy: a possible novel mechanism for cardiovascular protection of high-density lipoprotein. Lipids Health Dis 13:163. https://doi.org/10.1186/1476-511X-13-163
90. Wang P, Shen C, Diao L, Yang Z, Fan F, Wang C, Liu X, Sun X, Dong Z, Zhu H, Ma X, Cao Q, Zhao X, Ma D, Zou Y, Hu K, Sun A, Ge J (2015) Aberrant hypermethylation of aldehyde dehydrogenase 2 promoter upstream sequence in rats with experimental myocardial infarction. Biomed Res Int 2015:503692. https://doi.org/10.1155/2015/503692
91. Wang C, Fan F, Cao Q, Shen C, Zhu H, Wang P, Zhao X, Sun X, Dong Z, Ma X, Liu X, Han S, Wu C, Zou Y, Hu K, Ge J, Sun A (2016) Mitochondrial aldehyde dehydrogenase 2 deficiency aggravates energy metabolism disturbance and diastolic dysfunction in diabetic mice. J Mol Med (Berl) 94(11):1229–1240. https://doi.org/10.1007/s00109-016-1449-5
92. Wenzel P, Schuhmacher S, Kienhofer J, Muller J, Hortmann M, Oelze M, Schulz E, Treiber N, Kawamoto T, Scharffetter-Kochanek K, Munzel T, Burkle A, Bachschmid MM, Daiber A (2008) Manganese superoxide dismutase and aldehyde dehydrogenase deficiency increase mitochondrial oxidative stress and aggravate age-dependent vascular dysfunction. Cardiovasc Res 80(2):280–289. https://doi.org/10.1093/cvr/cvn182
93. Woods C, Shang C, Taghavi F, Downey P, Zalewski A, Rubio GR, Liu J, Homburger JR, Grunwald Z, Qi W, Bollensdorff C, Thanaporn P, Ali A, Riemer K, Kohl P, Mochly-Rosen D, Gerstenfeld E, Large S, Ali Z, Ashley E (2016) In vivo post-cardiac arrest myocardial dysfunction is supported by Ca2+/Calmodulin-dependent protein kinase II-mediated calcium long-term potentiation and mitigated by Alda-1, an agonist of aldehyde dehydrogenase type 2. Circulation 134(13):961–977. https://doi.org/10.1161/CIRCULATIONAHA.116.021618
94. Wu B, Yu L, Wang Y, Wang H, Li C, Yin Y, Yang J, Wang Z, Zheng Q, Ma H (2016) Aldehyde dehydrogenase 2 activation in aged heart improves the autophagy by reducing the carbonyl modification on SIRT1. Oncotarget 7(3):2175–2188. https://doi.org/10.18632/oncotarget.6814
95. Xia G, Fan F, Liu M, Wang S, Wu J, Shen C, Han S, Wang C, Jia J, Zou Y, Hu K, Ge J, Sun A (2016) Aldehyde dehydrogenase 2 deficiency blunts compensatory cardiac hypertrophy through modulating Akt phosphorylation early after transverse aorta constriction in mice. Biochim Biophys Acta 1862(9):1587–1593. https://doi.org/10.1016/j.bbadis.2016.05.019
96. Xiao Y, Yang Z, Wu QQ, Jiang XH, Yuan Y, Chang W, Bian ZY, Zhu JX, Tang QZ (2017) Cucurbitacin B protects against pressure overload induced cardiac hypertrophy. J Cell Biochem 118(11):3899–3910. https://doi.org/10.1002/jcb.26041
97. Xu T, Liu S, Ma T, Jia Z, Zhang Z, Wang A (2017) Aldehyde dehydrogenase 2 protects against oxidative stress associated with pulmonary arterial hypertension. Redox Biol 11:286–296. https://doi.org/10.1016/j.redox.2016.12.019
98. Xue R, Zeng J, Chen Y, Chen C, Tan W, Zhao J, Dong B, Sun Y, Dong Y, Liu C (2017) Sestrin 1 ameliorates cardiac hypertrophy via autophagy activation. J Cell Mol Med 21(6):1193–1205. https://doi.org/10.1111/jcmm.13052
99. Yang X, Sreejayan N, Ren J (2005) Views from within and beyond: narratives of cardiac contractile dysfunction under senescence. Endocrine 26(2):127–137. https://doi.org/10.1385/ENDO:26:2:127
100. Yu YH, Liao PR, Guo CJ, Chen CH, Mochly-Rosen D, Chuang LM (2016) PKC-ALDH2 pathway plays a novel role in adipocyte differentiation. PLoS One 11(8):e161993. https://doi.org/10.1371/journal.pone.0161993
101. Zhang Y, Ren J (2011) ALDH2 in alcoholic heart diseases: molecular mechanism and clinical implications. Pharmacol Ther 132(1):86–95. https://doi.org/10.1016/j.pharmthera.2011.05.008

102. Zhang P, Xu D, Wang S, Fu H, Wang K, Zou Y, Sun A, Ge J (2011) Inhibition of aldehyde dehydrogenase 2 activity enhances antimycin-induced rat cardiomyocytes apoptosis through activation of MAPK signaling pathway. Biomed Pharmacother 65(8):590–593. https://doi.org/10.1016/j.biopha.2009.12.001
103. Zhang H, Gong DX, Zhang YJ, Li SJ, Hu S (2012a) Effect of mitochondrial aldehyde dehydrogenase-2 genotype on cardioprotection in patients with congenital heart disease. Eur Heart J 33(13):1606–1614. https://doi.org/10.1093/eurheartj/ehs061
104. Zhang Y, Babcock SA, Hu N, Maris JR, Wang H, Ren J (2012b) Mitochondrial aldehyde dehydrogenase (ALDH2) protects against streptozotocin-induced diabetic cardiomyopathy: role of GSK3beta and mitochondrial function. BMC Med 10:40. https://doi.org/10.1186/1741-7015-10-40
105. Zhang Y, Mi SL, Hu N, Doser TA, Sun A, Ge J, Ren J (2014) Mitochondrial aldehyde dehydrogenase 2 accentuates aging-induced cardiac remodeling and contractile dysfunction: role of AMPK, Sirt1, and mitochondrial function. Free Radic Biol Med 71:208–220. https://doi.org/10.1016/j.freeradbiomed.2014.03.018
106. Zhang Y, Wang C, Zhou J, Sun A, Hueckstaedt LK, Ge J, Ren J (2017) Complex inhibition of autophagy by mitochondrial aldehyde dehydrogenase shortens lifespan and exacerbates cardiac aging. Biochim Biophys Acta 1863(8):1919–1932. https://doi.org/10.1016/j.bbadis.2017.03.016
107. Zhu H, Sun A, Zhu H, Li Z, Huang Z, Zhang S, Ma X, Zou Y, Hu K, Ge J (2014) Aldehyde dehydrogenase-2 is a host factor required for effective bone marrow mesenchymal stem cell therapy. Arterioscler Thromb Vasc Biol 34(4):894–901. https://doi.org/10.1161/ATVBAHA.114.303241
108. Zhu Y, Zhang D, Zhou D, Li Z, Li Z, Fang L, Yang M, Shan Z, Li H, Chen J, Zhou X, Ye W, Yu S, Li H, Cai L, Liu C, Zhang J, Wang L, Lai Y, Ruan L, Sun Z, Zhang S, Wang H, Liu Y, Xu Y, Ling J, Xu C, Zhang Y, Lv D, Yuan Z, Zhang J, Zhang Y, Shi Y, Lai M (2017) Susceptibility loci for metabolic syndrome and metabolic components identified in Han Chinese: a multi-stage genome-wide association study. J Cell Mol Med 21(6):1106–1116. https://doi.org/10.1111/jcmm.13042

Mitochondrial Aldehyde Dehydrogenase in Myocardial Ischemic and Ischemia-Reperfusion Injury

Jie Ding, Zheng Yang, Heng Ma, and Hao Zhang

Abstract Myocardial ischemia-reperfusion (IR) injury during acute myocardial infarction or open-heart surgery would promote oxidative stress, leading to the accumulation of reactive aldehydes that cause cardiac damage. It has been well demonstrated that aldehyde dehydrogenase (ALDH)-2 is an important cardioprotective enzyme for its central role in the detoxification of reactive aldehydes. ALDH2 activation by small molecule activators is a promising approach for cardioprotection for myocardial IR injury.

1 Introduction

Coronary artery disease (CAD), also known as ischemia heart disease (IHD), is referred to as the obstruction of coronary artery leading to the interruption of myocardium blood supply [1]. The most obvious manifestation and consequence of CAD is acute myocardial infarction (AMI), which results in leading cause of morbidity and mortality worldwide [2]. The most effective treatment for reducing myocardial infarction size is timely reperfusion. However, the reperfusion process then triggers myocardial damage, cardiomyocyte death, and cardiac dysfunction, the

Authors Jie Ding and Zheng Yang have been equally contributed to this chapter.

J. Ding
State Key Laboratory of Cardiovascular Disease, Anesthesia Department, Fuwai Hospital, National Center for Cardiovascular Diseases, Chinese Academy of Medical Science and Peking Union Medical College, Beijing, China

Z. Yang · H. Ma (✉)
Department of Physiology and Pathophysiology, Fourth Military Medical University, Xi'an, China
e-mail: hengma@fmmu.edu.cn

H. Zhang (✉)
Heart Center and Shanghai Institution of Pediatric Congenital Heart Disease, Shanghai Children's Medical Center, National Children's Medical Center, Shanghai Jiaotong University School of Medline, Shanghai, China

J. Ren et al. (eds.), *Aldehyde Dehydrogenases*, Advances in Experimental Medicine and Biology 1193, https://doi.org/10.1007/978-981-13-6260-6_6

so-called "ischemia-reperfusion injury" (IR injury) [3]. Unfortunately, there is no effective therapy against IR injury that is currently available.

Open-heart surgery with cardiopulmonary bypass (CPB) is a standard procedure for cardiac surgery. Myocardial IR injury is also unavoidable during cardiac arrest under CPB, which is characterized by the sharp production of superoxide, reactive oxygen species (ROS), and aldehyde accumulation [4, 5]. Although the hypothermia and cardioplegia infusion has become the routine approaches for protecting the heart from IR injury [6], many patients still are at high risk for complications following cardiac surgery, which mainly present as severe myocardial injury and worsened cardiac function [7].

It has been first reported by Mochly-Rosen's group that mitochondrial aldehyde dehydrogenase (ALDH)-2 has cardioprotective role in the onset and progression during myocardial IR injury [2]. They found that activation of ALDH2 could reduce ischemic cardiac damage. Using Alda-1, a small molecule ALDH2 activator, N-(1,3-benzodioxol-5-ylmethyl)-2,6-dichlorobenzamide, prior to ischemia could reduce infarct size. Thus, the potential role of ALDH2 activation to limit IR injury has drawn broad attention recently.

2 Mitochondrial Aldehyde Dehydrogenase in Coronary Artery Disease

CAD is a leading cause of death worldwide [8]. CAD includes a group of diseases, such as stable angina, unstable angina, myocardial infarction (MI), and sudden cardiac death [9], which involves the pathologic process of coronary arteries atherosclerosis. MI usually occurs without prodromes and clinical signs following with severe complications, such as cardiac rupture and ventricular fibrillation. Despite the advances in diagnosis and management of MI, the morbidity of MI is still high. The most effective treatment for MI damage is timely reperfusion; however, IR injury triggers the following: myocardial injury, cardiomyocyte death, and cardiac dysfunction itself. ACE inhibitors, β-blockers, statins, and mineralocorticoid receptor antagonists are widely used in post-MI management [10]; however, there are still a number of patients that ultimately develop heart failure or other complications. An amount of studies suggested the cardioprotective role of ALDH2 in ischemia and/or reperfusion injury. ALDH2 activation or overexpression could attenuate myocardial infarct size and restore cardiac function [11–14]. Thus, it is urgent to investigate the cardioprotection in the ischemia-reperfusion injury.

2.1 *ALDH2 Genetic Polymorphism and CAD*

2.1.1 ALDH2 Genetic Polymorphism Related with Increased Risks of CAD

ALDH2*2 mutation with compromised ALDH2 activity has been reported as an independent risk factor of acute coronary syndrome in epidemiological research [15]. Compared with ALDH2*1 carriers, ALDH2*2 carriers displayed a significantly higher incidence of acute coronary syndrome and ST-segment elevation myocardial infarction (STEMI) among Chinese male [16]. And an increased risk linked with the development of acute coronary syndrome was found in a case control study in Korean male [17]. ALDH2*2 carriers showed elevated inflammatory markers in response to acute myocardial infarction, like high C-sensitive protein, hs-CRP, which indicated severe damages following ischemic injury [18].

Genetic susceptibility, environmental triggers, and their interactions play a crucial role in the etiology of CAD [19, 20]. Alcohol consumption has been regarded as a major risk factor in the development of CAD [21, 22]. The associations between CAD and alcohol consumption show inconsistency in different distribution of genetic polymorphisms in drinking-related enzymes, especially ALDH2 [23], which has been considered to be the primary enzyme referring to acetaldehyde oxidation [24]. The SNP (rs671) allele [i.e., with 504glu encoded by glutamine (G)] is the most common form and is usually referred to as ALDH2*1, On the other hand, the ALDH2*2 (504lys) allele denotes ALDH2 mutation resulting in the expression of inactive catalytic isozyme [25]. The variant 504lys allele with elevated blood acetaldehyde levels after alcohol consumption is commonly present in decedents from Northeast Asia [17]. High acetaldehyde concentration contributes the development of CAD through a degraded detoxification of 4-hydroxy-2-nonenal (4-HNE) and asymmetric dimethylarginine (ADMA) [26]. Recent studies also reported that ALDH2 Glu504lys polymorphism could be associated with increased risk of CAD [17, 27]. ALDH2 rs671 polymorphism related with increased risks of CAD has also been reported [28].

2.1.2 ALDH2 Genetic Polymorphism Related with Drug Efficacy Treating CAD

Another aspect relevant to CAD and ALDH2 relates to nitroglycerin metabolism. Nitroglycerin has been widely used as a vasodilator to treat ischemic and congestive cardiac diseases, including angina pectoris, myocardial infarction, and heart failure

[29, 30]. Chen et al. [31] demonstrated that ALDH2 plays an essential role for the bioconversion of nitroglycerin to NO, to achieve its vasodilating effects. However, nitroglycerin is a potent inhibitor for ALDH2, and consecutive treatment of nitroglycerin will lead to insensitivity to the vasodilating effect in patients with CAD [32]. Yet nitroglycerin use in East Asia for patients with acute myocardial infarction and in congestive heart failure is very common. As expected, nitroglycerin is less cardiovascular protective in East Asians carrying ALDH2*2 mutation compared with those carriers with wild-type ALDH2 [24, 33, 34].

2.2 *Modified ALDH2 Activity Mouse Models in Myocardial Ischemic Injury*

ALDH2 is a tetrameric allosteric enzyme localized widely in organs including the heart and plays a main role in ethanol metabolism through oxidation. Ethanol is firstly catalyzed by alcohol dehydrogenase (ADH) to acetaldehyde, then catalyzed to acetic acid by ALDH2. ALDH2 could also eliminate toxic aldehydes produced during lipid peroxidation, like 4-hydroxy-2-nonenal (4-HNE) [35, 36].

Considering to its genetic polymorphism and enzymatic activity, ALDH2*1 is referred to as wild-type genotype with full enzymatic activity. A common human polymorphism ALDH2*2 is caused by a single nucleotide substitution, with glutamate being transferred to lysine at amino acid 487 (E487K) to yield compromised enzymatic activity [37]. ALDH2 enzymatic activity can be severely influenced by this change in any of its subunits. Heterozygotes ALDH2*1/2 display only 30%–40% of wild-type ALDH2 enzymatic activity, and the enzymatic activity of the mutant homozygotes ALDH2*2/2 allele only has <1–4% of wild-type activity. Up to 40% of the East Asian carry ALDH2*2 with compromised enzymatic activity. So in order to recapitulate the human ALDH2 phenotypes and relative diseases or physiological reactions, transgenic animal models were generated.

It manifests defects in alcohol metabolism, accumulation of aldehydic adducts, and susceptibility to myocardial IR injury [29]. Unlike ALDH2 knockout mice, ALDH2*2 transgenic mice keep some residual ALDH2 enzyme activity, and ALDH2 activity can be reversed by correcting the structural defect [38]. ALDH2 overexpressing mice manifest augmented ALDH2 enzymatic activity, which ameliorated deleterious effects of acetaldehyde, oxidative stress, and 4-HNE in various organs [39–41]. ALDH2*2 knock-in mice was developed with the homologous recombination of wild-type ALDH2 allele and E487K substituted ALDH2 allele. In vitro assays observed that, when activated by Alda-1, mouse ALDH2*2 recombination enzyme showed similar activation kinetics and potency with human, which is considered as the best model for human ALDH2*2 mutation phenotype study [42].

2.3 Possible Mechanisms of ALDH2 Cardioprotection

The protective effects in myocardial IR injury are involved in many mechanisms. The initial discovery by Mochly-Rosen's group suggested that ALDH2 is a downstream player of protein kinase C type ε (εPKC). There are several targets of εPKC, including ROS production, mitochondrial kATP channel regulation, mitochondrial permeability transition pore (MPTP) opening, etc. Accumulation of reactive aldehydes, 4-HNE, is a key process during IR injury; εPKC-mediating ALDH2 activation increases detoxification of 4-HNE [43]. ROS generation, with its peroxide ions, oxidize cell membranes and organelle lipids which lead to elevated reactive aldehydes. Reactive aldehydes, in turn, trigger augmented production of ROS. ALDH2 inhibits this malignant feedback by diminish reactive aldehydes, which provides a more specific therapeutic targets in cardioprotection.

Apoptosis contributes cardiomyocytes loss during IR injury. JNK/c-Jun signaling can be triggered by ROS through both intrinsic and extrinsic pathways. Cardiomyocytes with ALDH2 deficiency were found with accumulation of ROS, activation of JNK signaling, and increased expression of c-Jun [44]. It was also reported that downregulation of ALDH2 elevates 4-HNE, which inhibits HSP70 and triggers JNK/p53, resulting in increased cardiomyocyte apoptosis [45]. During IR, autophagy is upregulated; it can lead to loss of cardiomyocytes despite its myocardial protection effect [46–48]. ALDH2 plays an important role in regulating IR-induced autophagy. It has been found that, on the one hand, ALDH2 activates LKB1/AMPK/mTOR pathway during ischemia phase. On the other hand, ALDH2 suppresses excessive autophagy via PTEN/Akt/mTOR pathway during reperfusion [49].

Lipid peroxidation is a consequence of increased oxidative stress induced by IR injury [50]. Protein carbonylation by reactive carbonyl species (RCS) is irreversible and irreparable modification which may changes the conformation of polypeptide chain then partially or totally inactivates proteins. Myocardium protein carbonylation shows a time-dependent increase in ischemia and decrease in reperfusion. Activating ALDH2 by Alda-1 can eliminate RCS and improve the cardioprotection effects [51].

2.4 ALDH2 Protects the Heart Against Vulnerability to Ischemic Injury

ALDH2 protects the aging heart against ischemic injury. The free radical theory of aging was put up by Denham Harman in 1956. Getting rid of mitochondrial oxygen species increases life span, and it has been proved by many studies. Suppression of mitochondrial lipid peroxidative aldehydes is considered to be one of the underlying mechanisms [52, 53].

The ability of ALDH2 to remove 4-HNE declines with aging. Thus, loss of ALDH2 activation plays an important role in impaired cardiac tolerance to ischemic stress in aging. Recent research also found that ALDH2 ablation induced both carbonyl stress and autophagy decline attribute to heart senescence. Our group found that blunted SIRT1 activity due to carbonyl stress triggers chronic inflammation and cell arrest [54].

Decreased SIRT1 also leads to impairment in autophagy flux. Both of these contribute to age-related ischemic tolerance [55] [56].

Pain is not an isolated pathological process, it does impact on many organs. ALDH2 activation also provides a therapeutic strategy in pain, including peripheral neuropathy pain and inflammatory pain. It has been reported that transient peripheral nociception precondition was observed as a cardioprotective phenomenon similar to ischemic preconditioning (IPC). Chronic pain is deemed a long-term malignant stress with significant organ consequences. Epidemic study revealed high relevance between chronic pain and cardiovascular disease [57]. Clinical observations showed the increased mortality and morbidity of ischemic heart with chronic pain. Our group observed that chronic pain increases carbonyl stress after IR injury. SIRT1 carbonyl modification by 4-HNE induces the block of LKB1-AMPK interaction, consequently AMPK activation impairs. ALDH2 activation effectively prevents SIRT1 carbonylation and reverses the interaction of LKB1-AMPK, denoting its therapeutic promises in the management of chronic pain-related IR injury [58].

Diabetes mellitus is a metabolic disorder. The mortality and morbidity increases in diabetic patients, and the heart failure happens after myocardial infarction among diabetics is 2–3 times compared with nondiabetic patients [11]. Under diabetes condition, basal oxidative stress increases; ischemic conditioning impair is impaired. After IR injury, PI3K/Akt/GSK3-β signaling pathway and ERK1/2 phosphorylation are inhibited. Decreased generation and release of nitric oxide and CGRP and dysfunction of KATP channels lead to elevated oxidative stress [3]. Research showed that myocardial ALDH2 expression decreases with the progression of diabetes. Activating ALDH2 can decrease the ratio of Bax/Bcl2, activates PI3K/Akt/GSK3-β signaling pathway, and inhibits MPTP opening, which plays an antiapoptotic role in diabetic IR injury [59].

3 Mitochondrial Aldehyde Dehydrogenase in Myocardial Ischemia-Reperfusion (IR) Injury During Perioperative Period of Cardiac Surgery

Evidence from animal studies has shown that ALDH2 activation during cardioplegic arrest enhances the cardioprotection against myocardial ischemia-reperfusion injury. What's more, a clinical research indicated that ALDH2*2 carriers with cyanotic congenital heart disease were associated with an induced metabolic remodeling phenotype and a compensatory myocardium glutathione (GSH) pool. When

ALDH2 activity was impaired during open-heart surgery, this larger GSH pool could lead to unexpectedly better cardioprotection. This may aid in the prediction of cardioprotection outcomes and identification of individualized cardioprotective strategies.

Furthermore, ALDH2 may participate in the respiratory homeostasis. Lung IR injury often results in respiratory insufficiency following open-heart surgery such as cardiopulmonary bypass procedures. Cellular and animal studies indicated that activation of ALDH2 could attenuate lung IR injury, possibly through accelerating the clearance of 4-hydroxy-2-nonenal in human pulmonary alveolar epithelial cells.

3.1 Activity of ALDH2 to Prevent Myocardial IR Injury During Cardiac Surgery

During open-heart surgery, myocardial IR insult, caused by aortic cross-clamping and removal of aortic cross-clamping, remains the most common cause of cardiac morbidity and mortality despite major advances in cardiovascular medicine [60]. Therefore, it is of great importance to identify novel cardioprotective strategies to minimize myocardial IR injury.

When performing open-heart surgery, the mismatch between cardiac oxygen demand and supply will result in a myocardial oxidative stress state in which excessive ROS are accumulated. Most biological ROS are highly reactive and short-lived; thus, ROS-induced injury is restricted to their site of origin with limited significance. However, toxic aldehydes generated from the peroxidation of unsaturated lipids are abundant and more stable than ROS. Therefore, they are thought to be secondary products that amplify and propagate ROS-initiated IR damage [4]. A large amount of evidence has shown that these cytotoxic aldehydes play a deleterious role in the progression of IR injury [61]. Aldehydes can modify the proteasomes and decline their peptidase activities [62]; they directly inhibit NADH-linked mitochondrial respiration following both cardiac ischemia and reperfusion [63]; the activity of cardiac glycolytic enzymes can also be diminished by aldehydes following ischemia and reperfusion [64]; and significant associations between aldehydes and the onset of arrhythmias and heart failure have also been reported [65]. What's more, excessive aldehydes can damage cardiomyocytes and impair heart function through disruption of actin-myosin interactions, enzymatic functions, calcium sensitivity, and efficiency of cross-bridge cycling [66].

Cardioplegic solutions are used to alleviate IR injury during open-heart surgery. Although well developed, cardioplegic solutions cannot completely block IR injury. Histidine-tryptophan-ketoglutarate (HTK) solution is a widely used cardioprotective solution in open-heart surgery to alleviate ischemic damage [67]. Although it contains antioxidants, HTK cannot block aldehyde formation completely, and no specific agents are included in HTK solution to remove toxic aldehydes. Therefore, aldehyde accumulation is inevitable during cardiopulmonary bypass surgery.

Hypothermia is another powerful method used during open-heart surgery to reduce IR injury, and ALDH2 is a bio-enzyme whose activity can be easily inhibited by hypothermia. Previous studies have proved that during open-heart surgeries, ALDH2 activities can be inhibited to a very lower level after cardioplegia and ischemia under hypothermic conditions [68]. The inhibition of ALDH2 activity would significantly impair the clearance of active aldehydes and result in excessive aldehyde accumulation. On the other hand, aldehydes are both substrates for ALDH2 and direct inhibitors of this enzyme [69], ALDH2 can also be inhibited by aldehydes via adduct formation [70]. Therefore, excessive aldehydes in turn inhibit ALDH2 activity to form a vicious cycle in the hypothermic ischemic heart.

Evidence from animal studies [51] has shown that Alda-1 supplementation can significantly improve the cardioprotection effect of cardioplegia solution under hypothermic conditions, possibly through activation of ALDH2, to remove toxic aldehydes, and in turn attenuated IR-induced elevation in creatine kinase isoenzyme MB leakage and protein carbonyl formation. The Alda-1 addition also obtained higher glutathione/oxidized glutathione ratios and alleviated IR-induced cardiomyocyte contractile function impairment as evidenced by improved maximal velocity of pressure development and decline, left ventricular developed pressure, and heart rate.

3.2 *Effect of Gene Polymorphism of ALDH2 in Myocardial IR Injury During Cardiac Surgery*

At least 540 million of the world population carry a common single nucleotide polymorphism (rs671) in ALDH2 [71], where a glutamate at amino acid 504 is replaced by a lysine (E504K) to form an ALDH2 loss-of-function allele (ALDH2*2 allele) [37]. ALDH2*2 carriers display dramatically reduced ALDH2 activities [72]. Since ALDH2*2 carriers exhibit impaired ALDH2 activity in the detoxification of reactive aldehydes, it is warranted that cardioprotective outcomes should be reevaluated in these individuals.

However, Endo et al. [73] found that carrying an ALDH2 loss-of-function allele (ALDH2*2 allele) also has cardioprotective effects. Using ALDH2*2 transgenic mice, they found that myocardial oxidative stress increased the expression of AFT4 and 3-phosphoglycerate dehydrogenase (PHGDH) in ALDH2*2 transgenic hearts, which further led to metabolic remodeling to produce more GSH. Thereby, the increase in the intrinsic myocardial GSH pool made ALDH2*2 transgenic mice have a greater tolerance to IR damage.

Our group [68] investigated the effect of the ALDH2*2 allele on human cardioprotection after open-heart surgery. Young patients with congenital heart diseases were selected, and we found that cyanosis inhibited ALDH2 activity and led to accumulation of aldehyde in ALDH2*2 carriers. This accumulation was associated with greater expression of activating factor (ATF)-4 and larger myocardium GSH pools. Inhibited ALDH2 activity during surgery and lower intrinsic myocardium GSH levels could result in more aldehydes accumulated in the heart of ALDH2*2 non-carriers, and these excessive toxic aldehydes in turn caused deleterious cascades through ALDH2 adducts formation on a variety of macromolecules to impair heart function. Consequently, ALDH2*2 non-carriers exhibited poor tolerance to IR injury caused by cardiopulmonary bypass and manifested worse clinical outcomes.

3.3 ALDH2 Activation Attenuates Pulmonary IR Injury During Cardiac Surgery

Lung IR injury is the most common cause of respiratory insufficiency after open-heart surgery [74] and is inevitable during a cardiopulmonary bypass [75, 76]. Previous pulmonary protective strategies are effective but may not be sufficient to prevent lung IR injury [77]. Complicated postoperative courses caused by LIRI are not rare in clinical practice [74]. ALDH2 is abundantly expressed in the lung [78, 79].

A study about lung IR injury has demonstrated the potential protective effect of ALDH2[[80]. IR-induced pulmonary injury concomitantly induced aldehydes accumulation in human pulmonary alveolar epithelial cells and lung tissues, but not in human pulmonary microvascular endothelial cells. Moreover, Alda-1 pretreatment significantly elevated ALDH2 activity, increased surfactant-associated protein C, and attenuated elevation of 4-hydroxy-2-nonenal, apoptosis, intercellular adhesion molecule-1, inflammatory response, and the permeability of pulmonary alveolar capillary barrier, thus alleviated injury.

4 Conclusion and Perspective

As discussed here, ALDH2 is a well-known member of 19 different aldehyde dehydrogenases that are present in humans. Much of what we knew about ALDH2 stemmed from an interest in ethanol metabolism. However, an exciting body of more recent research indicates that this enzyme provides an important shield from oxidative stress and increased aldehydic load, thus contributing to better health. Myocardial IR injury induced by CAD and CPB during open-heart surgery is associated with increased aldehydic load.

In this review, we examined the evidence from animal studies and from human epidemiological studies. We believe that aldehydes are major contributors to the pathology of myocardial IR injury. ALDH2 is capable of attenuating myocardial IR injury through acetaldehyde detoxification. Allelic variations of ALDH2 show different relation with CAD. Several studies revealed ALDH2 cardioprotective role using genetically modified ALDH2 models in cardiac remodeling and myocardial dysfunction. Therefore, ALDH2 possesses crucial therapeutic potential against myocardial damage.

However, an intriguing mystery remains: How to explain the existence of widespread ALDH2*2 mutation in East Asians? Why could a single monogenetic mutation that occurred 2000–3000 years ago affect an estimated 8% of the human race? Could there be a selective advantage for having the mutation? A few hypotheses have been postulated, but none has been confirmed thus far [81].

References

1. Hanson MA, Fareed MT, Argenio SL, Agunwamba AO, Hanson TR (2013) Coronary artery disease. Prim Care 40:1–16. https://doi.org/10.1016/j.pop.2012.12.001
2. Mozaffarian D et al (2015) Heart disease and stroke statistics--2015 update: a report from the American Heart Association. Circulation 131:e29–e322. https://doi.org/10.1161/CIR.0000000000000152
3. Lejay A et al (2016) Ischemia reperfusion injury, ischemic conditioning and diabetes mellitus. J Mol Cell Cardiol 91:11–22. https://doi.org/10.1016/j.yjmcc.2015.12.020
4. Hill BG, Bhatnagar A (2009) Beyond reactive oxygen species: aldehydes as arbitrators of alarm and adaptation. Circ Res 105:1044–1046. https://doi.org/10.1161/CIRCRESAHA.109.209791
5. Milei J, Grana D (2007) R., Forcada, P. & Ambrosio, G. mitochondrial oxidative and structural damage in ischemia-reperfusion in human myocardium. Current knowledge and future directions. Front Biosci 12:1124–1130
6. White CW et al (2016) Avoidance of profound hypothermia during initial reperfusion improves the functional recovery of hearts donated after circulatory death. Am J Transplant 16:773–782. https://doi.org/10.1111/ajt.13574
7. Sepehri A et al (2014) The impact of frailty on outcomes after cardiac surgery: a systematic review. J Thorac Cardiovasc Surg 148:3110–3117. https://doi.org/10.1016/j.jtcvs.2014.07.087
8. Roger VL et al (2012) Executive summary: heart disease and stroke statistics--2012 update: a report from the American Heart Association. Circulation 125:188–197. https://doi.org/10.1161/CIR.0b013e3182456d46
9. Wong ND (2014) Epidemiological studies of CHD and the evolution of preventive cardiology. Nat Rev Cardiol 11:276–289. https://doi.org/10.1038/nrcardio.2014.26
10. Bonsu KO, Owusu IK, Buabeng KO, Reidpath DD, Kadirvelu A (2016) Review of novel therapeutic targets for improving heart failure treatment based on experimental and clinical studies. Therap Clin Risk Manag 12:887–906. https://doi.org/10.2147/tcrm.s106065
11. Zhang Y, Ren J (2011) ALDH2 in alcoholic heart diseases: molecular mechanism and clinical implications. Pharmacol Ther 132:86–95. https://doi.org/10.1016/j.pharmthera.2011.05.008
12. Yin G et al (2016) ALDH2 polymorphism is associated with fasting blood glucose through alcohol consumption in Japanese men. Nagoya J Med Sci 78:183–193
13. Yokoyama A et al (2013) Genetic polymorphisms of alcohol dehydrogenase-1B and aldehyde dehydrogenase-2 and liver cirrhosis, chronic calcific pancreatitis, diabetes mellitus, and hypertension among Japanese alcoholic men. Alcohol Clin Exp Res 37:1391–1401. https://doi.org/10.1111/acer.12108
14. Dakeishi M, Murata K, Sasaki M, Tamura A, Iwata T (2008) Association of alcohol dehydrogenase 2 and aldehyde dehydrogenase 2 genotypes with fasting plasma glucose levels in Japanese male and female workers. Alcohol Alcohol 43:143–147. https://doi.org/10.1093/alcalc/agm173
15. Takagi S et al (2002) Aldehyde dehydrogenase 2 gene is a risk factor for myocardial infarction in Japanese men. Hypertens Res 25:677–681

16. Xu F et al (2011) Role of aldehyde dehydrogenase 2 Glu504lys polymorphism in acute coronary syndrome. J Cell Mol Med 15:1955–1962. https://doi.org/10.1111/j.1582-4934.2010.01181.x
17. Jo SA et al (2007) A Glu487Lys polymorphism in the gene for mitochondrial aldehyde dehydrogenase 2 is associated with myocardial infarction in elderly Korean men. Clin Chim Acta 382:43–47. https://doi.org/10.1016/j.cca.2007.03.016
18. Bian Y et al (2010) The polymorphism in aldehyde dehydrogenase-2 gene is associated with elevated plasma levels of high-sensitivity C-reactive protein in the early phase of myocardial infarction. Tohoku J Exp Med 221:107–112
19. Broeckel U et al (2002) A comprehensive linkage analysis for myocardial infarction and its related risk factors. Nat Gen 30:210–214. https://doi.org/10.1038/ng827
20. Yamada Y et al (2002) Prediction of the risk of myocardial infarction from polymorphisms in candidate genes. N Engl J Med 347:1916–1923. https://doi.org/10.1056/NEJMoa021445
21. Hvidtfeldt UA et al (2010) Alcohol intake and risk of coronary heart disease in younger, middle-aged, and older adults. Circulation 121:1589–1597. https://doi.org/10.1161/circulationaha.109.887513
22. Huang B et al (2012) Adverse cardiovascular effects of concomitant use of proton pump inhibitors and clopidogrel in patients with coronary artery disease: a systematic review and meta-analysis. Arch Med Res 43:212–224. https://doi.org/10.1016/j.arcmed.2012.04.004
23. Edenberg HJ (2007) The genetics of alcohol metabolism: role of alcohol dehydrogenase and aldehyde dehydrogenase variants. Alcohol Res Health 30:5–13
24. Li Y et al (2006) Mitochondrial aldehyde dehydrogenase-2 (ALDH2) Glu504Lys polymorphism contributes to the variation in efficacy of sublingual nitroglycerin. J Clin Invest 116:506–511. https://doi.org/10.1172/JCI26564
25. Eng MY, Luczak SE, Wall TL (2007) ALDH2, ADH1B, and ADH1C genotypes in Asians: a literature review. Alcohol Res Health 30:22–27
26. Isse T, Matsuno K, Oyama T, Kitagawa K, Kawamoto T (2005) Aldehyde dehydrogenase 2 gene targeting mouse lacking enzyme activity shows high acetaldehyde level in blood, brain, and liver after ethanol gavages. Alcohol Clin Exp Res 29:1959–1964
27. Xu F et al (2007) The polymorphism in acetaldehyde dehydrogenase 2 gene, causing a substitution of Glu > Lys (504), is not associated with coronary atherosclerosis severity in Han Chinese. Tohoku J Exp Med 213:215–220
28. Han H et al (2013) Association of genetic polymorphisms in ADH and ALDH2 with risk of coronary artery disease and myocardial infarction: a meta-analysis. Gene 526:134–141. https://doi.org/10.1016/j.gene.2013.05.002
29. Fernandez E et al (2006) Monoamine metabolism and behavioral responses to ethanol in mitochondrial aldehyde dehydrogenase knockout mice. Alcohol Clin Exp Res 30:1650–1658. https://doi.org/10.1111/j.1530-0277.2006.00200.x
30. Parker JD, Parker JO (1998) Nitrate therapy for stable angina pectoris. N Engl J Med 338:520–531. https://doi.org/10.1056/NEJM199802193380807
31. Chen Z et al (2005) An essential role for mitochondrial aldehyde dehydrogenase in nitroglycerin bioactivation. Proc Natl Acad Sci U S A 102:12159–12164. https://doi.org/10.1073/pnas.0503723102
32. Ferreira JC, Mochly-Rosen D (2012) Nitroglycerin use in myocardial infarction patients. Circ J 76:15–21
33. Guo R et al (2008) Evidence for involvement of calcitonin gene-related peptide in nitroglycerin response and association with mitochondrial aldehyde dehydrogenase-2 (ALDH2) Glu504Lys polymorphism. J Am Coll Cardiol 52:953–960. https://doi.org/10.1016/j.jacc.2008.05.049
34. Zhang H et al (2007) The relationship between aldehyde dehydrogenase-2 gene polymorphisms and efficacy of nitroglycerin. Zhonghua Nei Ke Za Zhi 46:629–632
35. Chen CH et al (2008) Activation of aldehyde dehydrogenase-2 reduces ischemic damage to the heart. Science 321:1493–1495. https://doi.org/10.1126/science.1158554

36. Sun A et al (2014) Aldehyde dehydrogenase 2 ameliorates doxorubicin-induced myocardial dysfunction through detoxification of 4-HNE and suppression of autophagy. J Mol Cell Cardiol 71:92–104. https://doi.org/10.1016/j.yjmcc.2014.01.002
37. Goedde HW et al (1983) Population genetic studies on aldehyde dehydrogenase isozyme deficiency and alcohol sensitivity. Am J Human Gen 35:769–772
38. Hoshi H et al (2012) Aldehyde-stress resulting from Aldh2 mutation promotes osteoporosis due to impaired osteoblastogenesis. J Bone Miner Res 27:2015–2023. https://doi.org/10.1002/jbmr.1634
39. Doser TA et al (2009) Transgenic overexpression of aldehyde dehydrogenase-2 rescues chronic alcohol intake-induced myocardial hypertrophy and contractile dysfunction. Circulation 119:1941–1194. https://doi.org/10.1161/CIRCULATIONAHA.108.823799
40. Ge W, Guo R, Ren J (2011) AMP-dependent kinase and autophagic flux are involved in aldehyde dehydrogenase-2-induced protection against cardiac toxicity of ethanol. Free Radic Biol Med 51:1736–1748. https://doi.org/10.1016/j.freeradbiomed.2011.08.002
41. Ge W, Ren J (2012) mTOR-STAT3-notch signalling contributes to ALDH2-induced protection against cardiac contractile dysfunction and autophagy under alcoholism. J Cell Mol Med 16:616–626. https://doi.org/10.1111/j.1582-4934.2011.01347.x
42. Kim J, Chen CH, Yang J, Mochly-Rosen D (2017) Aldehyde dehydrogenase 2*2 knock-in mice show increased reactive oxygen species production in response to cisplatin treatment. J Biomed Sci 24:33. https://doi.org/10.1186/s12929-017-0338-8
43. Churchill EN, Disatnik MH, Mochly-Rosen D (2009) Time-dependent and ethanol-induced cardiac protection from ischemia mediated by mitochondrial translocation of varepsilonPKC and activation of aldehyde dehydrogenase 2. J Mol Cell Cardiol 46:278–284. https://doi.org/10.1016/j.yjmcc.2008.09.713
44. Ebert AD et al (2014) Characterization of the molecular mechanisms underlying increased ischemic damage in the aldehyde dehydrogenase 2 genetic polymorphism using a human induced pluripotent stem cell model system. Sci Transl Med 6:255 ra130. https://doi.org/10.1126/scitranslmed.3009027
45. Sun A et al (2014) Mitochondrial aldehyde dehydrogenase 2 plays protective roles in heart failure after myocardial infarction via suppression of the cytosolic JNK/p53 pathway in mice. J Am Heart Assoc 3:e000779. https://doi.org/10.1161/JAHA.113.000779
46. Cooper C, Campion G, Melton L (1992) J., 3rd. Hip fractures in the elderly: a world-wide projection. Osteoporos Int 2:285–289
47. Cui L et al (2010) Sublethal total body irradiation leads to early cerebellar damage and oxidative stress. Curr Neurovasc Res 7:125–135
48. Dandre F, Cassaigne A, Iron A (1995) The frequency of the mitochondrial aldehyde dehydrogenase I2 (atypical) allele in Caucasian, oriental and African black populations determined by the restriction profile of PCR-amplified DNA. Mol Cell Probes 9:189–193. https://doi.org/10.1006/mcpr.1995.0030
49. Ma H, Guo R, Yu L, Zhang Y, Ren J (2011) Aldehyde dehydrogenase 2 (ALDH2) rescues myocardial ischaemia/reperfusion injury: role of autophagy paradox and toxic aldehyde. Eur Heart J 32:1025–1038. https://doi.org/10.1093/eurheartj/ehq253
50. Becker LB (2004) New concepts in reactive oxygen species and cardiovascular reperfusion physiology. Cardiovasc Res 61:461–470. https://doi.org/10.1016/j.cardiores.2003.10.025
51. Gong D et al (2012) Aldehyde dehydrogenase-2 activation during cardioplegic arrest enhances the cardioprotection against myocardial ischemia-reperfusion injury. Cardiovasc Toxicol 12:350–358. https://doi.org/10.1007/s12012-012-9179-6
52. Ohsawa I et al (2008) Age-dependent neurodegeneration accompanying memory loss in transgenic mice defective in mitochondrial aldehyde dehydrogenase 2 activity. J Neurosci 28:6239–6249. https://doi.org/10.1523/JNEUROSCI.4956-07
53. Yu BP (2005) Membrane alteration as a basis of aging and the protective effects of calorie restriction. Mech Ageing Dev 126:1003–1010. https://doi.org/10.1016/j.mad.2005.03.020

54. Ota H et al (2006) Sirt1 inhibitor, Sirtinol, induces senescence-like growth arrest with attenuated Ras-MAPK signaling in human cancer cells. Oncogene 25:176–185. https://doi.org/10.1038/sj.onc.1209049
55. Wu B et al (2016) Aldehyde dehydrogenase 2 activation in aged heart improves the autophagy by reducing the carbonyl modification on SIRT1. Oncotarget 7:2175–2188. https://doi.org/10.18632/oncotarget.6814
56. Gu C et al (2013) Impaired cardiac SIRT1 activity by carbonyl stress contributes to aging-related ischemic intolerance. PLoS One 8:e74050. https://doi.org/10.1371/journal.pone.0074050
57. McBeth J et al (2009) Musculoskeletal pain is associated with a long-term increased risk of cancer and cardiovascular-related mortality. Rheumatology 48:74–77. https://doi.org/10.1093/rheumatology/ken424
58. Li C et al (2018) Targeting ALDH2 for therapeutic interventions in chronic pain-related myocardial ischemic susceptibility. Theranostics 8:1027–1041. https://doi.org/10.7150/thno.22414
59. Kang PF et al (2016) Activation of ALDH2 with low concentration of ethanol attenuates myocardial ischemia/reperfusion injury in diabetes rat model. Oxidative Med Cell Longev 2016:6190504. https://doi.org/10.1155/2016/6190504
60. Roger VL et al (2011) Heart disease and stroke statistics--2011 update: a report from the American Heart Association. Circulation 123:e18–e209. https://doi.org/10.1161/CIR.0b013e3182009701
61. Fukai M et al (2005) Lipid peroxidation during ischemia depends on ischemia time in warm ischemia and reperfusion of rat liver. Free Radic Biol Med 38:1372–1381. https://doi.org/10.1016/j.freeradbiomed.2005.02.004
62. Farout L, Mary J, Vinh J, Szweda LI, Friguet B (2006) Inactivation of the proteasome by 4-hydroxy-2-nonenal is site specific and dependant on 20S proteasome subtypes. Arch Biochem Biophys 453:135–142. https://doi.org/10.1016/j.abb.2006.02.003
63. Chen J, Henderson GI, Freeman GL (2001) Role of 4-hydroxynonenal in modification of cytochrome c oxidase in ischemia/reperfused rat heart. J Mol Cell Cardiol 33:1919–1927. https://doi.org/10.1006/jmcc.2001.1454
64. Lucas DT, Szweda LI (1999) Declines in mitochondrial respiration during cardiac reperfusion: age-dependent inactivation of alpha-ketoglutarate dehydrogenase. Proc Natl Acad Sci U S A 96:6689–6693
65. Bhatnagar A (2006) Environmental cardiology: studying mechanistic links between pollution and heart disease. Circ Res 99:692–705. https://doi.org/10.1161/01.RES.0000243586.99701.cf
66. Luo J et al (2007) Mechanisms of acrolein-induced myocardial dysfunction: implications for environmental and endogenous aldehyde exposure. Am J Physiol Heart Circ Physiol 293:H3673–H3684. https://doi.org/10.1152/ajpheart.00284.2007
67. Bellamy CA, Nicely B, Mattice BJ, Teaster R (2008) Comparative analysis of clinical efficacy and cost between University of Wisconsin solution and histidine-tryptophan-ketoglutarate. Prog Transplant 18:166–171
68. Zhang H, Gong DX, Zhang YJ, Li SJ, Hu S (2012) Effect of mitochondrial aldehyde dehydrogenase-2 genotype on cardioprotection in patients with congenital heart disease. Eur Heart J 33:1606–1614. https://doi.org/10.1093/eurheartj/ehs061
69. Luckey SW, Tjalkens RB, Petersen DR (1999) Mechanism of inhibition of rat liver class 2 ALDH by 4-hydroxynonenal. Adv Exp Med Biol 463:71–77
70. Petersen DR, Doorn JA (2004) Reactions of 4-hydroxynonenal with proteins and cellular targets. Free Radic Biol Med 37:937–945. https://doi.org/10.1016/j.freeradbiomed.2004.06.012
71. Brooks PJ, Enoch MA, Goldman D, Li TK, Yokoyama A (2009) The alcohol flushing response: an unrecognized risk factor for esophageal cancer from alcohol consumption. PLoS Med 6:e50. https://doi.org/10.1371/journal.pmed.1000050
72. Enomoto N, Takase S, Yasuhara M, Takada A (1991) Acetaldehyde metabolism in different aldehyde dehydrogenase-2 genotypes. Alcohol Clin Exp Res 15:141–144

73. Endo J et al (2009) Metabolic remodeling induced by mitochondrial aldehyde stress stimulates tolerance to oxidative stress in the heart. Circ Res 105:1118–1127. https://doi.org/10.1161/CIRCRESAHA.109.206607
74. Weyker PD, Webb CA, Kiamanesh D, Flynn BC (2013) Lung ischemia reperfusion injury: a bench-to-bedside review. Semin Cardiothorac Vasc Anesth 17:28–43. https://doi.org/10.1177/1089253212458329
75. Apostolakis E, Filos KS, Koletsis E, Dougenis D (2010) Lung dysfunction following cardiopulmonary bypass. J Card Surg 25:47–55. https://doi.org/10.1111/j.1540-8191.2009.00823.x
76. Schlensak C et al (2001) Bronchial artery perfusion during cardiopulmonary bypass does not prevent ischemia of the lung in piglets: assessment of bronchial artery blood flow with fluorescent microspheres. Eur J Cardiothorac Surg 19:326–331
77. Apostolakis EE, Koletsis EN, Baikoussis NG, Siminelakis SN, Papadopoulos GS (2010) Strategies to prevent intraoperative lung injury during cardiopulmonary bypass. J Cardiothorac Surg 5:1. https://doi.org/10.1186/1749-8090-5-1
78. Yoon M, Madden MC, Barton HA (2006) Developmental expression of aldehyde dehydrogenase in rat: a comparison of liver and lung development. Toxicol Sci 89:386–398. https://doi.org/10.1093/toxsci/kfj045
79. Stewart MJ, Malek K, Crabb DW (1996) Distribution of messenger RNAs for aldehyde dehydrogenase 1, aldehyde dehydrogenase 2, and aldehyde dehydrogenase 5 in human tissues. J Investig Med 44:42–46
80. Ding J et al (2016) Alda-1 attenuates lung ischemia-reperfusion injury by reducing 4-Hydroxy-2-Nonenal in alveolar epithelial cells. Crit Care Med 44:e544–e552. https://doi.org/10.1097/CCM.0000000000001563
81. Lin YP, Cheng TJ (2002) Why can't Chinese Han drink alcohol? Hepatitis B virus infection and the evolution of acetaldehyde dehydrogenase deficiency. Med Hypotheses 59:204–207

Association of East Asian Variant Aldehyde Dehydrogenase 2 Genotype (*ALDH2*2*) with Coronary Spasm and Acute Myocardial Infarction

Hirofumi Yasue, Yuji Mizuno, and Eisaku Harada

Abstract Coronary spasm plays an important role in the pathogenesis of ischemic heart disease, including angina pectoris, acute myocardial infarction (AMI), silent myocardial ischemia, and sudden death. The prevalence of coronary spasm is higher among East Asians probably due to genetic as well as environmental factors. ALDH2 eliminates toxic aldehydes including 4-hydroxy-2-nonenal (4-HNE) derived from lipid peroxidation and acrolein in tobacco smoking as well as ethanol-derived acetaldehyde and thereby protects tissues and cells from oxidative damage. Deficient variant *ALDH2*2* genotype is prevalent among East Asians and is a significant risk factor for both coronary spasm and AMI through accumulation of toxic aldehydes, thereby contributing to oxidative stress, endothelial damage, vasoconstriction, and thrombosis. Toxic aldehydes are thus identified as risk factors to be targeted for the treatment of coronary spasm and AMI.

Keywords Acute myocardial infarction (AMI) · Aldehyde dehydrogenase 2 (*ALDH2*) · Coronary spasm · East Asians · Reactive oxygen species

Coronary (artery) spasm is defined as an abnormal contraction of an epicardial coronary artery resulting in transient myocardial ischemia and plays an important role in the pathogenesis of ischemic heart disease in general including angina pectoris, silent myocardial ischemia, acute myocardial infarction (AMI), and sudden death [21, 26]. Angina pectoris caused by coronary spasm is called as "coronary spastic angina (CSA)" or "vasospastic angina (VSA)" and occurs most often from midnight to early morning and is usually not induced by exercise in the daytime [2, 21, 76]. The attack of CSA is associated with either ST segment elevation or depression or negative U wave on ECG and is rapidly relieved within minutes by sublingual administration or oral sprays of nitroglycerin. Ca channel blockers (CCBs) but not beta-blockers are the mainstay for the treatment and prevention of CSA.

H. Yasue (✉) · Y. Mizuno · E. Harada
Division of Cardiovascular Medicine, Kumamoto Kinoh Hospital,
Kumamoto Aging Research Institute, Kumamoto City, Japan
e-mail: yasue@juryo.or.jp

J. Ren et al. (eds.), *Aldehyde Dehydrogenases*, Advances in Experimental Medicine and Biology 1193, https://doi.org/10.1007/978-981-13-6260-6_7

Table 2 The multivariable logistic regression analysis for CSA

	OR	Std. Err	z	$p > \|z\|$	95% CI	
Age	1.342	0.435	0.91	0.363	0.712	2.532
Gender (male)	1.845	0.744	1.52	0.129	0.837	4.066
ALDH2 *2	3.607	1.205	3.84	0.000	1.874	6.943
Smoking	2.320	0.859	2.27	0.023	1.123	4.794
HDL-C	0.860	0.283	−0.46	0.648	0.452	1.639
Uric acid	1.271	0.170	1.79	0.074	0.977	1.652

*ALDH2**2* indicates aldehyde dehydrogenase 2 variant genotype; CI, confidence interval; CSA, coronary spastic angina; HDL-C, high-density lipoprotein cholesterol; OR, odds ratio; and Std. Err, Standard Error. (Reproduced with permission from Ref. [44])

ALDH2 is a 517 amino acid polypeptide encoded by a nuclear gene located at chromosome 12q24 and plays a crucial role in metabolizing not only ethanol-derived acetaldehyde but also endogenous toxic aldehydes such as 4-HNE derived from lipid peroxidation and environmental aldehydes such as acrolein in tobacco smoking. *ALDH2*2* with a substitution of glutamate to lysine at position 504 (Glu504Lys) is present in 30–50% of East Asians but is virtually absent in other populations of the world. This polymorphism exerts a dominant negative effect over wild-type homozygote *ALDH2*1/*1*, with heterozygote *ALDH2*1/*2* showing a severely reduced and homozygote *ALDH2*2/*2* negligible ALDH2 activity [9, 33]. The carriers of *ALDH2*2* thus manifest the characteristic AFS caused by accumulation of ethanol-derived acetaldehyde.

The present study reveals that *ALDH2*2* genotype is the most significant risk factors for CSA on the multiple logistic regression analysis. The study thus shows that *ALDH2*2* specifically present among East Asians is significantly associated with CSA as well as AFS and may thereby explain at least partially why CSA is prevalent among East Asians. Seki and his coworkers also reported the almost similar findings [63]. The previous studies reported that CSA was also associated with the genetic polymorphisms of endothelial nitric oxide synthase [54], paraoxonase [20], NADH/NADPH oxidase p22 phox, stromelysin-1, interleukin-6 [51], or phospholipase–δ 1 [53] as well as environmental factors such as tobacco smoking [31, 76]. It is thus likely that *ALDH2*2* may not be an indispensable player in the development of CSA. However, *ALDH2*2* genotype is unique in that it is specifically prevalent in East Asians among whom CSA is also prevalent. *ALDH2*2* was highly significantly associated with AFS in line with the previous studies [34, 77] and may thus be useful as a convenient marker for *ALDH2*2* in the absence of genotyping [44, 77].

Biological lipid membranes exposed to reactive oxygen species (ROS) (lipid peroxidation) generate numerous reactive aldehydes which are more stable and diffuse greater distances than ROS and thereby propagate oxidative damage [18, 64]. ALDH2 eliminates toxic aldehydes such as 4-hydroxy-2-nonenal (4-HNE) and malondialdehyde (MDA) derived from lipid peroxidation under oxidative stress as well as alcohol-derived acetaldehyde and protects thereby tissues and cells from oxidative damage [6, 18, 40, 64].

Tobacco smoking is the only established environmental risk factor for CSA as shown in this study as well as previous studies [21, 49, 76]. A variety of toxic aldehydes as well as various oxidants are present in tobacco smoking [28, 29, 42, 52], and tobacco smoking also generates endogenous aldehydes including 4-HNE by lipid peroxidation [61]. It is therefore likely that tobacco smoking interacts with *ALDH2*2* and exacerbates the risk for CSA by accumulating toxic aldehydes. Indeed, tobacco smoking synergistically amplifies the *ALDH2*2* risk for CSA more than the additive or multiplicative effects of each considered separately [46]. Conversely, tobacco smoking in turn impairs ALDH2 activity and thereby further increases reactive aldehydes, leading to vicious cycle [42, 45, 46]. Smoking rate is higher among East Asians compared to Westerns [72]. Our study therefore identifies deficient ALDH2 activity and hence increased reactive aldehydes as causative risk factors to be targeted and intervened for the treatment and prevention of CSA [44–46]. Indeed, reactive aldehydes such as 4-HNE cause vascular damage including endothelial dysfunction, smooth muscle proliferation, low-grade inflammation, oxidative stress, and thrombogenity, all of which have been shown to be associated with CSA [21, 76].

ALDH2 also plays a key role in the bioactivation of nitroglycerin widely used for the treatment of ischemic heart disease [10, 56]. However, continued administration of nitroglycerin leads to the tolerance or even cardiac events through inactivation of ALDH2 enzyme and increased ROS [10]. Accordingly, the carriers of *ALDH2*2* genotypes are less responsive to nitroglycerin and are more susceptible to nitroglycerin tolerance, ROS [4, 33, 39, 47], and cardiac events [27, 68]. Thus, new drugs which enhance the activity of ALDH2 are needed for the treatment of CSA, particularly in the patients with *ALDH2*2* [4, 5, 6].

2 Association of *ALDH2*2* with AMI

*ALDH2*2* genotypes are reported to be a risk factor for AMI in East Asians including Japanese, Chinese, and Koreans [12, 13, 23, 36, 48, 49, 67, 73, 74]. Takeuchi and coworkers recently identified the genetic locus of *ALDH2*2* (*rs*671) as the strongest predictor for AMI on a genome-wide association study (GWAS) in Japanese [69]. Coronary spasm is demonstrated in AMI patients [24, 71], and *ALDH2*2* genotype is a significant risk factor for CSA and acute coronary syndrome [44, 74]. We therefore examined the relationship between AMI and *ALDH2*2* genotype.

2.1 *Prevalence of* ALDH2*2 *in AMI Patients*

We examined the frequency of *ALDH2*2* in the 202 Japanese patients with acute ST segment elevation myocardial infarction (156 men and 46 women with a mean age of 67.3 ± 12.0) who underwent successful primary PCI with stent implantation within 12 hours after symptoms [45, 46]. The frequencies of *ALDH2*1/*1*, *ALDH2*1/*2*, and *ALDH2*2/*2* genotypes were 49.0%, 42.6%, and 8.4%, respectively. Accordingly, the frequency of *ALDH2*2* allele was 37.7% in the AMI patients, which is definitively higher as compared with that of the general population in Japan 24.3% in Tokyo, Japan) [34]. Thus, the findings indicate that *ALDH2*2* is significantly associated with AMI (OR = 1.88–1.98, $P < 0.001$) in agreement with those of previous reports [12, 13, 23, 36, 48, 49, 67, 73, 74]. Accordingly, 51.0% (103/202) or more than half of the AMI patients had variant *ALDH2*2*. Because the genotype is assigned randomly at conception independently of the possible confounding factors [32], the finding implies that *ALDH2*2* is causally associated with AMI and identified deficient ALDH2 activity and hence increased reactive aldehydes as a causative risk factor for AMI as well as for CSA [44–46].

2.2 *Association of* ALDH2*2 *with Coronary Spasm in AMI Patients*

There were no differences in the clinical features including age, gender, coronary risk factors, and other variables between the *ALDH2*2* group and the *ALDH2*1/*1* group, except the higher frequencies of AFS and CSA ($P < 0.001$ and $P = 0.001$, respectively) and the lower prevalence of alcohol habit ($P < 0.001$) in the *ALDH2*2* group than the *ALDH2*1/*1* group. Thus, *ALDH2*2* was significantly associated with CSA and AFS ($P < 0.001$, respectively) also in AMI patients.

2.3 *Myocardial Ischemia/Reperfusion Injury in AMI Patients with* ALDH2*2

The plasma levels of peak CPK were significantly higher ($P = 0.002$) (Fig. 2), and those of LDH, cTnT, and B-type natriuretic peptide (BNP) tended to be higher and left ventricular ejection fraction (LVEF) to be lower in the *ALDH2*2* than the *ALDH2*1/*1* group. This suggested that myocardial ischemia/reperfusion injury was more severe in the AMI patients with *ALDH2*2* than in those with *ALDH2*1/*1*. Accordingly, *ALDH2*2* is not only a risk factor for AMI but also aggravates the severity of ischemia/reperfusion injury in AMI patients. These findings are in line with those of the previous studies showing that ALDH2 protects against myocardial

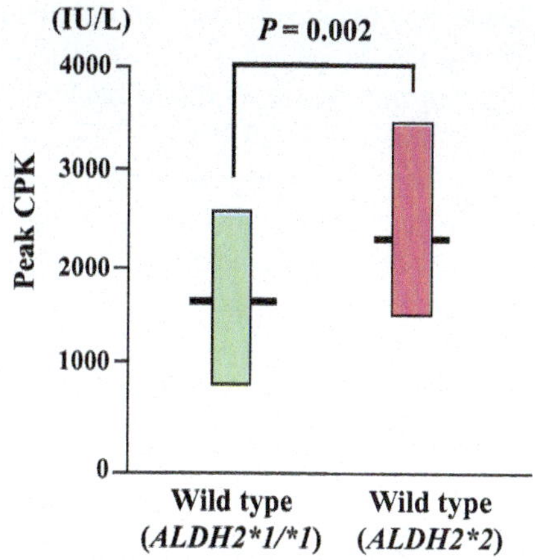

Fig. 2 Comparison of the levels of maximum CPK by *ALDH2*genotype. The peak CPK levels were higher in the *ALDH2*2/*2* vs the *ALDH2*1/*1* groups. *ALDH2*, aldehyde dehydrogenase 2; AFS, alcohol flush syndrome; CS, coronary spasm. (Reproduced with permission from Ref. [45])

ischemia/reperfusion injury by removing toxic aldehydes in the experimental AMI animal models [5, 14, 22, 38, 58, 66]. However, the number of study patients in our study was small, and further studies involving larger sample size are required to confirm this finding.

2.4 *Clinical Implications*

AMI is assumed to be triggered by plaque rupture or erosion resulting in coronary thrombotic occlusion [1, 37]. The present study revealed that *ALDH2*2* was causally associated with CSA in patients with AMI. CSA is associated with increased reactive oxygen species (ROS), endothelial impairment, and thrombogenicity [11, 21, 43, 50, 57, 76]. Accordingly, it is likely that coronary spasm may trigger plaque rupture or erosion and eventually lead to coronary occlusion, resulting in AMI, particularly in those with *ALDH2*2*. It is also to be noted that the episodes of CSA have a circadian variation with a peak incidence in the early morning, which corresponds to that of AMI [25, 76].

Timely reperfusion by direct PCI is mandatory to salvage ischemic myocardium from infarction, but reperfusion per se paradoxically contributes to myocardial injury [5, 14, 16, 17]. Toxic aldehydes such as 4-HNE are abundantly released in association with myocardial ischemia/reperfusion, and ALDH2 activity protects against the injury by eliminating toxic aldehydes in experimental models [22, 38, 66]. In the present study, the peak plasma CPK levels were significantly higher, and those of LDH, cTnT, and BNP tended to be higher and LVEF lower in the patients with *ALDH2*2* than in those with *ALDH2*1/*1*. These findings suggest that the carriers of variant *ALDH2*2* suffer more severe myocardial ischemia/reperfusion injury after AMI as compared to those of wild *ALDH2*1/*1*. It is intriguing in this connection to note that the in-hospital mortality rate for AMI is the highest in Japan among the developed countries of the OECD member nations [55] and that the adjusted in-hospital mortality rates for AMI are higher in Asian-Americans than White Americans [60]. There is a recent report that *ALDH2*2* could predict a worse prognosis of acute coronary syndrome [58]. Calcium channel blockers are the mainstay in the treatment of CSA at present [21, 76]. The results of our study thus suggest that calcium channel blockers also should be considered for the treatment of post-AMI patients with *ALDH2*2*.

3 Conclusions

Deficient variant *ALDH2*2* genotype is prevalent among East Asians and is a risk factor for both coronary spasm and AMI. ALDH2 eliminates not only alcohol-derived acetaldehyde but also other toxic aldehydes including 4-HNE derived from lipid peroxidation and acrolein in tobacco smoking and thereby protects tissues and cells from oxidative damage. Reactive aldehydes cause vascular damage including endothelial dysfunction, oxidative stress, low-grade inflammation, smooth muscle proliferation, and thrombogenicity, all of which are associated with coronary spasm. Bursts of toxic aldehydes are released on myocardial ischemia/reperfusion in patients with AMI, contributing to oxidative myocardial injury. Accordingly, ischemia/reperfusion injury is more severe in AMI patients with *ALDH2*2*, and hence toxic aldehydes are identified as the risk factors to be targeted and intervened for the treatment and prevention of coronary spasm and AMI (Fig. 3).

4 Perspective

No effective therapy currently exists for reducing ischemia/reperfusion injury in AMI patients except ischemic post- and remote conditionings, which reduce infarct size [14, 17] and are enhanced by ALDH2 activity in the animal models [8]. It is to be noted in this connection that volatile anesthetics such as isoflurane or sevoflurane, which have been widely used in clinical practice, have a protective effect on ischemia/reperfusion injury through enhancing ALDH2 activity in animal studies and cardiac surgery [30, 31]. Alpha-lipoic acid also is reported to protect the heart from ischemia/reperfusion injury through ALDH2 activation [15, 35]. Novel activator of ALDH2, Alda-1, enhanced the activity of ALDH2 and effectively restored the activity of *ALDH2*2* to wild-type levels in animal models [5, 6, 22, 38]. Accordingly, it is expected that these classes of drug or ALDH2 activators may serve as a novel treatment for AMI particularly in those with *ALDH2*2* in the future. The underlying mechanisms of coronary spasm involved in AMI remain to be further elucidated.

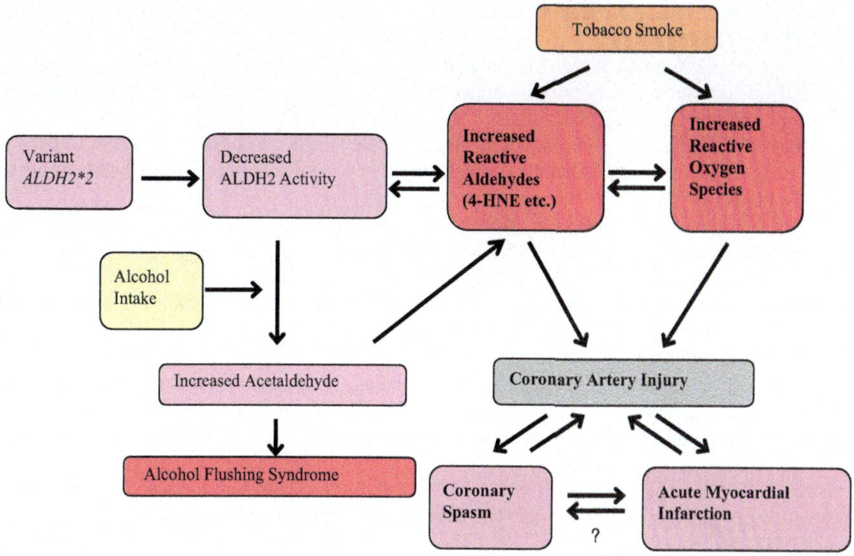

Fig. 3 Association of variant *ALDH2*2* with coronary spasm and acute myocardial infarction. Variant *ALDH2*2* with deficient ALDH2 activity leads to increased reactive aldehydes such as 4-HNE associated with increased reactive oxygen species. Increased reactive aldehydes and reactive oxygen species cause coronary artery injury, which may lead to coronary spasm and acute myocardial infarction. Carriers of variant *ALDH2*2* also have increased acetaldehyde due to deficient ALDH2 activity on alcohol intake and may thereby suffer from alcohol flushing syndrome. (Reproduced with permission from Ref. [45])

Acknowledgment This chapter was modified from the paper published by our group in Circulation [44] and J Am Heart Association [45]. The related contents are reused with the permission. We appreciate the secretarial assistance of Ms. Oda A. in the preparation of the manuscript.

References

1. American College of Emergency Physicians, Society for Cardiovascular Angiography and Interventions, O'Gara PT, Kushner FG, Ascheim DD, Casey DE Jr, Chung MK, de Lemos JA, Ettinger SM, Fang JC, Fesmire FM, Franklin BA, Granger CB, Krumholz HM, Linderbaum JA, Morrow DA, Newby LK, Ornato JP, Ou N, Radford MJ, Tamis-Holland JE, Tommaso CL, Tracy CM, Woo YJ, Zhao DX, Anderson JL, Jacobs AK, Halperin JL, Albert NM, Brindis RG, Creager MA, De Mets D, Guyton RA, Hochman JS, Kovacs RJ, Kushner FG, Ohman EM, Stevenson WG, Yancy CW (2013) ACCF/AHA guideline for the management of ST-elevation myocardial infarction: executive summary: a report of the American College of Cardiology Foundation/American Heart Association task force on practice guidelines. J Am Coll Cardiol 61:485–510
2. Beltrame JF, Crea F, Kaski JC, Ogawa H, Ong P, Sechtem U, Shimokawa H, Bairey Merz CN, Coronary Vasomotion Disorders International Study Group (COVADIS) (2017) International standardization of diagnostic criteria for vasospastic angina. Eur Heart J 38:2565–2568

3. Bentzon JF, Otsuka F, Virmani R, Falk E (2014) Mechanisms of plaque formation and rupture. Circ Res 114:1852–1866
4. Beretta M, Gorren AC, Wenzl MV, Weis R, Russwurm M, Koesling D, Schmidt K, Mayer B (2010) Characterization of the east Asian variant of aldehyde dehydrogenase-2: bioactivation of nitroglycerin and effects of Alda-1. J Biol Chem 285:943–952
5. Chen CH, Budas GR, Churchill EN, Disatnik MH, Hurley TD, Mochly-Rosen D (2008) Activation of aldehyde dehydrogenase-2 reduces ischemic damage to the heart. Science 321:1493–1495
6. Chen CH, Ferreira JC, Gross ER, Mochly-Rosen D (2014) Targeting aldehyde dehydrogenase 2: new therapeutic opportunities. Physiol Rev 94:1–34
7. Choi BG, Park SH, Rha SW, Ahn J, Choi SY, Byun JK, Li H, Mashaly A, Shim MS, Kang JH, Kim W, Choi JY, Park EJ, Lee S, Na JO, Choi CU, Lim HE, Kim EJ, Park CG, Seo HS, Oh DJ (2017) Three-year follow-up of patients with acetylcholine-induced coronary artery spasm combined with insignificant coronary stenosis. Int J Cardiol 238:66–71
8. Contractor H, Støttrup NB, Cunnington C, Manlhiot C, Diesch J, Ormerod JO, Jensen R, Bøtker HE, Redington A, Schmidt MR, Ashrafian H, Kharbanda RK (2013) Aldehyde dehydrogenase-2 inhibition blocks remote preconditioning in experimentaland human models. Basic Res Cardiol 108(3):343
9. Crabb DW, Edenberg HJ, Bosron WF, Li TK (1989) Genotypes for aldehyde dehydrogenase deficiency and alcohol sensitivity. The inactive ALDH2(2) allele is dominant. J Clin Invest 83:314–316
10. Daiber A, Münzel T (2015) Organic nitrate therapy, nitrate tolerance, and nitrate- induced endothelial dysfunction: emphasis on redox biology and oxidative stress. Antioxid Redox Signal 23(11):899–942
11. Fournier S, Taffé P, Radovanovic D, Von Elm E, Morawiec B, Stauffer JC, Erne P, Beggah A, Monney P, Pascale P, Iglesias JF, Eeckhout E, Muller O (2015) Myocardial infarct size and mortality depend on the time of day-a large multicenter study. PLoS One 10:e0119157
12. Gu JY, Li LW (2014) ALDH2 Glu504Lys polymorphism and susceptibility to coronary artery disease and myocardial infarction in east Asians: a meta-analysis. Arch Med Res 45(1):76–83
13. Han H, Wang H, Yin Z, Jiang H, Fang M, Han J (2013) Association of genetic polymorphisms in *ADH* and *ALDH2* with risk of coronary artery disease and myocardial infarction: a meta-analysis. Gene 526:134–141
14. Hausenloy DJ, Yellon DM (2013) Myocardial ischemia-reperfusion injury: a neglected therapeutic target. J Clin Invest 123:92–100
15. He L, Liu B, Dai Z, Zhang HF, Zhang YS, Luo XJ, Ma QL, Peng J (2012) Alpha lipoic acid protects heart against myocardial ischemia-reperfusion injury through a mechanism involving aldehyde dehydrogenase 2 activation. Eur J Pharmacol 678(1–3):32–38
16. Heusch G (2015) Molecular basis of cardioprotection: signal transduction in ischemic pre-, post-, and remote conditioning. Circ Res 116:674–699
17. Heusch G, Rassaf T (2016) Time to give up on Cardioprotection? A critical appraisal of clinical studies on ischemic pre-, post-, and remote conditioning. Circ Res 119:676–695
18. Hill GG, Bhatnagar A (2009) Beyond reactive oxygen species: aldehydes as arbitrators of alarm and adaption. Circ Res 105:1044–1046
19. Hung MJ, Hu P, Hung MY (2014) Coronary artery spasm: review and update. Int J Med Sci 11(11):1161–1171
20. Ito T, Yasue H, Yoshimura M, Nakamura S, Nakayama M, Shimasaki Y, Harada E, Mizuno Y, Kawano H, Ogawa H (2002) Paraoxonase gene Gln192Arg (Q192R) polymorphism is associated with coronary artery spasm. Hum Genet 110:89–94
21. JCS Joint Working Group (2014) Guidelines for diagnosis and treatment of patients with vasospastic angina (coronary spastic angina). Circ J 78:2779–2801
22. Ji W, Wei S, Hao P, Xing J, Yuan Q, Wang J, Xu F, Chen Y (2016) Aldehyde dehydrogenase 2 has Cardioprotective effects on myocardial/Ischaemia/reperfusion injury via suppressing Mitophagy. Front Pharmacol 7:101

23. Jo SA, Kim EK, Park MH, Han C, Park HY, Jang Y, Song BJ, Jo I (2007) A Glu487Lys polymorphism in the gene for mitochondrial aldehyde dehydrogenase 2 is associated with myocardial infarction in elderly Korean men. Clin Chim Acta 382:43–47
24. Katoh D, Mizuno Y, Harada E, Ito T, Morikawa Y, Nakagawa H, Saito Y, Yoshimura M, Yasue H (2012) High incidence of provoked coronary spasm in the presence of a stent after myocardial infarction: therapeutic and prognostic implications. Coron Artery Dis 23:141–145
25. Kim HO, Kim JM, Woo JS, Park CB, Cho JM, Lee SU, Kim CJ, Jeong MH, Kim W (2018) Korea acute myocardial infarction registry investigators. Circadian distribution of acute myocardial infarction in different age groups. Am J Cardiol 121:1279–1284
26. Kinlay S (2014) Coronary artery spasm as a cause of angina. Circulation 129:1717–1719
27. Kosugi M, Nakagomi A, Shibui T, Kato K, Kusama Y, Atarashi H, Mizuno K (2011) Effect of long-term nitrate treatment on cardiac events in patients with vasospastic angina. Circ J 75(9):2196–2205
28. Kugiyama K, Yasue H, Ohgushi M, Motoyama T, Kawano H, Inobe Y, Hiroshima O, Sugiyama S (1996a) Deficiency in nitric oxide bioactivity in epicardial coronary arteries of cigarette smokers. J Am Coll Cardiol 28:1161–1167
29. Kugiyama K, Yasue H, Okumura K, Ogawa H, Fujimoto K, Nakao K, Yoshimura M, Motoyama T, Inobe Y, Kawano H (1996b) Nitric oxide activity is deficient in spasm arteries of patients with coronary spastic angina. Circulation 94:266–271
30. Lang XE, Wang X, Zhang KR, Lv JY, Jin JH, Li QS (2013) Isoflurane preconditioning confers cardioprotection by activation of ALDH2. PLoS One 8:e52469
31. Lemoine S, Tritapepe L, Hanouz JL, Puddu PE (2016) The mechanisms of cardio-protective effects of desflurane and sevoflurane at the time of reperfusion: anaesthetic post-conditioning potentially translatable to humans? Br J Anaesth 116:456–475
32. Lewis SJ (2010) Mendelian randomization as applied to coronary heart disease, including recent advances incorporating new technology. Circ Cardiovasc Genet 3:109–117
33. Li Y, Zhang D, Jin W, Shao C, Yan P, Xu C, Sheng H, Liu Y, Yu J, Xie Y, Zhao Y, Lu D, Nebert DW, Harrison DC, Huang W, Jin L (2006) Mitochondrial aldehyde dehydrogenase-2 (ALDH2) Glu504Lys polymorphism contributes to the variation in efficacy of sublingual nitroglycerin. J Clin Invest 116:506–511
34. Li H, Borinskaya S, Yoshimura K, Kal'ina N, Marusin A, Stepanov VA, Qin Z, Khaliq S, Lee MY, Yang Y, Mohyuddin A, Gurwitz D, Mehdi SQ, Rogaev E, Jin L, Yankovsky NK, Kidd JR, Kidd KK (2009) Refined geographic distribution of the oriental *ALDH2**504Lys (nee 487Lys) variant. Ann Hum Genet 73:335–345
35. Li RJ, Ji WQ, Pang JJ, Wang JL, Chen YG, Zhang Y (2013) Alpha-lipoic acid ameliorates oxidative stress by increasing aldehyde dehydrogenase-2 activity in patients with acute coronary syndrome. Tohoku J Exp Med 229(1):45–51
36. Li YY, Wang H, Wu JJ, Kim HJ, Yang XX, Geng HY, Gong G (2018) ALDH2 gene G487A polymorphism and coronary artery disease: a meta-analysis including 5644 participants. J Cell Mol Med 22(3):1666–1674
37. Libby P (2013) Mechanisms of acute coronary syndromes and their implications for therapy. N Engl J Med 368:2004–2013
38. Ma H, Guo R, Yu L, Zhang Y, Ren J (2011) Aldehyde dehydrogenase 2 (ALDH2) rescues myocardial ischaemia/reperfusion injury: role of autophagy paradox and toxic aldehyde. Eur Heart J 32:1025–1038
39. Mackenzie IS, Maki-Petaja KM, McEniery CM, Bao YP, Wallace SM, Cheriyan J, Monteith S, Brown MJ, Wilkinson IB (2005) Aldehyde dehydrogenase 2 plays a role in the bioactivation of nitroglycerin in humans. Arterioscler Thromb Vasc Biol 25(9):1891–1895
40. Mali VR, Palaniyandi SS (2014) Regulation and therapeutic strategies of 4-hydroxy-2-nonenal metabolism in heart disease. Free Radic Res 48:251–263
41. Maseri A, Davies G, Hackett D, Kaski JC (1990) Coronary artery spasm and vasoconstriction. The case for a distinction. Circulation 81:1983–1991

42. Messner B, Bernhard D (2014) Smoking and cardiovascular disease: mechanisms of endothelial dysfunction and early atherogenesis. Arterioscler Thromb Vasc Biol 34:509–515
43. Miyamoto S, Kawano H, Sakamoto T, Soejima H, Kajiwara I, Hokamaki J, Hirai N, Sugiyama S, Yoshimura M, Yasue H, Nakamura H, Yodoi J, Ogawa H (2004) Increased plasma levels of thioredoxin in patients with coronary spastic angina. Antioxid Redox Signal 6:75–80
44. Mizuno Y, Harada E, Morita S, Kinoshita K, Hayashida M, Shono M, Morikawa Y, Murohara T, Nakayama M, Yoshimura M, Yasue H (2015) East Asian variant of aldehyde dehydrogenase 2 is associated with coronary spastic angina: possible roles of reactive aldehydes and implications of alcohol flushing syndrome. Circulation 131:1665–1673
45. Mizuno Y, Hokimoto S, Harada E, Kinoshita K, Nakagawa K, Yoshimura M, Ogawa H, Yasue H (2016a) Variant aldehyde dehydrogenase 2 (ALDH2*2) is a risk factor for coronary spasm and ST-segment elevation myocardial infarction. J Am Heart Assoc 5:pii: e003247
46. Mizuno Y, Hokimoto S, Harada E, Kinoshita K, Yoshimura M, Yasue H (2016b) Variant aldehyde dehydrogenase 2 (ALDH2*2) in east Asians interactively exacerbates tobacco smoking risk for coronary spasm - possible role of reactive aldehydes. Circ J 81:96–102
47. Morikawa Y, Mizuno Y, Harada E, Kuboyama O, Yoshimura M, Yasue H (2010) Nitrate tolerance as a possible cause of multidrug-resistant coronary artery spasm. Int Heart J 51:211–213
48. Morita K, Miyazaki H, Saruwatari J, Oniki K, Kumagae N, Tanaka T, Kajiwara A, Otake K, Ogata Y, Arima Y, Hokimoto S, Ogawa H, Nakagawa K (2014a) Combined effects of current-smoking and the aldehyde dehydrogenase 2*2 allele on the risk of myocardial infarction in Japanese patients. Toxicol Lett 232:221–225
49. Morita S, Mizuno Y, Harada E, Nakagawa H, Morikawa Y, Saito Y, Katoh D, Kashiwagi Y, Yoshimura M, Murohara T, Yasue H (2014b) Differences and interactions between risk factors for coronary spasm and atherosclerosis--smoking, aging, inflammation, and blood pressure. Intern Med 53:2663–2670
50. Motoyama T, Kawano H, Kugiyama K, Hirashima O, Ohgushi M, Tsunoda R, Moriyama Y, Miyao Y, Yoshimura M, Ogawa H, Yasue H (1998) Vitamin E administration improves impairment of endothelium-dependent vasodilation in patients with coronary spastic angina. J Am Coll Cardiol 32:1672–1679
51. Murase Y, Yamada Y, Hirashiki A, Ichihara S, Kanda H, Watarai M, Takatsu F, Murohara T, Yokota M (2004) Genetic risk and gene-environment interaction in coronary artery spasm in Japanese men and women. Eur Heart J 25:970–977
52. Murohara T, Kugiyama K, Ohgushi M, Sugiyama S, Yasue H (1994) Cigarette smoke extract contracts isolated porcine coronary arteries by superoxide anion-mediated degradation of EDRF. Am J Phys 266:H874–H880
53. Nakano T, Osanai T, Tomita H, Sekimata M, Homma Y, Okumura K (2002) Enhanced activity of variant phospholipase C-delta1 protein (R257H) detected in patients with coronary artery spasm. Circulation 105:2024–2029
54. Nakayama M, Yasue H, Yoshimura M, Shimasaki Y, Kugiyama K, Ogawa H, Motoyama T, Saito Y, Ogawa Y, Miyamoto Y, Nakao K (1999) T-786-C mutation in the 5′-flanking region of the endothelial nitric oxide synthase gene is associated with coronary spasm. Circulation 99:2864–2870
55. OECD (2013) Mortality following acute myocardial infarction (AMI). In health at a glance 2013. OECD Indicators, OECD Publishing, Paris, pp 112–113
56. Opelt M, Eroglu E, Waldeck-Weiermair M, Russwurm M, Koesling D, Malli R, Graier WF, Fassett JT, Schrammel A, Mayer B (2016) Formation of nitric oxide by aldehyde Dehydrogenase-2 is necessary and sufficient for vascular bioactivation of nitroglycerin. J Biol Chem 291:24076–24084
57. Oshima S, Yasue H, Ogawa H, Okumura K, Matsuyama K (1990) Fibrinopeptide a is released into the coronary circulation after coronary spasm. Circulation 82:2222–2225
58. Pan C, Zhao Y, Bian Y, Shang R, Wang JL, Xue L, Wei SJ, Zhang H, Chen YG, Xu F (2018) Aldehyde dehydrogenase 2 Glu504Lys variant predicts a worse prognosis of acute coronary syndrome patients. J Cell Mol Med 22(4):2518–2522

59. Pristipino C, Beltrame JF, Finocchiaro ML, Hattori R, Fujita M, Mongiardo R, Cianflone D, Sanna T, Sasayama S, Maseri A (2000) Major racial differences in coronary constrictor response between Japanese and Caucasians with recent myocardial infarction. Circulation 101:1102–1108
60. Qian F, Ling FS, Deedwania P, Hernandez AF, Fonarow GC, Cannon CP, Peterson ED, Peacock WF, Kaltenbach LA, Laskey WK, Schwamm LH, Bhatt DL, Get With The Guidelines Steering Committee and Investigators (2012) Care and outcomes of Asian-American acute myocardial infarction patients: findings from the American Heart Association get with the guidelines-coronary artery disease program. Circ Cardiovasc Qual Outcomes 5:126–133
61. Reilly M, Delanty N, Lawson JA, FitzGerald GA (1996) Modulation of oxidant stress in vivo in chronic cigarette smokers. Circulation 94:19–25
62. Schuckit MA (2011) Ethanol and methanol. In: Brnton L, Chabner B, Knollmann BC (eds) Goodman & Gilman's pharmacological basis of therapeutics, 12th edn. McGraw Hill, New York
63. Seki T, Okayama H, Isoyama S, Kagaya Y, Shirato K, Munakata K, Kanazawa M, Tamaki K, Hiramoto T, Okayama M, Kasahara S (1999) The role of alcohol dehydrogenase 2 and aldehyde dehydrogenase 2 genotypes in alcohol-induced vasospastic angina. Tohoku J Exp Med 187:311–322
64. Singh S, Brocker C, Koppaka V, Chen Y, Jackson BC, Matsumoto A, Thompson DC, Vasiliou V (2013) Aldehyde dehydrogenases in cellular responses to oxidative/electrophilic stress. Free Radic Biol Med 56:89–101
65. Sohn SM, Choi BG, Choi SY, Byun JK, Mashaly A, Park Y, Jang WY, Kim W, Choi JY, Park EJ, Na JO, Choi CU, Lim HE, Kim EJ, Park CG, Seo HS, Oh DJ, Rha SW (2018) Impact of alcohol drinking on acetylcholine-induced coronary artery spasm in Korean populations. Atherosclerosis 268:163–169
66. Sun A, Zou Y, Wang P, Xu D, Gong H, Wang S, Qin Y, Zhang P, Chen Y, Harada M, Isse T, Kawamoto T, Fan H, Yang P, Akazawa H, Nagai T, Takano H, Ping P, Komuro I, Ge J (2014) Mitochondrial aldehyde dehydrogenase 2 plays protective roles in heart failure after myocardial infarction via suppression of the cytosolic JNK/p53 pathway in mice. J Am Heart Assoc 3:e000779
67. Takagi S, Iwai N, Yamauchi R, Kojima S, Yasuno S, Baba T, Terashima M, Tsutsumi Y, Suzuki S, Morii I, Hanai S, Ono K, Baba S, Tomoike H, Kawamura A, Miyazaki S, Nonogi H, Goto Y (2002) Aldehyde dehydrogenase 2 gene is a risk factor for myocardial infarction in Japanese men. Hypertens Res 25:677–681
68. Takahashi J, Nihei T, Takagi Y, Miyata S, Odaka Y, Tsunoda R, Seki A, Sumiyoshi T, Matsui M, Goto T, Tanabe Y, Sueda S, Momomura S, Yasuda S, Ogawa H, Shimokawa H, Japanese Coronary Spasm Association (2015) Prognostic impact of chronic nitrate therapy in patients with vasospastic angina: multicentre registry study of the Japanese coronary spasm association. Eur Heart J 36:228–237
69. Takeuchi F, Yokota M, Yamamoto K, Nakashima E, Katsuya T, Asano H, Isono M, Nabika T, Sugiyama T, Fujioka A, Awata N, Ohnaka K, Nakatochi M, Kitajima H, Rakugi H, Nakamura J, Ohkubo T, Imai Y, Shimamoto K, Yamori Y, Yamaguchi S, Kobayashi S, Takayanagi R, Ogihara T, Kato N (2012) Genome-wide association study of coronary artery disease in the Japanese. Eur J Hum Genet 20:333–340
70. Takizawa A, Yasue H, Omote S, Nagao M, Hyon H, Nishida S, Horie M (1984) Variant angina induced by alcohol ingestion. Am Heart J 107:25–27
71. Wakabayashi K, Suzuki H, Honda Y, Wakatsuki D, Kawachi K, Ota K, Koba S, Shimizu N, Asano F, Sato T, Takeyama Y (2008) Provoked coronary spasm predicts adverse outcome in patients with acute myocardial infarction. a novel predictor of prognosis after acute myocardial infarction J Am Coll Cardiol 52:518–522
72. World Health Organization (2008) WHO Report on the global tobacco epidemic. 2008; 67176. http://www.who.int/tobacco/global_report/summary/en/#. Accessed 16 Feb 2016

73. Xu F, Chen YG, Xue L, Li RJ, Zhang H, Bian Y, Zhang C, Lv RJ, Feng JB, Zhang Y (2011) Role of aldehyde dehydrogenase 2 Glu504lys polymorphism in acute coronary syndrome. J Cell Mol Med 15:1955–1962
74. Xu F, Sun Y, Shang R, Li M, Cui L, Cui Z, Chen Y (2014) The Glu504Lys polymorphism of aldehyde dehydrogenase 2 contributes to development of coronary artery disease. Tohoku J Exp Med 234:143–150
75. Yasue H, Sasayama S, Kikuchi K, Okumura K, Matsubara T, Miwa K, Gotoh Y, Shimokawa H, Yokoyama M, Yokoyama H, Nakamura Y, Kawada M, Inoue K, Nakamura K (2000) The study on the role of CAS in ischemic heart disease. In: Annual report of the research on cardiovascular diseases 2000. National Cardiovascular Center, Osaka, pp 96–97. (in Japanese)
76. Yasue H, Nakagawa H, Itoh T, Harada E, Mizuno Y (2008) Coronary artery spasm—clinical features, diagnosis, pathogenesis, and treatment. J Cardiol 51:2–17
77. Yokoyama A, Omori T, Yokoyama T (2010) Alcohol and aldehyde dehydrogenase polymorphisms and a new strategy for prevention and screening for cancer in the upper aerodigestive tract in East Asians. Keio J Med 59:115–130

Aldehyde Dehydrogenases Genetic Polymorphism and Obesity: From Genomics to Behavior and Health

Cheng Hu

Abstract Obesity is multifactorial and complex. Remarkable progress has been made recently in search for polygenic obesity through genome-wide association study (GWAS), but biology of polygenic effects on obesity is largely poor. This review summarizes the available evidence and provides an overview of the links between *ALDH2* variants and adiposity, which were firstly and mainly derived from studies of polygenic obesity and also indirectly investigated by using cell lines and mice. The genetic association studies have observed consistent associations of *ALDH2* variants with obesity-related traits including BMI, waist circumference (WC), waist-to-hip ratio (WHR), and visceral fat accumulation. In consideration of *ALDH2* variants with enzyme activity and alcohol consumption behavior in physiological mechanism studies, we proposed a model by which the physiological and behavioral consequences of alcohol consumption serve as an intermediary process between polymorphisms in *ALDH2* and obesity.

Keywords ALDH2 · Obesity · BMI · Central obesity · Variant

Abbreviations

4-HNE	4-hydroxykenals
ADH	alcohol dehydrogenase
ALDHs	aldehyde dehydrogenases
BMI	body mass index
FADE	FAt Distribution and diseasE

C. Hu (✉)
Shanghai Diabetes Institute, Shanghai Key Laboratory of Diabetes Mellitus, Shanghai Clinical Center for Diabetes, Shanghai Jiao Tong University Affiliated Sixth People's Hospital, Shanghai, China

Institute for Metabolic Diseases, Shanghai Jiao Tong University Affiliated Sixth People's Hospital South Campus, Shanghai, China
e-mail: alfredhc@sjtu.edu.cn

J. Ren et al. (eds.), *Aldehyde Dehydrogenases*, Advances in Experimental Medicine and Biology 1193, https://doi.org/10.1007/978-981-13-6260-6_8

GABA	γ-aminobutyric acid
GWAS	genome-wide association studies
IDF	International Diabetes Federation
PPAR-γ	peroxisome proliferator-activated receptor γ
ROS	reactive oxygen species
SFA	subcutaneous fat area
SGWAS	Shanghai Genome-Wide Association Studies
VFA	visceral fat area
WC	waist circumference
WHO	World Health Organization
WHR	waist-to-hip ratio

1 Introduction

Obesity, defined as an excessive fat accumulation, has increased at an alarming pace around the world. For the years 2013–2014, more than one-third (37.7%) of US adults have obesity, and the number is projected to 51% in 2030 [1]. Not only that, obesity is also a major heath concern in developing countries, where almost two in three of the world's people live with obesity. The rate of obesity has tripled since 1980 in the Middle East, the Pacific Islands, and China [2]. Obesity-related diseases such as type 2 diabetes, stroke, coronary heart disease, and some cancers have been the leading cause of disability and death. Clinically, overweight and obesity are quantified through the surrogate measure of body mass index (BMI), calculated as weight divided by the square of height. Individuals with a BMI of ≥28 kg/m^2 are considered obese and those with a BMI of 25 kg/m^2 to 29.9 kg/m^2 considered overweight according to World Health Organization (WHO) criteria [3]. The given BMI values indicate a higher percentage of body fat and metabolic disease risk for Asian populations than European populations; therefore, some Asian countries have revised the definition of obesity to adjust for ethnic differences [4]. In China, overweight and obesity are, respectively, identified as 24 kg/m^2≤ BMI <28 kg/m^2 and BMI ≥28 kg/m^2 [5]. Central obesity (also known as visceral, android, apple-shaped or upper body obesity, abdominal obesity) can be clinically assessed by waist circumference (WC) which is highly associated with intra-abdominal fat. International Diabetes Federation (IDF) consensus defined central obesity in Europeans as a WC of ≥94 cm in males and ≥ 80 cm in females. Lower WC cutoffs are proposed for some ethnic groups (e.g., Chinese, Southeast Asians) [6]. Some indices corresponding to fat distribution, including waist-to-hip ratio (WHR), visceral fat are also used to assess central obesity.

Obesity is a product of the interaction between genetic factors and lifestyle risk factors (like smoking, drinking, lack of physical activity, sedentary, and Western diet). Even though surrounded by "obesogenic environment," a certain number of individuals could maintain their weight, suggesting genetic factors play a critical role in regulation of obesity. The advent of genome technology innovation (e.g., microarray and sequencing) has allowed far more detailed investigation of genetic factors than previously known. Approximately 150 variants linked to obesity (e.g., BMI, WC, and WHR) have been identified through genome-wide association studies (GWAS) [7–24]. To be frustrated, little is known about the underlying mechanism by which those surprisingly high numbers of obesity-related variants imposed risk to obesity or body fat distribution. The interpretation of genomics to clinical care and public health is not enough, which requires the knowledge of genetic basis for obesity and interactions with health behaviors. Population research on the gene-environment interaction, which means that the sensitivity to environmental influences is regulated modulated by genetic factors, was still lacking.

The p.Glu504Lys (c.1510G>A, rs671) *ALDH2* missense variant is among numerous variants linked to obesity which reached the genome-wide significance. Aldehyde dehydrogenase 2 encoded by *ALDH2* is well known for most efficient enzyme of alcohol oxidation in the liver and other organs. Since alcohol consumption is also a public health concern, we focused on aldehyde dehydrogenases (ALDHs) genetic polymorphisms in this review and elucidated a genetic link to a common behavior and health.

2 Aldehyde Dehydrogenases

Human aldehyde dehydrogenases superfamily is a group of oxidizing enzymes responsible for the metabolism of endogenous and exogenous aldehydes. There are 19 functional isoforms with a wide range of tissue distribution, and some of them display specific subscapular compartments [25, 26]. Amino acid sequence similarities are about 40% between families and 60% or higher between the subfamilies. While some aldehydes functioned by ALDHs play key roles in normal physiological processes including vision, embryonic development, and neurotransmission, most of aldehydes can lead to cytotoxic damage. In this case, ALDHs are regarded as detoxification enzymes and serve as an important shield from the cytotoxic damage of aldehydes by converting them to their respective carboxylic acids.

Mitochondrial ALDH2 emerges as a particularly important enzyme for the oxidation of acetaldehyde in vivo, an immediate metabolite of alcohol. Alcohol-detoxifying pathway, specifically, consists of two-step enzymatic reaction. The first step is catalyzed by enzyme alcohol dehydrogenase (ADH), which converts alcohol to acetaldehyde. The second step is mainly catalyzed by ALDH2, dehydrogenating acetaldehyde into acetate to keep low levels of circulating acetaldehyde under normal condition. Any abnormal endogenous (i.e., mutations or liver disease) and exogenous (i.e., drugs) conditions which influence the enzyme activity could lead to the accumulation of acetaldehyde, manifested by a variety of unpleasant effects such as alcohol-flushing responses, nausea, vomiting, hypotension, or rapid heart rate. These unpleasant effects may prevent individuals from consuming more alcohol. Besides, it can also serve to remove lipid peroxidation-derived aldehydes and other reactive aldehydes to protect from the damage of excessive oxidative stress. Compelling evidence indicates that ALDH2 is a key mediator of endogenous cytoprotection against ischemia injury [27, 28], gastrointestinal cancers [29, 30], late-onset Alzheimer's disease [31, 32], and a variety of human diseases.

The importance of ALDH2 proteins in physiological or pathological processes might be best evidenced by the associations between *ALDH2* functional variants and distinct disease phenotypes in humans. The p.Glu504Lys (c.1510G>A, rs671) *ALDH2* is the most common single point mutation in humans. This single point mutation occurs 35–45% among East Asians (approximately 560 million) but very rare in European populations [33]. The *ALDH2**2 (termed A allele) carriers have a lower ALDH2 enzymatic activity. *ALDH2*1/*2* heterozygotes are expected to have dramatically lower than 50% of the wild-type's enzymatic activity and *ALDH2*2/*2* homozygotes have <1–4% of the wild-type activity. Several large meta-analyses identified that carriers of the highly active *ALDH2*1* allele (or G allele) had an increased risk of alcoholism. rs886205 is another variant in the promoter region of *ALDH2* gene and was linked to ALDH2 activity through changing transcriptional activity in European populations.

3 The Associations Between Obesity and *ALDH2* Variants

3.1 *What's for the GWAS of Obesity?*

Multiple measurements including BMI, WC, WHR, and visceral fat area (VFA) are applied to quantify the degree of obesity. BMI, is a simple but standard measurement for overall obesity. Great advances in identification of variants linked to obesity can be largely attributed to the strategy of GWAS. The attempts to identify BMI-related variants are considered to facilitate some patterns of discovery for neuronal regulation in overall obesity [12, 16]. WC and WHR are considered as simple and commonly used markers for central obesity. Besides, other indices corresponding to fat distribution imaged by MRI or CT technology are superior to WC and WHR in terms of distinguishing between visceral fat and abdominal subcutaneous fat. The important thing for the central obesity-related variants has proven vital to elucidate the signals either shared with overall obesity or specific to central obesity.

3.2 *The Link Between Obesity and* ALDH2 *Variants*

The involvement of *ALDH2* in obesity and fat distribution was first suggested by GWAS in East Asian populations. So far, a total of three large-scale GWAS analyses and one replication study were performed among East Asian populations (as shown in Table 1) [21, 22, 34, 35]. The first GWAS analysis, published in *Nature Genetics* in 2009, uncovered a novel locus rs2074356 affecting WHR that reached

Table 1 The associations of *ALDH2* variants with obesity-related traits

SNP	Nearest gene	Chr	Position	Allele	EAF (%)	Traits	Effect sizes (SE)	*P* values	Date	References
rs2074356	*ALDH2-HECTD4*	12	112,207,597	C/T	0.85	WHRnoBMI	0.006 (0.001)	**7.8×10^{-12}**	2009	*Nature Genetics*
rs671	*ALDH2*	12	111,803,962	G/A	0.76	BMI	0.0378 (0.0057)	**3.4×10^{-11}**	2014	*Human Molecular Genetics*
rs12229654	*MYL2*	12	110,976,657	T/G	0.80	BMI	0.0341 (0.0058)	**4.56×10^{-9}**		
rs671	*ALDH2*	12	111,803,962	G/A	0.78	WCnoBMI	0.0482 (0.0089)	**6.73×10^{-8}**	2016	*Scientific Reports*
						WCadjBMI	0.0119 (0.005)	**0.0165**		
						WHRnoBMI	0.0255 (0.0089)	**0.0042**		
						WHRadjBMI	0.0062 (0.0078)	0.4220		
rs12229654	*MYL2*	12	110,976,657	T/G	0.83	WCnoBMI	0.0468 (0.0087)	**7.94×10^{-8}**		
						WCadjBMI	0.0174 (0.005)	**5.25×10^{-4}**		
						WHRnoBMI	0.0366 (0.0085)	**1.80×10^{-5}**		
						WHRadjBMI	0.0216 (0.0074)	**0.0035**		
rs671	*ALDH2*	12	111,803,962	G/A	0.78	BMI	0.2932 (0.1052)	**0.0053**	2016	*Scientific Reports*
						WCnoBMI	0.0075 (0.0015)	**4.05×10^{-7}**		
						WCadjBMI	0.0042 (0.0009)	**1.96×10^{-6}**		
						WHRnoBMI	0.002 (0.0009)	**0.0220**		
						WHRadjBMI	0.001 (0.0008)	0.2143		
						VFAnoBMI	0.036 (0.0075)	**1.94×10^{-6}**		
						VFAadjBMI	0.0224 (0.0057)	**9.64×10^{-5}**		
						VFA/SFAnoBMI	0.0177 (0.0069)	**0.0104**		
						VFA/SFAadjBMI	0.0164 (0.0068)	**0.0155**		
						SFAnoBMI	0.0042 (0.0061)	0.4953		
						SFAadjBMI	0.0006 (0.0043)	0.8827		

(continued)

Table 1 (continued)

SNP	Nearest gene	Chr	Position	Allele	EAF (%)	Traits	Effect sizes (SE)	*P* values	Date	References
rs2074356	*ALDH2-HECTD4*	12	112,207,597	C/T	0.85	BMI	0.1324 (0.1217)	0.2766		
						WCnoBMI	0.0035 (0.0017)	**0.0418**		
						WCadjBMI	0.002 (0.001)	**0.0477**		
						WHRnoBMI	0.0012 (0.001)	0.2157		
						WHRadjBMI	0.0008 (0.0009)	0.3860		
						VFAnoBMI	0.0219 (0.0087)	**0.0124**		
						VFAadjBMI	0.0158 (0.0066)	**0.0175**		
						VFA/SFAnoBMI	0.0229 (0.0059)	**1.19×10^{-4}**		
						VFA/SFAadjBMI	0.0199 (0.0058)	**6.54×10^{-4}**		
						SFAnoBMI	0.0131 (0.0053)	**0.0132**		
						SFAadjBMI	0.0025 (0.0037)	0.4966		

Traits were adjusted for age and sex in the additive genetic model
SNP single nucleotide polymorphism, *Chr* chromosome, *Allele* minor allele/major allele, *EAF* effect allele frequency
P values<0.05 are shown in bold

genome-wide significance in Korean populations (8842 and 7861 samples in stages 1 and 2, respectively) [34]. This locus is mapped to chromosome 12q24 in the 24th intron of the *C12orf51* and in moderate linkage disequilibrium with the rs671 at *ALDH2* in East Asian populations ($r^2 = 0.58$ in JPT and CHB). Then, two enlarged GWAS of obesity among East Asian populations were performed. The latter GWAS was conducted by Wen et al. to test the association of BMI with 2.5 million genotyped or imputed SNPs in Asian population in 2014 [21]. The significant associations of the two related SNPs in 12q24 region (rs671 at *ALDH2*, rs12229654 at *MYL2*, $r^2 = 0.58$) with BMI has been primarily identified in a population of 86,757 individuals and replicated in an independent sample of 11,233 and 23,454 individuals, respectively. The carriers of G allele (namely, highly active *ALDH2*1* allele) conferred higher BMI compared with non-carriers. There was substantial overlapping between overall obesity and central obesity and previously reported locus near *ALDH2* for WHR was not adjusted for BMI. Therefore, Wen et al. conducted a new round of meta-analyses to test the associations of WC and WHR with 2.5 million SNPs among individuals of East Asian ancestry in 2016 ($n = 53{,}052$ and 48,312 for WC and WHR, respectively) [22]. They confirmed the effects of rs671 in *ALDH2* on WC and WHR before or after adjusting for BMI. Even though WC and WHR are considered good markers for central obesity, they cannot distinguish between visceral fat and abdominal subcutaneous fat directly. The genetic study for more accurate proxy of central obesity may reveal novel variants that are not necessarily discovered when WC and WHR are used as the outcomes. The study by Wang et al. in 2016, consisted of 2958 subjects in FAt Distribution and diseasE (FADE) cohort from Chinese Han populations with refined visceral fat area (VFA) and subcutaneous fat area (SFA) imaged by MRI, explored whether *ALDH2* variants directly imposed effects on visceral fat or subcutaneous fat deposit [35]. They demonstrated that rs671 at *ALDH2* was associated with visceral fat accumulation specifically. The carriers of G allele (highly active *ALDH2*1* allele) confers more visceral fat accumulation compared with non-carriers. All of studies mentioned above indicated that *ALDH2* variants have substantial influence on obesity, especially for visceral fat accumulation.

3.3 *Subgroup Analysis Stratified by Alcohol Composition*

rs671 in *ALDH2* was previously demonstrated a robust association with alcohol consumption in genetic and functional studies. To evaluate the underlying effect of alcohol consumption on the association between *ALDH2* variants and obesity-related traits, the subgroup analyses were conducted but produced inconsistent results (as shown in Table 2). Two studies mentioned above by Wen et al. mainly involved a total of 6918 Chinese individuals with data of alcohol consumption available from Shanghai Genome-Wide Association Studies (SGWAS). The information about alcohol consumption was collected using a standard questionnaire. The association of rs671 with BMI was mainly observed among nondrinkers in SGWAS

Table 2 Subgroup analysis of *ALDH2* variant and obesity-related traits stratified by alcohol composition

Traits	Drinkers			Nondrinkers					
	N	β (SE)	P values	N	β (SE)	P values	P for interaction	Date	References
SGWAS cohort									
BMI	643	−0.0174 (0.0979)	0.8586	8617	0.0512 (0.0188)	**0.0065**	0.5656	2014	*Human Molecular Genetics*
WCnoBMI	643	0.0190 (0.0947)	0.8413	8616	0.0466 (0.0185)	**0.0119**	0.5950	2016	*Scientific Reports*
WCadjBMI	643	0.0327 (0.0555)	0.5568	8616	0.0084 (0.0121)	0.4843	0.7639		
WHRnoBMI	643	−0.0582 (0.0954)	0.5422	8615	−0.0003 (0.0175)	0.9861	0.0745		
WHRadjBMI	643	−0.0497 (0.0827)	0.5483	8615	−0.0195 (0.0160)	0.2218	0.0787		
FADE cohort									
BMI	1192	0.3165 (3.495)	**8.46 × 10^{-5}**	1696	0.1104 (−0.8466)	0.4000	**0.0071**	2016	*Scientific Reports*
WCnoBMI	1211	0.0139 (0.0028)	**9.6 × 10^{-7}**	1726	0.0047 (0.0019)	**0.0112**	**0.0188**		
WCadjBMI	1211	0.005 (0.0017)	**0.0024**	1726	0.0034 (0.0011)	**0.0029**	0.6301		
WHRnoBMI	1211	0.0033 (0.0016)	**0.0412**	1726	0.0004 (0.0011)	0.7063	0.2344		
WHRadjBMI	1211	0.0005 (0.0015)	0.7556	1726	0 (0.001)	0.9985	0.8903		
VFAnoBMI	1211	0.0846 (0.0154)	**5.18 × 10^{-8}**	1726	0.009 (0.009)	0.3197	**1.00 × 10^{-4}**		
VFAadjBMI	1211	0.0451 (0.0118)	**0.0001**	1726	0.0042 (0.0068)	0.5416	**0.0055**		
VFA/SFAnoBMI	1211	0.0421 (0.0119)	**0.0004**	1726	0.0058 (0.0072)	0.4216	**0.0316**		
VFA/SFAadjBMI	1211	0.0335 (0.0118)	**0.0047**	1726	0.0049 (0.0071)	0.4880	0.0973		
SFAnoBMI	1211	0.0425 (0.0104)	**4.48 × 10^{-5}**	1726	0.0033 (0.0066)	0.6209	**2.00 × 10^{-3}**		
SFAadjBMI	1211	0.0116 (0.0069)	0.0906	1726	−0.0007 (0.0048)	0.8909	0.0946		

Traits were adjusted for age and sex in the additive genetic model

P values<0.05 are shown in bold

cohort, suggesting an antagonistic effect of alcohol consumption on the *ALDH2*-BMI association. Nonetheless, WC-increasing effect or WHR-increasing effect conferred by rs671 G allele failed to be replicated among drinkers or nondrinkers separately. In FADE cohort, Wang et al. had access to individual data of alcohol consumption among a total of 2937 Chinese Han individuals. Drinkers were defined if subjects had ever consumed alcohol in their lifetime including chance drinkers (less than three times in every week) and regular drinkers (equal or more than three times in every week), whereas nondrinkers were those who never drank in their lifetime. The effects of rs671 on BMI, WC, and WHR with or without adjustment of BMI were mainly observed among drinkers. More importantly, the effect of *ALDH2* rs671 on VFA and VFA/SFA was also observed among drinkers, indicating association of rs671 with obesity-related traits mediated by alcohol consumption. Wang et al. also conducted subgroup analysis which draws a distinction between chance drinkers and regular drinkers to strengthen their findings. The results showed that the nominal associations of *ALDH2* variant with WC and VFA after adjustment of BMI were restricted to regular drinkers specifically but did not observe associations in chance drinkers. The mixed results between SGWAS and FADE study might be explained by the difference in study design. Although the definition of alcohol consumption in SGWAS and FADE study was approximately similar, the proportion of alcohol consumption was different (drinkers vs nondrinkers = 1:13 in SGWAS and 2:3 in FADE study). The possibility that BMI- or WC- or WHR-increasing effects conferred by rs671 G allele were not observed because of small sample size of drinkers in SGWAS cannot be excluded (Table 3).

3.4 Gender Difference

Generally, males have more visceral fat deposit, whereas females have more subcutaneous fat deposit before menopause. Taking into account this heterogeneity in fat distribution in both genders, the subgroup analyses in males and females were performed separately. The results showed more pronounced associations in males than in females, and there is evident heterogeneity in both gender for obesity-related traits (Table 4). Additionally, Wang et al. also tested sex difference among drinkers and nondrinkers separately. The nominal associations between the *ALDH2* variant and visceral fat accumulation were restricted to male drinkers specifically, and the effects of rs671 in *ALDH2* on VFA and SFA revealed a borderline sex-related significance among overall drinkers. Note that the male to female ratio was not balanced between drinkers and nondrinkers, that is, both of SGWAS and FADE study have more male drinkers than female drinkers. We believe that the statistical power of rs671 with central obesity-related traits in female drinkers was inadequate due to the relatively small sample size of female drinkers. The studies with comparable amount of males and females regarding to alcohol consumption to test whether alcohol consumption affect obesity or visceral fat accumulation in a sex-dependent manner are warranted.

Table 3 The analysis of rs671 and obesity-related traits in chance drinkers and regular drinkers in FADE cohort

Traits	Nondrinkers			Chance drinkers			Regular drinkers			P for interaction	Date	Reference
	N	β (SE)	P	N	β (SE)	P	N	β (SE)	P			
BMI	1696	0.1104 (−0.3466)	0.4000	625	0.8656 (0.2653)	**0.0012**	567	0.8678 (0.3533)	**0.0143**	**0.0099**	2016	*Scientific reports*
WCnoBMI	1726	0.0047 (0.0019)	**0.0112**	620	0.0127 (0.0036)	**0.0004**	565	0.0161 (0.0048)	**0.0009**	**0.0145**		
WCadjBMI	1726	0.0034 (0.0011)	**0.0029**	620	0.0036 (0.0022)	0.1071	565	0.0064 (0.0027)	**0.0174**	0.4514		
WHRnoBMI	1726	0.0004 (0.0011)	0.7063	620	0.0011 (0.0021)	0.5879	564	0.0052 (0.0027)	0.0599	0.1664		
WHRadjBMI	1726	0 (0.001)	0.9985	620	−0.0019 (0.0019)	0.3219	564	0.002 (0.0024)	0.4059	0.6961		
VFAnoBMI	1726	0.009 (0.009)	0.3197	625	0.0676 (0.0197)	**0.0007**	567	0.1056 (0.0264)	**0.0001**	**0.0001**		
VFAadjBMI	1726	0.0042 (0.0068)	0.5416	625	0.0236 (0.0146)	0.1048	567	0.0655 (0.0209)	**0.0018**	**0.0029**		
VFA/SFAnoBMI	1726	0.0058 (0.0072)	0.4216	625	0.0221 (0.0153)	0.1487	567	0.0572 (0.0202)	**0.0048**	**0.0164**		
VFA/SFAadjBMI	1726	0.0049 (0.0071)	0.4880	625	0.0102 (0.015)	0.4973	567	0.0509 (0.0201)	**0.0118**	0.0527		
SFAnoBMI	1726	0.0033 (0.0066)	0.6209	625	0.0453 (0.0133)	**0.0007**	567	0.0487 (0.0177)	**0.0062**	**0.0035**		
SFAadjBMI	1726	−0.0007 (0.0048)	0.8909	625	0.0133 (0.009)	0.1425	567	0.0149 (0.0112)	0.1860	0.1216		

Traits were adjusted for age and sex in the additive genetic model
P values<0.05 are shown in bold

Table 4 Gender difference in association of *ALDH2* variants with obesity-related traits

SNP	Nearest Gene	Allele	Traits	Males		Females		P for difference	Date	Reference
				Beta (SE)	P values	Beta (SE)	P values			
rs2074356	*ALDH2-HECTD4*	C/T	WHR	–	–	–	–	–	2009	*Nature Genetics*
rs671	*ALDH2*	G/A	BMI	–	–	–	–	–	2014	*Human Molecular Genetics*
rs12229654	*MYL2*	T/G	BMI	–	–	–	–	–		
rs671	*ALDH2*	G/A	WCnoBMI	0.1089 (0.0167)	**7.78×10^{-11}**	0.0297 (0.0106)	**0.0050**	**6.30×10^{-5}**	2016	*Scientific reports*
			WCadjBMI	0.0292 (0.0098)	**0.0029**	0.0072 (0.0064)	0.2581	0.0604	2016	
			WHRncBMI	0.1236 (0.0183)	**1.34×10^{-11}**	−0.0017 (0.0107)	0.8727	**3.29×10^{-9}**	2016	
			WHRadjBMI	0.065 (0.0146)	**8.76×10^{-6}**	−0.0151 (0.0097)	0.1194	**4.96×10^{-6}**	2016	
rs12229654	*MYL2*	T/G	WHRncBMI	0.1109 (0.0167)	**3.16×10^{-11}**	0.0139 (0.01)	0.1646	**6.49×10^{-7}**	2016	
			WHRadjBMI	0.0707 (0.0134)	**1.33×10^{-7}**	0.0046 (0.009)	0.6112	**4.28×10^{-5}**	2016	
			WCnoBMI	0.0969 (0.0162)	**2.45×10^{-9}**	0.0316 (0.0103)	**0.0022**	**6.85×10^{-4}**	2016	
			WCadjBMI	0.0272 (0.0085)	**0.0013**	0.0123 (0.006)	**0.0408**	0.1506	2016	
rs671	*ALDH2*	G/A	BMI	0.4053 (0.1539)	**0.0085**	0.2047 (0.1438)	0.15	0.3409	2016	*Scientific reports*
			WCnoBMI	0.01 (0.0022)	**4.05×10^{-6}**	0.0055 (0.002)	**0.0062**	0.1301	2016	

(continued)

Table 4 (continued)

SNP	Nearest Gene	Allele	Traits	Males		Females		P for difference	Date	Reference
				Beta (SE)	P values	Beta (SE)	P values			
			WCadjBMI	0.0053 (0.0012)	**7.06×10^{-6}**	0.0033 (0.0013)	**0.0104**	0.2583	2016	
			WHRnoBMI	0.0041 (0.0011)	**3.58×10^{-4}**	0.0002 (0.0012)	0.8418	**0.0166**	2016	
			WHRadjBMI	0.0026 (0.001)	**0.0087**	−0.0004 (0.0012)	0.7340	0.0548	2016	
			VFAnoBMI	0.0748 (0.0125)	**2.56×10^{-9}**	0.0057 (0.009)	0.5294	**7.25×10^{-6}**	2016	
			VFAadjBMI	0.0529 (0.0093)	**1.75×10^{-8}**	−0.0025 (0.0069)	0.7173	**1.72×10^{-6}**	2016	
			VFA/SFAnoBMI	0.0539 (0.0091)	**3.48×10^{-9}**	−0.0011 (0.0078)	0.8872	**4.46×10^{-6}**	2016	
			VFA/SFAadjBMI	0.049 (0.0089)	**4.43×10^{-8}**	−0.0027 (0.0077)	0.7241	**1.12×10^{-5}**	2016	
			SFAnoBMI	0.0209 (0.0082)	**0.0110**	0.0069 (0.0069)	0.3182	0.1914	2016	
			SFAadjBMI	0.004 (0.0051)	0.4345	0.0003 (0.0051)	0.9544	0.6080	2016	
rs2074356	*ALDH2-HECTD4*	C/T	WCnoBMI	0.0037 (0.0025)	0.1491	0.0038 (0.0023)	0.1031	0.9765	2016	
			WCadjBMI	0.0012 (0.0014)	0.3759	0.0029 (0.0015)	**0.0473**	0.4074	2016	
			WHRnoBMI	0.0023 (0.0013)	0.0877	0.0007 (0.0014)	0.6049	0.4023	2016	
			WHRadjBMI	0.0015 (0.0011)	0.1858	0.0005 (0.0013)	0.7238	0.5571	2016	

	VFAnoBMI	0.0499 (0.0147)	**6.69×10^{-4}**	0.0028 (0.0104)	0.7857	**0.0089**	2016	
	VFAadjBMI	0.0384 (0.0109)	**4.56×10^{-4}**	−0.0003 (0.0079)	0.9699	**0.0040**	2016	
	VFA/ SFAnoBMI	0.0423 (0.0106)	**7.33×10^{-5}**	0.0005 (0.009)	0.9594	**0.0026**	2016	
	VFA/ SFAadjBMI	0.0396 (0.0104)	**1.45×10^{-4}**	−0.0001 (0.0088)	0.9949	**0.0036**	2016	
	SFAnoBMI	0.0075 (0.0096)	0.4341	0.0025 (0.0079)	0.7561	0.6876	2016	
	SFAadjBMI	−0.0014 (0.006)	0.8180	−0.0002 (0.0059)	0.9788	0.8866	2016	

Traits were adjusted for age and sex in the additive genetic model
SNP, single nucleotide polymorphism; Chr, chromosome
P values<0.05 are shown in bold

3.5 Ethnicity

Varied ancestry populations differed in fat distribution and underlying genetic background [36, 37]. For a given amount of BMI, Asian populations seem to be prone to the accumulation of visceral fat compared to European populations [38]. An indisputable fact is that large-scale obesity GWAS that include Asian and other non-European populations are more likely to provide insight into different genetic architectures and identify evidence for specific causal genes [19, 20]. 12q24 region is polymorphic only in East Asians, and it is still unknown to what extent *ALDH2* gene contributed to the risk of visceral fat accumulation among East Asian populations exactly.

4 A Possible Model for ALDH2-Induced Obesity

The *ALDH2* gene encodes the mitochondrial ALDH2, a critical enzyme not only for ethanol oxidation during alcohol ingestion but also for several endogenous aldehydes such as propionaldehyde, butyraldehyde, and 4-hydroxykenals (4-HNE) originated from mitochondrial production of reactive oxygen species (ROS). Therefore, two plausible mechanisms have been postulated for the link between *ALDH2* variants and obesity including ethanol/acetaldehyde metabolism and endogenous bioactive aldehyde metabolism.

4.1 Proposed Mechanism One: Ethanol/Acetaldehyde Metabolism

ALDH2 encodes a functionally ALDH enzyme subunit that leads to impaired the removal of acetaldehyde, a toxic byproduct of ethanol metabolism. The A allele (also *ALDH2**2 allele), common in 30–50% of individuals of northeast Asian descent, is associated with a significantly reduced likelihood of heavy drinking and alcohol dependence due to a unpleasant symptoms like flushing, tachycardia, and nausea. The physiological and behavioral consequences of alcohol consumption serve as an intermediary process between *ALDH2* genetics and obesity. In other words, rs671 in *ALDH2* may influence the obesity and visceral fat accumulation by affecting alcohol consumption behavior, with A allele carriers having lower BMI and visceral fat depots due to lower alcohol consumption. A similar conclusion was drawn from a Mendelian randomization analysis among Korean population regarding rs671 in *ALDH2* as an instrumental variable, which indicated the marked positive effects of alcohol intake (as indexed by the absence of alcohol flushing and the *ALDH2* rs671 GG genotype) on blood pressure [39].

Recent evidence indicated that excess drinking was consistently associated with weight gain or increased waist circumference, whereas light-to-moderate alcohol consumption is not linked to adiposity gain [40–42]. Further, Molenaar et al. [43] reported that intake of large amounts of alcohol was associated with decrease of subcutaneous adiposity in females and increase of visceral adiposity in males from Framingham Offspring Study. A common trend appears to be that it is multifactorial and involves cross talk among various organs and tissues.

In food, the energy content is derived from macronutrients (carbohydrate, fat, protein, and alcohol). Both of carbohydrate and protein provide 4 kcal per gram and fat provides 9 kcal per gram. Alcohol is a calorically dense substance and produces 7.1 kcal (29 kJ) per gram [44], which should theoretically play a critical role in energy balance. The findings about the effects of alcohol-derived energy on body mass and fat deposit are debatable. A line of evidence indicates that alcohol-derived calorie consumption seems to supplement food-derived energy [45, 46], and individuals with excessive drinking appear to increase adiposity indices among varied populations [47–50]. A diet recall study in 951 healthy males from Koreans, which recorded the dietary intake of energy from food and alcohol, showed that total energy intake increased with higher alcohol consumption and further observed that there was an increase in visceral fat accumulation with either decrease or no change in subcutaneous fat accumulation [51]. Another study, however, found that chronic and moderate alcohol intake was likely to lead to a decrease in macronutrients intake to compensate for ethanol calories [52]. A certain number of studies have examined the short-term effect of alcohol consumption on appetite control and feeding behavior. In these studies, alcohol may amplify individuals' perception of appetite in response to food stimuli but fail to produce sufficient signals on satiety or enhance the rewarding effects [53–55]. Besides, several neurotransmitters including γ-aminobutyric acid (GABA), opioid and dopaminergic system were considered to be vital for motivational effect of alcohol on stimulation of appetite [56, 57]. Genomic-based evidence showed that compared to individuals with the GG genotype (*ALDH2*1/*1* homozygotes), those with the inactive A allele (*ALDH2*2*) reported greater negative alcohol expectancies, and lower risk of alcohol abuse, indicating that differences in alcohol metabolism were reported to influence how drinking events are experienced, interpreted, and stored in memory in central nervous system [58].

Abundance of data showed that higher amounts of daily alcohol intake were positively associated with visceral adiposity [43, 59, 60]. The correlation between alcohol intake and fat distribution was likely mediated by plasma androgens at least in part or fatty liver, which can result in hepatic insulin resistance and subsequent weight gain. Several studies, however, have been rather inconsistent, reporting no association between alcohol consumption and visceral fat [61, 62]. Genomic-based evidence indicated that *ALDH2* variants strongly correlated with obesity, especially for visceral fat accumulation. Further studies are required to validate the association and get understanding of the mechanism process of ALDH2 which manifested as visceral adiposity.

Moreover, increasing studies showed that excess alcohol consumption were often associated with chronic systemic inflammation status and high circulating proinflammatory cytokine levels [63–65] as well as high circulating cortisol levels [66, 67]. Alcohol intake may enhance cortisol secretion which changed the pattern of fat distribution, together with an increase in abdominal and hepatic fat deposition and subcutaneous adiposity lipolysis [66].

Obesity is a combination of genetic and environmental factors. These factors may act independently or they may interact with each other. Gene-environment interaction, in a statistical sense, refers to a situation in which the impacts of genes depend on the environment or the impacts of the environment depend on genotype [68]. For example, a study found that *GNB3* variation interacts with physical activity to influence obesity. They reported that carriers of 825 T allele in physical active group had a 20% lower prevalence of obesity for each additional T allele, while those with the same genotype who were not physically active had a 23% greater prevalence of obesity [69]. The integration of gene-environment information is crucial to move genomic discoveries in obesity to actual behavioral interventions or medications that reduce the burden of obesity. Previous genomic-based evidence indicated the association of *ALDH2* variants with obesity-related indices, which were mediated with alcohol consumption. However, the knowledge on how *ALDH2* and environment interact at a biological level is crucial in fully understanding the processes of obesity or visceral fat accumulation but remains unclear currently. Research on the advance of a wide range of biological responses (e.g., energy intake, appetite control, systemic inflammation, or some hormones) after alcohol consumption in the internal environment among individuals with certain genotype is needed.

4.2 Proposed Mechanism Two: Endogenous Bioactive Aldehyde Metabolism

In the body, several degradation reactions are known to form endogenous acetaldehyde during the oxidative stress, many of which are highly reactive and toxic. Apart from alcohol metabolism, ALDH2 is also responsible for oxidizing several bioactive aldehydes (i.e., propionaldehyde, butyraldehyde, and 4-HNE). It is suggested that ALDH2 could protect against oxidative stress-related diseases such as atherosclerosis, tumors, diabetes, and acute lung injury and pulmonary arterial hypertension [70–72]. Recent evidence indicated that reactive oxygen species (ROS) balance was required for the physiology adipocyte function and differentiation. Yu et al. reported that ALDH2 overexpression or ALDH2 agonist Alda1 was correlated with adipocyte differentiation, mediated by signaling pathways downstream of peroxisome proliferator-activated receptor (PPAR)-γ [72]. As oxidative stress is consolidated in obesity complications such as cardiomyopathy, Wang et al. reported that ALDH2 can help preserve high-fat diet-induced obesity cardiomyopathy through a mechanism related to modulation of autophagy and SUV39H-Sirt1-dependent PGC-1α deacetylation [73].

5 Ways Ahead and Conclusions

Great advance has been made through GWAS in different ethnic groups for the discovery the susceptibility of obesity. Findings of this research in large-scale population-based studies have proven significant to advancing our knowledge of the pathways by which obesity development is modulated. Currently, evidence on quantifying the role of gene-environment interactions in the development of obesity, which are crucial for clinical or public health practice, is still lacking. Along this review, we illustrated the available evidence on *ALDH2* variants and obesity and then proposed a model by which the physiological and behavioral consequences of alcohol consumption are considered to be an intermediary process between *ALDH2* genetics and obesity. This integration of *ALDH2* variants and environment or personal behavior is of great value: (1) shedding new light on the role of aldehyde dehydrogenases in biology process in humans, (2) improving the integration of currently uncertain data on alcohol consumption in etiology of obesity, and (3) regarded as an important way to understand the functional diversity of the numerous genetic polymorphisms for obesity and any other serious, chronic pathologies and ultimately to improve population health.

However, to explore the interpretation of genetic findings to environmental factors for obesity or fat distribution is currently in the early stage, and much of that should be validated. First, to obtain the better assessments of gene-environment interactions, alcohol type as well as consumption pattern including frequency and amount of alcohol intake should be taken into account in quantitative models. Measurement of environmental factors including behavioral and lifestyle factors appears to be less certain and complete rather than measurement of the genomics in epidemiological studies. Therefore, the advances in genomics on intervention and management of obesity will depend on the reliability of evidence obtained from epidemiological studies. Second, whether alcohol-derived metabolites per se or other endogenous reactive dehydrates play a key role in regulation of obesity should be studied since the cytoprotection of ALDH2 is considered to be a prominent function for a variety of human diseases. Another concern is that the usefulness of *ALDH2*-variants findings should be determined through longitudinal studies or intervention studies. Such longitudinal studies in which *ALDH2* variants could be of predictive value or serve as markers to identify individuals who are at high risk of obesity are warranted. Additional research on behavioral interventions targeted to subgroups with varied genotypes is needed to understand behavioral responses to genetic information. For instance, researchers might as well examine whether health guidance based on genetic testing would be more beneficial in framing behavior of certain genotype subgroup.

References

1. Flegal KM et al (2016) Trends in obesity among adults in the United States, 2005 to 2014. JAMA 315(21):2284–2291
2. Ellulu M et al (2014) Epidemiology of obesity in developing countries: challenges and prevention. Global Epidem Obes 2(1):2
3. James WP (2008) The epidemiology of obesity: the size of the problem. J Intern Med 263(4):336–352
4. Misra A et al (2009) Consensus statement for diagnosis of obesity, abdominal obesity and the metabolic syndrome for Asian Indians and recommendations for physical activity, medical and surgical management. J Assoc Physicians India 57:163–170
5. Zhou BF (2002) Predictive values of body mass index and waist circumference for risk factors of certain related diseases in Chinese adults—study on optimal cut-off points of body mass index and waist circumference in Chinese adults. Biomed Environ Sci 15(1):83–96
6. Alberti KG, Zimmet P, Shaw J (2005) The metabolic syndrome—a new worldwide definition. Lancet 366(9491):1059–1062
7. Frayling TM et al (2007) A common variant in the FTO gene is associated with body mass index and predisposes to childhood and adult obesity. Science 316(5826):889–894
8. Loos RJ et al (2008) Common variants near MC4R are associated with fat mass, weight and risk of obesity. Nat Genet 40(6):768–775
9. Heard-Costa NL et al (2009) NRXN3 is a novel locus for waist circumference: a genome-wide association study from the CHARGE Consortium. PLoS Genet 5(6):e1000539
10. Lindgren CM et al (2009) Genome-wide association scan meta-analysis identifies three loci influencing adiposity and fat distribution. PLoS Genet 5(6):e1000508
11. Thorleifsson G et al (2009) Genome-wide association yields new sequence variants at seven loci that associate with measures of obesity. Nat Genet 41(1):18–24
12. Willer CJ et al (2009) Six new loci associated with body mass index highlight a neuronal influence on body weight regulation. Nat Genet 41(1):25–34
13. Heid IM et al (2010) Meta-analysis identifies 13 new loci associated with waist-hip ratio and reveals sexual dimorphism in the genetic basis of fat distribution. Nat Genet 42(11):949–960
14. Speliotes EK et al (2010) Association analyses of 249,796 individuals reveal 18 new loci associated with body mass index. Nat Genet 42(11):937–948
15. Fox CS et al (2012) Genome-wide association for abdominal subcutaneous and visceral adipose reveals a novel locus for visceral fat in women. PLoS Genet 8(5):e1002695
16. Locke AE et al (2015) Genetic studies of body mass index yield new insights for obesity biology. Nature 518(7538):197–206
17. Shungin D et al (2015) New genetic loci link adipose and insulin biology to body fat distribution. Nature 518(7538):187–196
18. Chambers JC et al (2008) Common genetic variation near MC4R is associated with waist circumference and insulin resistance. Nat Genet 40(6):716–718
19. Okada Y et al (2012) Common variants at CDKAL1 and KLF9 are associated with body mass index in east Asian populations. Nat Genet 44(3):302–306
20. Wen W et al (2012) Meta-analysis identifies common variants associated with body mass index in east Asians. Nat Genet 44(3):307–311
21. Wen W et al (2014) Meta-analysis of genome-wide association studies in East Asian-ancestry populations identifies four new loci for body mass index. Hum Mol Genet 23(20):5492–5504
22. Wen W et al (2016) Genome-wide association studies in East Asians identify new loci for waist-hip ratio and waist circumference. Sci Rep 6:17958
23. Ng MC et al (2012) Genome-wide association of BMI in African Americans. Obesity (Silver Spring) 20(3):622–627
24. Monda KL et al (2013) A meta-analysis identifies new loci associated with body mass index in individuals of African ancestry. Nat Genet 45(6):690–696

25. Marchitti SA et al (2008) Non-P450 aldehyde oxidizing enzymes: the aldehyde dehydrogenase superfamily. Expert Opin Drug Metab Toxicol 4(6):697–720
26. Vasiliou V et al (2013) Aldehyde dehydrogenases: from eye crystallins to metabolic disease and cancer stem cells. Chem Biol Interact 202(1–3):2–10
27. Zhang Y, Ren J (2010) Autophagy in ALDH2-elicited cardioprotection against ischemic heart disease: slayer or savior? Autophagy 6(8):1212–1213
28. Sun L, Ferreira JC, Mochly-Rosen D (2011) ALDH2 activator inhibits increased myocardial infarction injury by nitroglycerin tolerance. Sci Transl Med 3(107):107–111
29. Lee CH et al (2009) Genetic modulation of ADH1B and ALDH2 polymorphisms with regard to alcohol and tobacco consumption for younger aged esophageal squamous cell carcinoma diagnosis. Int J Cancer 125(5):1134–1142
30. Toh Y et al (2010) Alcohol drinking, cigarette smoking, and the development of squamous cell carcinoma of the esophagus: molecular mechanisms of carcinogenesis. Int J Clin Oncol 15(2):135–144
31. Hao PP et al (2011) Meta-analysis of aldehyde dehydrogenase 2 gene polymorphism and Alzheimer's disease in East Asians. Can J Neurol Sci 38(3):500–506
32. Wang B et al (2008) The association of mitochondrial aldehyde dehydrogenase gene (ALDH2) polymorphism with susceptibility to late-onset Alzheimer's disease in Chinese. J Neurol Sci 268(1–2):172–175
33. Brooks PJ et al (2009) The alcohol flushing response: an unrecognized risk factor for esophageal cancer from alcohol consumption. PLoS Med 6(3):e50
34. Cho YS et al (2009) A large-scale genome-wide association study of Asian populations uncovers genetic factors influencing eight quantitative traits. Nat Genet 41(5):527–534
35. Wang T et al (2016) Effects of obesity related genetic variations on visceral and subcutaneous fat distribution in a Chinese population. Sci Rep 6:20691
36. Lear SA et al (2009) Ethnic variation in fat and lean body mass and the association with insulin resistance. J Clin Endocrinol Metab 94(12):4696–4702
37. Lear SA et al (2007) Visceral adipose tissue accumulation differs according to ethnic background: results of the Multicultural Community Health Assessment Trial (M-CHAT). Am J Clin Nutr 86(2):353–359
38. Deurenberg P, Deurenberg-Yap M, Guricci S (2002) Asians are different from Caucasians and from each other in their body mass index/body fat per cent relationship. Obes Rev 3(3):141–146
39. Cho Y et al (2018) Exploring the utility of alcohol flushing as an instrumental variable for alcohol intake in Koreans. Sci Rep 8(1):458
40. MacInnis RJ et al (2014) Predictors of increased body weight and waist circumference for middle-aged adults. Public Health Nutr 17(5):1087–1097
41. Sayon-Orea C et al (2011) Type of alcoholic beverage and incidence of overweight/obesity in a Mediterranean cohort: the SUN project. Nutrition 27(7–8):802–808
42. Schutze M et al (2009) Beer consumption and the 'beer belly': scientific basis or common belief? Eur J Clin Nutr 63(9):1143–1149
43. Molenaar EA et al (2009) Association of lifestyle factors with abdominal subcutaneous and visceral adiposity: the Framingham Heart Study. Diabetes Care 32(3):505–510
44. Yeomans MR (2010) Alcohol, appetite and energy balance: is alcohol intake a risk factor for obesity? Physiol Behav 100(1):82–89
45. Austin GL, Krueger PM (2013) Increasing the percentage of energy from dietary sugar, fats, and alcohol in adults is associated with increased energy intake but has minimal association with biomarkers of cardiovascular risk. J Nutr 143(10):1651–1658
46. Yeomans MR (2004) Effects of alcohol on food and energy intake in human subjects: evidence for passive and active over-consumption of energy. Br J Nutr 92(Suppl 1):S31–S34
47. Coulson CE et al (2013) Alcohol consumption and body composition in a population-based sample of elderly Australian men. Aging Clin Exp Res 25(2):183–192
48. French MT et al (2010) Alcohol consumption and body weight. Health Econ 19(7):814–832

49. Mozaffarian D et al (2011) Changes in diet and lifestyle and long-term weight gain in women and men. N Engl J Med 364(25):2392–2404
50. Wannamethee SG, Shaper AG, Whincup PH (2005) Alcohol and adiposity: effects of quantity and type of drink and time relation with meals. Int J Obes 29(12):1436–1444
51. Kim KH et al (2012) Alcohol consumption and its relation to visceral and subcutaneous adipose tissues in healthy male Koreans. Ann Nutr Metab 60(1):52–61
52. Cornier MA, Gayles EC, Bessesen DH (2002) The effects of chronic ethanol consumption on energy balance in rats. Metabolism 51(6):787–791
53. Yeomans MR, Phillips MF (2002) Failure to reduce short-term appetite following alcohol is independent of beliefs about the presence of alcohol. Nutr Neurosci 5(2):131–139
54. Yeomans MR (2010) Short term effects of alcohol on appetite in humans. Effects of context and restrained eating. Appetite 55(3):565–573
55. Jerlhag E et al (2009) Requirement of central ghrelin signaling for alcohol reward. Proc Natl Acad Sci U S A 106(27):11318–11323
56. Yeomans MR, Caton S, Hetherington MM (2003) Alcohol and food intake. Curr Opin Clin Nutr Metab Care 6(6):639–644
57. Rodrigue FD, Covenas R (2011) Targeting opioid and neurokinin-1 receptors to treat alcoholism. Curr Med Chem 18(28):4321–4334
58. Hendershot CS et al (2009) ALDH2, ADH1B and alcohol expectancies: integrating genetic and learning perspectives. Psychol Addict Behav 23(3):452–463
59. Komiya H et al (2006) Smoking as a risk factor for visceral fat accumulation in Japanese men. Tohoku J Exp Med 208(2):123–132
60. Cigolini M et al (1996) Moderate alcohol consumption and its relation to visceral fat and plasma androgens in healthy women. Int J Obes Relat Metab Disord 20(3):206–212
61. Halkjaer J et al (2004) Food and drinking patterns as predictors of 6-year BMI-adjusted changes in waist circumference. Br J Nutr 92(4):735–748
62. Koppes LL et al (2005) Cross-sectional and longitudinal relationships between alcohol consumption and lipids, blood pressure and body weight indices. J Stud Alcohol 66(6):713–721
63. Gonzalez-Reimers E et al (2014) Alcoholism: a systemic proinflammatory condition. World J Gastroenterol 20(40):14660–14671
64. Kang L et al (2007) Chronic ethanol-induced insulin resistance is associated with macrophage infiltration into adipose tissue and altered expression of adipocytokines. Alcohol Clin Exp Res 31(9):1581–1588
65. Sun X et al (2012) Activation of peroxisome proliferator-activated receptor-gamma by rosiglitazone improves lipid homeostasis at the adipose tissue-liver axis in ethanol-fed mice. Am J Physiol Gastrointest Liver Physiol 302(5):G548–G557
66. Leggio L et al (2009) Is cortisol involved in the alcohol-related fat mass impairment? A longitudinal clinical study. Alcohol Alcohol 44(2):211–215
67. Brkic S, Soderpalm B, Gordh AS (2016) High cortisol responders to stress show increased sedation to alcohol compared to low cortisol responders: An alcohol dose-response study. Pharmacol Biochem Behav 143:65–72
68. Dick DM (2011) Gene-environment interaction in psychological traits and disorders. Annu Rev Clin Psychol 7:383–409
69. Grove ML et al (2007) Gene-environment interaction and the GNB3 gene in the Atherosclerosis Risk in Communities study. Int J Obes 31(6):919–926
70. Yin H, Xu L, Porter NA (2011) Free radical lipid peroxidation: mechanisms and analysis. Chem Rev 111(10):5944–5972
71. Xu T et al (2017) Aldehyde dehydrogenase 2 protects against oxidative stress associated with pulmonary arterial hypertension. Redox Biol 11:286–296
72. Voulgaridou GP et al (2011) DNA damage induced by endogenous aldehydes: current state of knowledge. Mutat Res 711(1–2):13–27
73. Wang S et al (2018) ALDH2 protects against high fat diet-induced obesity cardiomyopathy and defective autophagy: role of CaM kinase II, histone H3K9 methyltransferase SUV39H, Sirt1, and PGC-1alpha deacetylation. Int J Obes 42(5):1073–1087

Aldehyde Dehydrogenase (ALDH) 2 in Diabetic Heart Diseases

Srikar Munukutla, Guodong Pan, and Suresh S. Palaniyandi

Abstract A major pathophysiological mechanism behind the development of diabetic heart diseases is oxidative stress mediated by toxic reactive aldehydes such as 4-hydroxynonenal (4HNE). Aldehyde dehydrogenase (ALDH) 2 is a mitochondrial enzyme that has been found to detoxify these deleterious aldehydes and thereby mitigate cardiac damage. Furthermore, its protective role in cellular signaling reverses aberrations caused by hyperglycemia, thereby protecting cardiac function. This chapter assesses the role of ALDH2 in diabetic heart diseases by examining preclinical studies where ALDH2 activity is perturbed in both decreased and increased directions. In doing so, issues in improving ALDH2 activity in select human populations are elucidated, and further research directions are discussed.

1 Introduction

The development of diabetes mellitus (DM) is rooted in either insulin depletion (type 1) or insulin insensitivity (type 2). Pathophysiological alterations in DM including dysregulation of energy (glucose, amino acid, and fatty acid) metabolism, insulin resistance, oxidative stress, as well as inflammation lead to diverse complications thus resulting in end-organ damage to the heart, kidney, liver, and eyes [17]. Particularly, DM is especially implicated in the pathogeneses of various heart diseases, which include chronic conditions such as atherosclerosis [3] and congestive heart failure (CHF) [56] as well as acute events such as myocardial infarction (MI) and unstable angina [15]. Moreover, cardiovascular diseases (CVD) such as

S. Munukutla · G. Pan
Division of Hypertension and Vascular Research, Department of Internal Medicine, Henry Ford Health System, Detroit, MI, USA

S. S. Palaniyandi (✉)
Division of Hypertension and Vascular Research, Department of Internal Medicine, Henry Ford Health System, Detroit, MI, USA

Department of Physiology, Wayne State University, Detroit, MI, USA
e-mail: spalani2@hfhs.org

J. Ren et al. (eds.), *Aldehyde Dehydrogenases*, Advances in Experimental Medicine and Biology 1193, https://doi.org/10.1007/978-981-13-6260-6_9

cerebrovascular disease, peripheral vascular disease, and coronary heart disease (CHD) are all considerable contributors to mortality in those afflicted with DM [57]. Furthermore, a unique development is diabetic cardiomyopathy, which itself creates cardiac functional and structural aberrations when other comorbidities such as CHD and contributive conditions like dyslipidemia and hypertension are not present [27]. Taken altogether, these diabetic heart diseases are detrimental to the health of those afflicted. The American Heart Association (AHA) reports that diabetics (as opposed to nondiabetics) have a two- to four-fold increase in risk of death from cardiac complications [1]. In fact, 68% of deaths in diabetic patients who are at least 65 years of age were associated with CVD alone [1].

There are a variety of mechanisms integral to the pathogenesis of diabetic complications; however, reactive oxygen species (ROS)-mediated oxidative stress is a particularly key component [26]. In normal conditions, ROS is needed to sustain homeostatic cellular signaling, as well as instigate antioxidant responses to stress [26]. However, when ROS levels become excessive, as present in diabetes, deleterious consequences result [26]. In DM, increased and sustained levels of ROS trigger direct consequences to mitochondrial integrity and cellular signaling [26]. Furthermore, increase in ROS also triggers excessive generation of reactive aldehydes and carbonyl compounds [8]. Normally, aldehydes are ubiquitously generated in routine cellular metabolic processes, which include carbohydrate and lipid metabolism [8]. However, the accumulation of particularly harmful aldehydes such as 4-hydroxynonenal (4HNE), malondialdehyde (MDA), and acrolein is deleterious in diabetic tissues [26]. In particular, the reactivity of 4HNE, generated by lipid peroxidation, is of major concern in the pathogenesis of diabetic complications and especially in DM-related heart diseases [47]. The toxic effects are reduced by the detoxification of reactive aldehydes via enzyme systems such as aldehyde dehydrogenase (ALDH) [68]. In this chapter, we will discuss the role of ALDH2 in diabetic heart diseases. Initially, we will outline 4HNE and its role in inducing cardiac damage during DM. After elaborating on the functions of myocardial ALDH2, the consequences of low ALDH2 activity, especially in unique patient populations, will be detailed. Subsequently, the benefits of augmenting ADH2 activity will also be reported. Finally, future research directions regarding new treatments as well as the repurposing of old ones will be outlined.

2 4HNE-Mediated Aldehydic Stress in Diabetic Heart Diseases

The diverse etiopathogeneses of diabetic heart diseases are representative of the various determinants – including genetics, lifestyle, and comorbidities – that are unique to each patient and disease type. However, as oxidative stress is a significant and unifying mediator of the development of these heart diseases, we strategized to target this pathophysiological mechanism. In oxidative stress-mediated effects, we

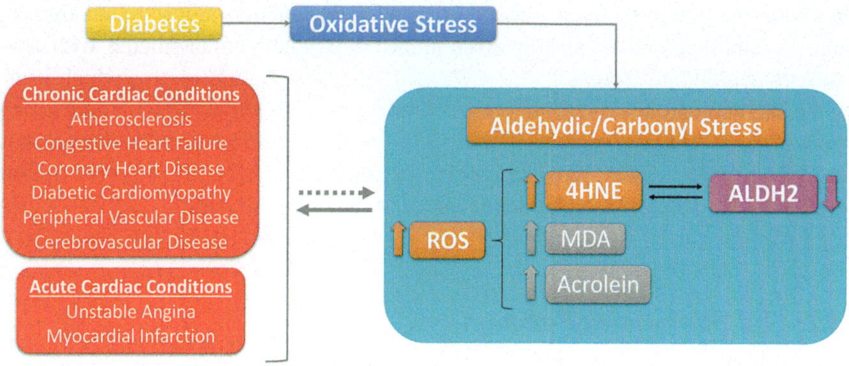

Fig. 1 In diabetes mellitus (DM), hyperglycemia-induced oxidative stress mediated by reactive oxygen species (ROS) results in the accumulation of toxic aldehydes such as 4-hydroxynonenal (4HNE). In the diabetic heart, toxic aldehyde production may impair myocardial function and contribute to both acute and chronic cardiac diseases. Interestingly, these cardiac conditions alone may also induce increases in aldehyde generation and stress

particularly focus on the deleterious role of aldehydic stress, derived significantly from 4HNE accumulation (Fig. 1).

4HNE adduction in the diabetic heart has been known to induce endothelial cell dysfunction [73] as well as cardiomyocyte hypertrophy [48]. Specifically, 4HNE creates adducts with significant intracellular enzymes like 26S proteasome [8, 16], succinate dehydrogenase (SDH) [33], and glyceraldehyde 3-phosphate dehydrogenase [8, 72], thereby restricting their function in the myocardium. As 4HNE-modified proteins accumulate due to low ALDH2 enzymatic activity, 4HNE exacerbates this condition by damaging cardiac mitochondria structurally and functionally [8]. 4HNE mediates increased mitochondrial pore opening, thereby reducing the structural stability of mitochondria [8, 32].

3 ALDH2 in Myocardium and Cardiac Metabolism

3.1 Mitigation of Oxidative Stress

The existence of NAD(P)+-reliant ALDH enzymes is significant in many flora and fauna [68]. Particularly, in the human genome, this family of enzymes encoded by 19 genes plays a key role in the body's ability to combat aldehydic stress [68]. As ROS production instigates toxic aldehyde generation and accumulation of 4HNE-like aldehydes can cause ROS production, a detrimental cycle sets in [65]. Thus, ALDH enzymes are essential in oxidizing the reactive aldehydes to nonreactive carboxylic acids, prevent damage to key intracellular macromolecules, and avert the pathogeneses of various diseases [68]. In this chapter, ALDH2 is discussed with respect to its role in salvaging the myocardium from DM-induced oxidative stress.

It should be noted, however, that other machineries such as aldose reductase [28] and glutathiones [68] are also involved in the detoxification of aldehydes and therefore myocardial protection.

In the myocardium, ALDH2 prevents the formation of toxic protein adducts by neutralizing aldehydes such as 4HNE [43]. Furthermore, it is also involved in key signaling pathways that regulate cardiac metabolism [23, 24, 31, 83]. We outline two such circumstances below with respect to the diabetic heart.

3.2 *Cardiomyocyte Function*

A key signaling role of ALDH2 involves peroxisome proliferator-activated receptor-γ coactivator (PGC-1α) (Fig. 2) [24]. Diabetes-associated insulin resistance promotes increased levels of acetylated PGC-1α, a coactivator that plays a key role in oxidative phosphorylation, biosynthesis of mitochondria, and cardiac development [24, 39]. When acetylated PGC-1α accumulates to an elevated state, regulation of intracellular Ca^{2+} is impaired, ROS levels are markedly increased, and cardiac contractility is damaged [24]. In response, mitochondrial ALDH2 encourages PGC-1α deacetylation by Sirt3, a deacetylating enzyme [24]. A report by Kong

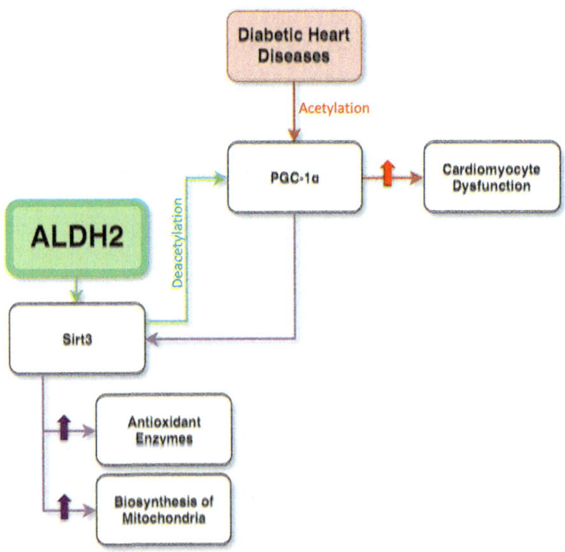

Fig. 2 In DM-related heart diseases, acetylated peroxisome proliferator-activated receptor-γ coactivator (PGC-1α) levels are elevated, which can lead to cardiomyocyte dysfunction. Aldehyde dehydrogenase (ALDH2) uses Sirt3 as a mediator to deacetylate PGC-1α. Additionally, PGC-1α increases the expression of Sirt3 by activating the estrogen-related receptor (ERRE) binding element on Sirt3 promoter. This cross talk is vital as Sirt3 is involved in the generation of antioxidant enzymes such as SOD2 and GPx1 as well as biosynthesis of mitochondria

et al. demonstrated that PGC-1α utilized Sirt3 as a key mediator in regulating metabolism in the mitochondria [31]. PGC-1α increased the expression of Sirt3 by binding to and activating the estrogen-related receptor binding element (ERRE) on Sirt3 promoter [31]. Subsequently, Sirt3 production elevated generation of SOD2 and GPx1, the enzymes that inactivate ROS [31]. Additionally, not only did Sirt3 aid PGC-1α in mitigating ROS generation; it was also directly involved in biosynthesis of mitochondria as well [31]. Therefore the cross talk between PGC-1α and Sirt3, which is vitally mediated by ALDH2, is essential in mitigating a cascade of mitochondrial and cardiac damage in DM [24, 31].

3.3 Autophagy Regulation

5' adenosine monophosphate-activated protein kinase (AMPK) is another important regulatory molecule that is involved in cardiac metabolism, autophagy, and contractility (Fig. 3) [23]. Guo et al. reported that in DM-associated hyperglycemia, phosphorylation of AMPK is lowered, thereby damaging cardiac function and particularly impairing autophagy, a homeostatic cellular clearing process [23]. Due to the inactivity of AMPK, activity of Unc-51-like kinase (ULK1), a key signaling molecule used to commence autophagy, is also diminished [23]. Furthermore, another signaling molecule involved in autophagy is forkhead box class O 3a (Foxo3a), whose phosphorylation and activation in DM also prevent the occurrence of autophagy [23]. Taken altogether, as AMPK and ULK1 activity is stymied while Foxo3a activity is increased, these adjustments impair autophagy in cardiomyocytes, contributing to impairment in intracellular Ca^{2+} management, cell death, and cardiac functional aberrations [23]. However, though the mechanism is still unclear, ALDH2 increased the phosphorylation of AMPK and ULK1 while favoring Foxo3a dephosphorylation [23]. Therefore, the interventional role of ALDH2 in cellular signaling involving the activation of AMPK1 and ULK1 as well as the inactivation of Foxo3a is important for the occurrence of autophagy and maintenance of cardiac homeostasis in the diabetic myocardium [23].

3.4 Energy Metabolism

Current research has not demonstrated a direct role of ALDH2 in the regulation of glucose metabolism, an important factor in diabetes and its complications. However, altered glucose metabolism due to DM has been demonstrated in ALDH2-deficient knockout mice [78]. Wang et al. reported in these ALDH2-deficient mice, induction of DM led to elevated adenosine monophosphate/adenosine triphosphate (AMP/ATP) and adenosine diphosphate/adenosine triphosphate (ADP/ATP) ratios as well reduced phosphocreatine (PCr)/ATP ratio, all indices of hampered energy

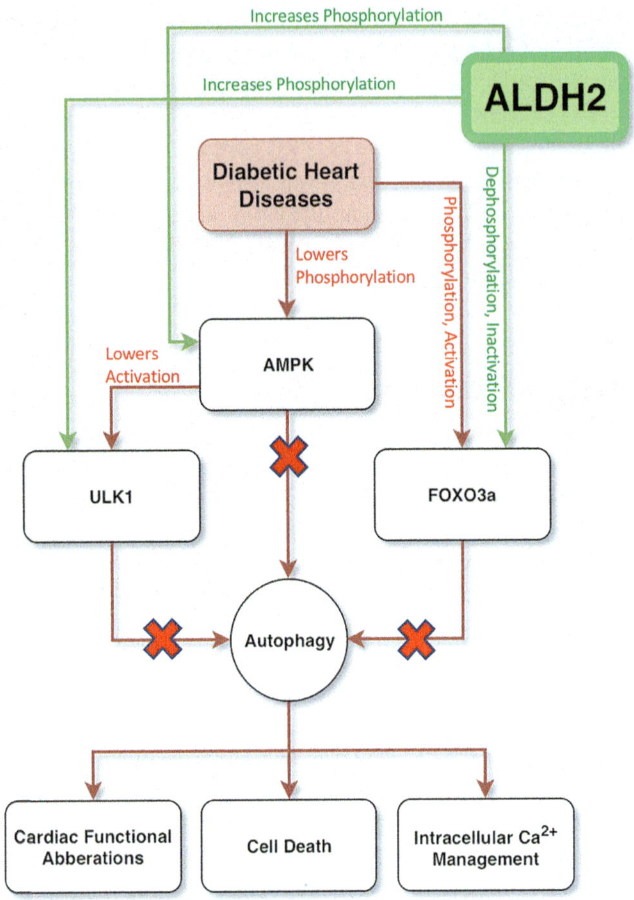

Fig. 3 In DM-related heart diseases, phosphorylation of 5′ adenosine monophosphate-activated protein kinase (AMPK) is decreased, which also leads to inactivation of Unc-51-like kinase (ULK1). Furthermore, forkhead box class O 3a (Foxo3a) phosphorylation is increased. Together, these DM-induced alterations prevent the homeostatic process of autophagy, which then leads to cardiac damage. ALDH2 increases both the phosphorylation of AMPK and ULK1 and promotes Foxo3a dephosphorylation, which allows for proper induction of autophagy

metabolism. Furthermore, in both DM and ALDH2 deficiencies, lowered hypoxanthine and inosine as well as elevated glycocholic acid were present, indicative of perturbed glucose metabolism [78]. In another abnormal metabolic state such as myocardial ischemia, ALDH2 deficiency has been shown to modify glucose metabolism [14, 70]. Endo et al. reported that in ischemia, pentose phosphate shunt activity was markedly increased in the presence of low ALDH2 activity [14]. Specifically, metabolomic analysis revealed that pentose phosphate shunt metabolites such as ribose 5-phosphate and 6-phosphogluconate as well as other relevant amino acids in the pathway such as homocysteine, glycine, and cystathionine were increased in the presence of low ALDH2 activity [14]. Furthermore, a heightened ratio of NADPH to NADP+ further supported these findings [14]. Further investigation is needed to elucidate the precise involvement of ALDH2 in glucose metabolism in DM.

4 Low ALDH2 Activity in Diabetic Cardiac Complications

Numerous studies have investigated the consequences of low ALDH2 activity. This section entails studies that used a variety of methods to inhibit ALDH2 such as pharmacological inhibition, siRNA transfection, mutation, and knockout mouse models. Furthermore, human studies will be discussed, with respect to both those suffering from the ALDH2*2 mutation and diabetic patients without the mutation. Particularly, in polymorphic diabetic patients, the potential toxicity from alcohol consumption will also be examined.

4.1 ALDH2 Deficiency in Animal Models

4.1.1 Pharmacological Inhibition

By binding to sulfhydryl groups, daidzin reversibly inhibits ALDH2 function [29]. In hyperglycemic conditions with daidzin induction, oxidative stress is increased, and mitochondrial membrane potential is decreased, leading to the mishandling of Ca^{2+} as well as lowering of ATP levels [76]. Eventually, myocardial damage occurs, and cardiac contractility is adversely affected [76]. Disulfiram, another inhibitor of ALDH2, operates by creating permanent adducts on a key cysteine (namely, Cys302) [41]. Due to this adduction, disulfiram irreversibly inhibits the active site on ALDH2 [41]. Compared to counterpart cells under normal equimolar conditions, H9C2 cardiomyocytes under high glucose conditions exhibited lowered ALDH2 activity with disulfiram treatment [60]. In the presence of normal conditions (no hyperglycemia), disulfiram alone stimulated increased levels of ROS [60]. Also, in hyperglycemic conditions, disulfiram potentiated the lowering of mitochondrial membrane potential, which eventually leads to apoptosis [60]. Pan et al. reported that H9C2 cardiomyocytes treated with disulfiram and high glucose experienced more than a 4.5-fold increase in apoptosis compared to those treated with high glucose alone [60]. Taken altogether, ALDH2 inhibition not only leads to increased mitochondrial deterioration by itself; it potentiates cell injury and apoptosis in hyperglycemic conditions [60, 76]. Finally, other ALDH2 inhibitors such as cyanimide [25] and CVT-1026 [9] are available; however, their effects on ALDH2 in diabetic conditions have not been elucidated.

4.1.2 siRNA Transfection

siRNA transfection against ALDH2 mRNA results in the targeted lessening of ALDH2 expression specifically and thus its activity [10]. Liao et al. demonstrated that with transfection and subsequent reduction in ALDH2 activity, ER stress is significantly heightened after prior induction of stress [40]. As this increased stress adversely affected proper protein folding and development in the ER, serious complications resulted such as decreases in cardiac fractional shortening and ejection fraction, both indices of impaired cardiac contractility [40]. Furthermore, ALDH2 siRNA transfection has also been shown to increase both mitochondrial dysfunction and cardiomyocyte apoptosis [49].

4.1.3 Knockout

In diabetic ALDH2 knockout mice, key metabolic alterations resulted in the exacerbation of diastolic dysfunction, which is normally also found in DM [78]. Wang et al. demonstrated that ALDH2 deficiency potentiated the inability of the diabetic myocardium to utilize glucose [78]. Also, reserved energy, as denoted by the ratio between phosphocreatine and ATP, was lowered in ALDH2-deficient diabetic mice [78]. Furthermore, after acute events such as ischemia and reperfusion, ALDH2 knockout mice displayed exacerbated endothelial dysfunction as well as elevated ROS levels, both evidences of worsened cardiac stress [9]. Finally, Shen et al. also revealed increased cardiac impairment caused by transverse aortic constriction (TAC) (aka pressure overload dysfunction) in mice that were ALDH2 null [67]. Mitochondrial injury and reduced autophagy due to loss of ALDH2 activity were key alterations that caused this cardiac dysfunction [67]. These studies using the knockout strategy established a definitive role of ALDH2 in cardiac metabolism particularly in diabetes, ischemia, and heart failure.

4.1.4 Mutation

The E487K mutation, an ALDH2*2 point mutation in humans, results in the modification of Glu^{487} to Lys^{487}, leading to low ALDH2 enzymatic activity [82]. Globally, around 8% of people suffer from impaired ALDH2 activity as a result of this mutation [82]. In order to replicate the mutation in experimental animals, Mochly-Rosen's group used pPNT, a mutagenic genetic vector, to modify the ALDH2 gene fragment and create the required point mutation [82]. Eventually, this gene fragment is incorporated into a larger plasmid and then implanted into embryonic stem cells [82]. Finally, these embryonic cells are used to impregnate mice and eventually result in ALDH2*2 knock-in mutant mice [82]. Thus, these ALDH2*2 knock-in mutant mice can be used as a clinically relevant model to study the effect of ALDH2*2. These mice display significantly lowered ALDH2 activity and exhibit increased oxidative stress due to elevated 4HNE levels [see Table 2 in [9]]. However, in contrast, Endo et al. have demonstrated that these mutant mice may become accustomed to the increased stress, in part due to a key metabolic change, i.e., the overexpression of the glutathione synthesis pathway [14]. This discrepancy needs to be studied further as to whether it is due to the method of knock-in mice generation, diet/housing conditions, time period of experiment, and so on. Recently, Pan et al. demonstrated that ALDH2*2 mutant mice with high-fat diet-induced type 2 diabetes exhibited the heart failure with preserved ejection fraction (HFpEF) phenotype with preserved systolic function [61]. But when subjected to exercise stress, they displayed systolic functional impairments in % ejection fraction and % fractional shortening in conscious echocardiography [61]. Similar to ALDH2*2 mice, diabetic patients with ALDH2*2 mutation show preserved systolic cardiac function but diastolic cardiac dysfunction [61]. Thus cardiac research involving ALDH2*2 mutant diabetic mice will provide further insight into the pathophysiology of diabetic heart diseases in ALDH2*2 carriers.

5 Increasing ALDH2 Activity Rescues Diabetic Cardiac Complications

5.1 Overexpression

Whole-body overexpression of ALDH2 is performed in mice models using chicken β-actin promoter [83]. In the diabetic heart, ALDH2 overexpression mitigated key pathophysiological changes [23, 83]. Guo et al. reported increased cardiomyocyte size and reduced autophagy in diabetic heart [23]. Also, DM increased relengthening time and hypertrophy while lowering cardiomyocyte shortening, leading to contractile impairment and structural cardiac damage [23]. All of these alterations were ameliorated by whole-body overexpression of ALDH2 perhaps due to ALDH2-mediated phosphorylation of ULK1 and AMPK as well as dephosphorylation of FOXO3a [23]. Furthermore, Zhang et al. reported that ALDH2 overexpression reduced mitochondrial injury and apoptosis associated with DM in heart tissue [83]. They also demonstrated that deleterious diabetic cardiomyopathy-associated changes in cardiac architecture and function such as increased end-diastolic diameter and end-systolic diameter, lowered thickness in LV wall, and reduced cardiac contractility were all mitigated by ALDH2 overexpression [83]. Zhang et al. attributed these improvements to the role of ALDH2 in mitigating DM-induced decrease of PGC-1α and mitochondrial uncoupling protein 2 (UCP-2) [83]. Furthermore, DM-related deteriorations in membrane potential of mitochondria as well as activity of aconitase, a key Krebs cycle enzyme that is especially vulnerable to ROS, were also ameliorated by ALDH2 [83]. Possible cell signaling pathways in which ALDH2 is participating in to induce these improvements include Foxo3a inactivation, as well as activation of both glycogen synthase kinase 3 beta (GSK3β) and protein kinase B (Akt) [83]. During DM, these signaling molecules are damaged due to oxidative stress, leading to cardiomyocyte dysfunction and overall cardiac damage [83].

Finally, other benefits conferred by increased ALDH2 whole-body expression include lowered fibrosis in the myocardium and reduced mishandling of intracellular Ca^{2+} [12]. Further research is needed to understand the specific mechanisms by which ALDH2 elicits these improvements.

Cell-specific ALDH2 overexpression in the case of diabetic heart complications has not been studied. Further studies using specific cardiomyocyte promoters such as alpha-myosin heavy chain (α-MHC) and cardiac troponin T promoter (cTnT), as well as Tie-2 promoters for endothelial cells, will be valuable in understanding the exact mechanism by which ALDH2 activity affects cardiac diabetic pathology at the specific cardiac cell type level.

5.2 Pharmacological Activation

Alda-1, an agonist and allosteric activator of ALDH2, improves and restores enzymatic capability [64]. In ALDH2*2 mutants, Alda-1 roughly restored almost complete ALDH2 activity [64]. A number of studies have reported that pharmacological activation via Alda-1 attenuated cardiac damage in the diabetic condition [18, 20, 23, 59]. According to Pan et al., after ischemia and reperfusion in diabetic ALDH2*2 mice, Alda-1 administration both attenuated damage to mitochondrial oxidative phosphorylation by elevating the presence of complex V and lowered 4HNE generation [59]. Other benefits conferred by Alda-1 included lowered infarct size as well as elevated left ventricular pressure in ALDH2*2 mutant diabetic mice [59]. Gomes et al. echoed these benefits in the ischemic condition in their work [18]. They demonstrated that in MI-induced heart failure, Alda-1 diminished aldehyde levels, lowered oxidative stress, and altogether ameliorated mitochondrial injury [18]. In addition, Guo et al. reported that under hyperglycemic conditions, Alda-1 promoted autophagy by preventing both a DM-induced decline in AMPK phosphorylation as well as increase in FOXO3a phosphorylation, therefore preventing damage to cardiomyocytes [23]. Furthermore, Gu et al. has shown that Alda-1 attenuation of lowered ALDH2 activity improved function of cardiac fibroblasts [20]. In high glucose conditions, Alda-1 stymied oxidative stress by lowering 4HNE and ROS production [20]. Furthermore, collagen I and III production was decreased, thereby contributing to lowered cardiac fibrosis [20]. Fibroblast proliferation and apoptosis were also mitigated by Alda-1 induction [20]. Alda-1 is the first promising pharmacological agent which can activate both wild-type and mutant ALDH2 [64]. Future studies focusing on its long-term effects with different doses and routes would benefit its candidacy for human use.

5.3 *Other Strategies to Improve ALDH2 Activity in the Diabetic Heart*

Alpha-lipoic acid, a phytochemical antioxidant, has been shown to increase ALDH2 activity [38]. In doing so, Lee et al. demonstrated that intravenous alpha-lipoic acid administration restored AMPK phosphorylation and thereby mitigated oxidative stress and receptor for advanced glycation end products (RAGE) production perhaps via ALDH2, as well as diminished cardiac fibrosis in diabetic rat hearts [34].

Additional methods to aid ALDH2 activity can undertake an indirect approach. Upstream and downstream molecules, such as PKCε and PGC-1α, respectively, are directly involved in ALDH2 activation and activity. Therefore, targeting their signaling cascades may yield beneficial effects. In fact, Chen et al. has shown during ischemia, an agonist-induced PKCε activation elevated ALDH2 function and decreased infarct volume [7]. Additionally, Malhotra et al. has demonstrated that agonist-mediated PKCε activation localized in the heart can mitigate DM-induced ventricular dysfunction [46]. Although studies have yet to utilize PGC-1α activation to specifically increase ALDH2 activity, research has demonstrated improving PGC-1α activity via a plasmid expression vector in DM can mitigate both ROS generation and mitochondrial damage [22]. Thus, targeting close signaling partners of ALDH2 can both improve ALDH2 activity and preserve the integrity of cell signaling in the diabetic heart.

6 Future Directions

6.1 *iPSCs Approach to Increase the Effectiveness of ALDH2 in Diabetic Heart Research*

The induced pluripotent stem cells (iPSCs) approach is meant to investigate solely the genetic contributions to disease etiology, without considerations for epigenetics involved [54]. In the case of patient populations such as ALDH2*2 East Asians, iPSCs research can model the clinical situation and provide novel insights into the pathophysiology of diabetic heart diseases at the cellular level [4]. Although current research using iPSCs in this population is limited, the feasibility and significance of this method have been validated [4]. iPSC cardiomyocytes from participants with the ALDH2*2 mutation were obtained and subsequently treated with Alda-1 [13]. The Alda-1 treatment group displayed increased cellular viability [13]. Since diabetic patients die of CVD, we can use iPSC-induced cardiac cells and study potential disease mechanisms. Thus we propose to determine the severity of heart diseases by using the iPSC approach and thereby address how the ALDH2*2 mutation in each cardiac cell type affects its function and the overall pathogenesis (Fig. 4).

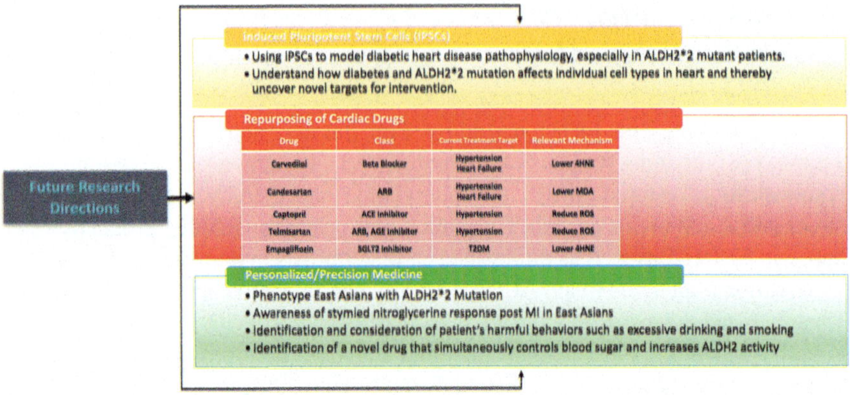

Fig. 4 Future research should focus on using induced pluripotent stem cells (IPSCs) to model diabetic heart pathophysiology, repurposing existing cardiac drugs to find novel ways to improve ALDH2 activity, as well as incorporating personalized/precision medicine to provide a holistic approach into the treatment of DM-related heart diseases, especially in unique patient populations

6.2 Repurposing Existing Cardiac Drugs to Activate ALDH2

Methods to activate and increase ALDH2 activity can involve the detoxification of 4HNE, which can also inhibit ALDH2 function (Fig. 4) [11]. Carvedilol, a beta-blocker and antioxidant-like drug used in diabetic heart failure as well as hypertension, has been shown to significantly lower the presence of 4HNE in the heart [55]. Testing the 4HNE-reducing efficacy of carvedilol in ALDH2*2 mutant diabetic patients is warranted. Candesartan, an angiotensin II receptor blocker (ARB) used to treat hypertension and heart failure, functions by elevating levels of antioxidant glutathiones as well as lowering levels of MDA [62]. Prioritizing the modifications of other such antioxidant-promoting drugs that can increase ALDH2 activity is a clinical necessity. Next, captopril, an angiotensin-converting enzyme (ACE) inhibitor normally used to treat high blood pressure, has been shown to reduce ROS [58]. Therefore, by mitigating ROS levels, it could potentially lead to a reduction in toxic aldehydes [58]. Additionally, telmisartan, an ARB as well as advanced glycation end-product (AGE) inhibitor used to treat hypertension, has been shown to scavenge ROS as well as mitigate oxidative stress [81]. Also, in a recent study, we found empagliflozin, a SGLT2 inhibitor, reduced 4HNE levels in the diabetic heart and improved cardiac function (manuscript in revision). Finally, another approach could be to use these drugs as prototypes and subsequently perform in silico structure-activity modeling using computerized methods in order to identify a suitable compound with increased ALDH2 activity. Thus, this drug could potentially be used in diabetic patients with multiple cardiac ailments and metabolic conditions.

7 Clinical Aspects

7.1 *ALDH2 Deficiency and Cardiac Impairments in Human Patients*

The E487K point mutation, which results in the ALDH2*2 allele, has been studied in East Asians. This mutation decreases ALDH2 activity and is associated with increased incidence of DM as well as myocardial stress [9]. Furthermore, for acute coronary syndrome (ACS), the mutation is considered as a unique precipitating factor in East Asians [9, 80]. More recent investigations have also implicated ALDH2*2 with increased incidence of coronary spastic angina [51], hypertension [44], ischemic stroke [71], and MI [79]. Since diabetic patients would have underlying oxidative stress in their hearts, they are more vulnerable to developing severe injury during ischemic events compared to nondiabetic patients. More notably, mutant diabetic patients would experience an even exacerbated state.

Problematically, these ALDH2*2 patients face a unique situation in the emergency treatment for myocardial ischemic diseases [9]. Nitroglycerin is a sublingual pill given during cardiac ischemic events for quick relief from spasm/ischemia [45]. It functions by inducing the release of nitric oxide (NO), a key vasodilator molecule, in which ALDH2 aids in the production of an intermediate molecule that is directly involved in NO generation [30]. Although tolerance to nitroglycerin happens naturally and gradually with use, patients with the ALDH2*2 mutation exhibit heightened resistance [45]. In fact, a number of studies have shown that in comparison to normal nonmutant patients who have taken nitroglycerin, ALDH2*2 East Asian counterparts demonstrated significant decrease in vasodilation [9, 21, 37]. When nitroglycerine dose is increased in the presence of low ALDH2 activity, infarct size after MI is increased [7] possibly due to heightened ROS generation. Thus, we anticipate ALDH2*2 mutant diabetic patients to be further disadvantaged as they are prone to ischemic events with DM-induced cellular stress.

Finally, even without the ALDH2*2 mutation, there is evidence of reduced ALDH2 activity in the presence of DM [76, 77]. This is perhaps due to increased 4HNE or other oxidative metabolites, which might affect ALDH2 activity or lower expression levels. While the pathophysiology of this condition is yet unclear, further studies may help elucidate this presentation.

7.2 *Alcohol Consumption in Different Polymorphic Diabetic Patients*

In the East Asian population, numerous studies have associated excessive consumption of alcohol with increased incidence of DM in groups such as Koreans [36], Chinese [42], and Japanese [66]. Particularly in ALDH2-deficient Asians in these regions, coexistent DM and heavy alcohol consumption may have synergistically harmful effects [74].

Peng et al. have reported that in ALDH2*2 mutant East Asians, even minimal alcohol consumption induced acetaldehyde buildup and contributed to facial flushing [63]. Alarmingly, the prevalence of this phenomenon termed "Asian Flush" in East Asian populations may be ~40% [35]. Other symptoms include palpitations, nausea/vomiting, as well as tachycardia [75]. In some cases, these aforementioned symptoms may dissuade East Asians from consuming alcohol [75]. For this reason, disulfiram is used as an anti-abuse medication for alcoholics in the general population too [69]. However, if alcohol consumption persists in ALDH2*2 carriers, severe consequences may result [6]. It has been a long-standing belief that alcohol consumption is beneficial to diabetic patients; however, we summarized that it could be detrimental to multiple organs such as the heart, liver, and kidney in both general and ALDH2*2 mutant diabetic patients [53].

7.3 Personalized/Precision Medicine Approach for Unique Patient Populations Such as East Asians with ALDH2*2 Mutation

For more than 500 million people globally who have the ALDH2*2 allele, clinical care necessitates personalized/precision medicine [19, 50]. Genotyping East Asians for the mutation will aid healthcare providers in offering the most precise treatment [19]. Several commercial toolkits in the market already offer this service; however, little is advertised about the benefits of early ALDH2*2 detection [19]. Other methods to detect this mutation include clinical options such as an ethanol patch test, flushing questionnaire [5], and a diagnostic ethanol breath test [2]. As the East Asian population increases, healthcare providers should target a diverse set of mediums – including reliable online sources – in order to educate and empower this growing body of people (Fig. 4) [19].

In clinical situations, physicians need to be cognizant of unique considerations in the treatment of ALDH2*2 mutant patients with cardiac afflictions. The administration of nitroglycerin post MI in East Asian patient populations is of special concern [50]. The ALDH2*2 allele creates tolerance toward nitroglycerin and initiates a stymied response [50]. In fact, there is a 40% decline in efficacy of vasodilation induced by nitroglycerin in relevant patients [45, 50]. Also, clinicians should be aware that East Asians with ALDH2*2 display high incidences of unique forms of MI, such as ST-segment elevation myocardial infarction (STEMI) [52]. Thus, East Asians afflicted with diabetes and cardiac diseases should particularly be screened for the mutation in order to be provided the most effective personalized care. As healthcare technology can support the ability to manage longitudinal care, one-time identification of the ALDH2*2 mutation may be indispensable for the entirety of a patient's lifetime medical history. Finally, in ALDH2*2 patients who consume alcohol, preventative surveillance for esophageal cancer should be a mandatory part of care [19]. Especially in patients who also smoke and report a family history of this cancer, screening is imperative and can be life-saving [19].

As DM is a long-standing disease, it is a hassle for patients to regulate their blood glucose levels. After a few decades, the complications of DM set in, affecting quality of life for many patients, especially ALDH2*2 mutant diabetic patients or non-mutant patients with low cardiac ALDH2 activity. Thus, by using the precision medicine approach, a selective drug that can both control blood sugar level and increase ALDH2 activity is ideal. Through this method, perhaps we can decrease the severity of diabetic complications in general patients and especially more so in ALDH2*2 mutant patients.

8 Conclusion

In this chapter we summarized how ALDH2, a key mitochondrial enzyme, detoxifies 4HNE and participates in a host of metabolic and cell signaling processes in the myocardium. DM lowers ALDH2 activity, which promotes aldehydic stress and impairs cellular signaling, both of which cause detrimental cardiac damage. Especially in ALDH2*2 mutant East Asians, the consequences of low ALDH2 activity may be severe. Complicating the situation are heavy drinking patterns, which may potentiate aldehydic stress. In animal models, increasing ALDH2 activity through overexpression and pharmacological activation has demonstrated improvements in cardiac function. Thus, activation of ALDH2 is a viable strategy in patients too. Continued research, however, should explore novel ways to activate ALDH2. Furthermore, in the clinical situation, more efforts are needed to diagnose aldehyde toxicity, which is elusive even now. As we approach a new age of personalized medicine, healthcare should also adapt to the unique considerations required in the treatment of ALDH2*2 mutant diabetic patients or other diabetic patients with low ALDH2 activity. Finally, comorbidities of DM that alter ALDH2 activity should be taken into consideration for the diagnosis and treatment of diabetic heart diseases.

References

1. AHA (2018) Cardiovascular disease & diabetes. 2018
2. Aoyama I, Ohashi S, Amanuma Y, Hirohashi K, Mizumoto A, Funakoshi M, Tsurumaki M, Nakai Y, Tanaka K, Hanada M, Uesaka A, Chiba T, Muto M (2017) Establishment of a quick and highly accurate breath test for ALDH2 genotyping. Clin Transl Gastroenterol 8(6):e96. https://doi.org/10.1038/ctg.2017.24
3. Beckman JA, Creager MA, Libby P (2002) Diabetes and atherosclerosis: epidemiology, pathophysiology, and management. JAMA 287(19):2570–2581

4. Brandao KO, Tabel VA, Atsma DE, Mummery CL, Davis RP (2017) Human pluripotent stem cell models of cardiac disease: from mechanisms to therapies. Dis Model Mech 10(9):1039–1059. https://doi.org/10.1242/dmm.030320
5. Brooks PJ, Enoch MA, Goldman D, Li TK, Yokoyama A (2009) The alcohol flushing response: an unrecognized risk factor for esophageal cancer from alcohol consumption. PLoS Med 6(3):e50. https://doi.org/10.1371/journal.pmed.1000050
6. Chang JS, Hsiao JR, Chen CH (2017) ALDH2 polymorphism and alcohol-related cancers in Asians: a public health perspective. J Biomed Sci 24(1):19. https://doi.org/10.1186/s12929-017-0327-y
7. Chen CH, Budas GR, Churchill EN, Disatnik MH, Hurley TD, Mochly-Rosen D (2008) Activation of aldehyde dehydrogenase-2 reduces ischemic damage to the heart. Science 321(5895):1493–1495. https://doi.org/10.1126/science.1158554
8. Chen CH, Sun L, Mochly-Rosen D (2010) Mitochondrial aldehyde dehydrogenase and cardiac diseases. Cardiovasc Res 88(1):51–57. https://doi.org/10.1093/cvr/cvq192
9. Chen CH, Ferreira JC, Gross ER, Mochly-Rosen D (2014) Targeting aldehyde dehydrogenase 2: new therapeutic opportunities. Physiol Rev 94(1):1–34. https://doi.org/10.1152/physrev.00017.2013
10. Cortinez G, Sapag A, Israel Y (2009) RNA interference against aldehyde dehydrogenase-2: development of tools for alcohol research. Alcohol 43(2):97–104. https://doi.org/10.1016/j.alcohol.2008.12.007
11. Doorn JA, Hurley TD, Petersen DR (2006) Inhibition of human mitochondrial aldehyde dehydrogenase by 4-hydroxynon-2-enal and 4-oxonon-2-enal. Chem Res Toxicol 19(1):102–110. https://doi.org/10.1021/tx0501839
12. Doser TA, Turdi S, Thomas DP, Epstein PN, Li SY, Ren J (2009) Transgenic overexpression of aldehyde dehydrogenase-2 rescues chronic alcohol intake-induced myocardial hypertrophy and contractile dysfunction. Circulation 119(14):1941–1949. https://doi.org/10.1161/CIRCULATIONAHA.108.823799
13. Ebert AD, Kodo K, Liang P, Wu H, Huber BC, Riegler J, Churko J, Lee J, de Almeida P, Lan F, Diecke S, Burridge PW, Gold JD, Mochly-Rosen D, Wu JC (2014) Characterization of the molecular mechanisms underlying increased ischemic damage in the aldehyde dehydrogenase 2 genetic polymorphism using a human induced pluripotent stem cell model system. Sci Transl Med 6(255):255ra130. https://doi.org/10.1126/scitranslmed.3009027
14. Endo J, Sano M, Katayama T, Hishiki T, Shinmura K, Morizane S, Matsuhashi T, Katsumata Y, Zhang Y, Ito H, Nagahata Y, Marchitti S, Nishimaki K, Wolf AM, Nakanishi H, Hattori F, Vasiliou V, Adachi T, Ohsawa I, Taguchi R, Hirabayashi Y, Ohta S, Suematsu M, Ogawa S, Fukuda K (2009) Metabolic remodeling induced by mitochondrial aldehyde stress stimulates tolerance to oxidative stress in the heart. Circ Res 105(11):1118–1127. https://doi.org/10.1161/CIRCRESAHA.109.206607
15. Fan W (2017) Epidemiology in diabetes mellitus and cardiovascular disease. Cardiovasc Endocrinol Metabol 6(1):8–16
16. Farout L, Mary J, Vinh J, Szweda LI, Friguet B (2006) Inactivation of the proteasome by 4-hydroxy-2-nonenal is site specific and dependant on 20S proteasome subtypes. Arch Biochem Biophys 453(1):135–142. https://doi.org/10.1016/j.abb.2006.02.003
17. Forbes JM, Cooper ME (2013) Mechanisms of diabetic complications. Physiol Rev 93(1):137–188. https://doi.org/10.1152/physrev.00045.2011
18. Gomes KM, Campos JC, Bechara LR, Queliconi B, Lima VM, Disatnik MH, Magno P, Chen CH, Brum PC, Kowaltowski AJ, Mochly-Rosen D, Ferreira JC (2014) Aldehyde dehydrogenase 2 activation in heart failure restores mitochondrial function and improves ventricular function and remodelling. Cardiovasc Res 103(4):498–508. https://doi.org/10.1093/cvr/cvu125
19. Gross ER, Zambelli VO, Small BA, Ferreira JC, Chen CH, Mochly-Rosen D (2015) A personalized medicine approach for Asian Americans with the aldehyde dehydrogenase

2*2 variant. Annu Rev Pharmacol Toxicol 55:107–127. https://doi.org/10.1146/annurev-pharmtox-010814-124915
20. Gu X, Fang T, Kang P, Hu J, Yu Y, Li Z, Cheng X, Gao Q (2017) Effect of ALDH2 on high glucose-induced cardiac fibroblast oxidative stress, apoptosis, and fibrosis. Oxidative Med Cell Longev 2017:9257967. https://doi.org/10.1155/2017/9257967
21. Guo R, Chen XP, Guo X, Chen L, Li D, Peng J, Li YJ (2008) Evidence for involvement of calcitonin gene-related peptide in nitroglycerin response and association with mitochondrial aldehyde dehydrogenase-2 (ALDH2) Glu504Lys polymorphism. J Am Coll Cardiol 52(11):953–960. https://doi.org/10.1016/j.jacc.2008.05.049
22. Guo K, Lu J, Huang Y, Wu M, Zhang L, Yu H, Zhang M, Bao Y, He JC, Chen H, Jia W (2015a) Protective role of PGC-1alpha in diabetic nephropathy is associated with the inhibition of ROS through mitochondrial dynamic remodeling. PLoS One 10(4):e0125176. https://doi.org/10.1371/journal.pone.0125176
23. Guo Y, Yu W, Sun D, Wang J, Li C, Zhang R, Babcock SA, Li Y, Liu M, Ma M, Shen M, Zeng C, Li N, He W, Zou Q, Zhang Y, Wang H (2015b) A novel protective mechanism for mitochondrial aldehyde dehydrogenase (ALDH2) in type i diabetes-induced cardiac dysfunction: role of AMPK-regulated autophagy. Biochim Biophys Acta 1852(2):319–331. https://doi.org/10.1016/j.bbadis.2014.05.017
24. Hu N, Ren J, Zhang Y (2016) Mitochondrial aldehyde dehydrogenase obliterates insulin resistance-induced cardiac dysfunction through deacetylation of PGC-1alpha. Oncotarget 7(47):76398–76414. https://doi.org/10.18632/oncotarget.11977
25. Ishibashi T, Nomura-Nakamoto M, Abe F, Masuoka T, Imaizumi N, Yoshida J, Kawada T, Nishio M (2013) Effects of cyanamide, an aldehyde dehydrogenase type 2 inhibitor, on glyceryl trinitrate- and isosorbide dinitrate-induced vasodilation in rabbit excised aorta and in anesthetized whole animal. Biol Pharm Bull 36(4):556–563
26. Jaganjac M, Tirosh O, Cohen G, Sasson S, Zarkovic N (2013) Reactive aldehydes – second messengers of free radicals in diabetes mellitus. Free Radic Res 47(Suppl 1):39–48. https://doi.org/10.3109/10715762.2013.789136
27. Jia G, Hill MA, Sowers JR (2018) Diabetic cardiomyopathy: an update of mechanisms contributing to this clinical entity. Circ Res 122(4):624–638. https://doi.org/10.1161/CIRCRESAHA.117.311586
28. Kaiserova K, Srivastava S, Hoetker JD, Awe SO, Tang XL, Cai J, Bhatnagar A (2006) Redox activation of aldose reductase in the ischemic heart. J Biol Chem 281(22):15110–15120. https://doi.org/10.1074/jbc.M600837200
29. Keung WM, Vallee BL (1993) Daidzin: a potent, selective inhibitor of human mitochondrial aldehyde dehydrogenase. Proc Natl Acad Sci U S A 90(4):1247–1251
30. Kollau A, Hofer A, Russwurm M, Koesling D, Keung WM, Schmidt K, Brunner F, Mayer B (2005) Contribution of aldehyde dehydrogenase to mitochondrial bioactivation of nitroglycerin: evidence for the activation of purified soluble guanylate cyclase through direct formation of nitric oxide. Biochem J 385. (Pt 3:769–777. https://doi.org/10.1042/BJ20041354
31. Kong X, Wang R, Xue Y, Liu X, Zhang H, Chen Y, Fang F, Chang Y (2010) Sirtuin 3, a new target of PGC-1alpha, plays an important role in the suppression of ROS and mitochondrial biogenesis. PLoS One 5(7):e11707. https://doi.org/10.1371/journal.pone.0011707
32. Kristal BS, Park BK, Yu BP (1996) 4-Hydroxyhexenal is a potent inducer of the mitochondrial permeability transition. J Biol Chem 271(11):6033–6038
33. Lashin OM, Szweda PA, Szweda LI, Romani AM (2006) Decreased complex II respiration and HNE-modified SDH subunit in diabetic heart. Free Radic Biol Med 40(5):886–896. https://doi.org/10.1016/j.freeradbiomed.2005.10.040
34. Lee JE, Yi CO, Jeon BT, Shin HJ, Kim SK, Jung TS, Choi JY, Roh GS (2012) Alpha-lipoic acid attenuates cardiac fibrosis in otsuka long-Evans Tokushima fatty rats. Cardiovasc Diabetol 11:111. https://doi.org/10.1186/1475-2840-11-111

35. Lee H, Kim SS, You KS, Park W, Yang JH, Kim M, Hayman LL (2014) Asian flushing: genetic and sociocultural factors of alcoholism among east asians. Gastroenterol Nurs 37(5):327–336. https://doi.org/10.1097/SGA.0000000000000062
36. Lee DY, Yoo MG, Kim HJ, Jang HB, Kim JH, Lee HJ, Park SI (2017) Association between alcohol consumption pattern and the incidence risk of type 2 diabetes in Korean men: a 12-years follow-up study. Sci Rep 7(1):7322. https://doi.org/10.1038/s41598-017-07549-2
37. Li Y, Zhang D, Jin W, Shao C, Yan P, Xu C, Sheng H, Liu Y, Yu J, Xie Y, Zhao Y, Lu D, Nebert DW, Harrison DC, Huang W, Jin L (2006) Mitochondrial aldehyde dehydrogenase-2 (ALDH2) Glu504Lys polymorphism contributes to the variation in efficacy of sublingual nitroglycerin. J Clin Invest 116(2):506–511. https://doi.org/10.1172/JCI26564
38. Li RJ, Ji WQ, Pang JJ, Wang JL, Chen YG, Zhang Y (2013) Alpha-lipoic acid ameliorates oxidative stress by increasing aldehyde dehydrogenase-2 activity in patients with acute coronary syndrome. Tohoku J Exp Med 229(1):45–51
39. Liang H, Ward WF (2006) PGC-1alpha: a key regulator of energy metabolism. Adv Physiol Educ 30(4):145–151. https://doi.org/10.1152/advan.00052.2006
40. Liao J, Sun A, Xie Y, Isse T, Kawamoto T, Zou Y, Ge J (2012) Aldehyde dehydrogenase-2 deficiency aggravates cardiac dysfunction elicited by endoplasmic reticulum stress induction. Mol Med 18:785–793. https://doi.org/10.2119/molmed.2011.00466
41. Lipsky JJ, Shen ML, Naylor S (2001) In vivo inhibition of aldehyde dehydrogenase by disulfiram. Chem Biol Interact 130-132(1–3):93–102
42. Liu C, Yu Z, Li H, Wang J, Sun L, Qi Q, Lin X (2010) Associations of alcohol consumption with diabetes mellitus and impaired fasting glycemia among middle-aged and elderly Chinese. BMC Public Health 10:713. https://doi.org/10.1186/1471-2458-10-713
43. Ma H, Guo R, Yu L, Zhang Y, Ren J (2011) Aldehyde dehydrogenase 2 (ALDH2) rescues myocardial ischaemia/reperfusion injury: role of autophagy paradox and toxic aldehyde. Eur Heart J 32(8):1025–1038. https://doi.org/10.1093/eurheartj/ehq253
44. Ma C, Yu B, Zhang W, Wang W, Zhang L, Zeng Q (2017) Associations between aldehyde dehydrogenase 2 (ALDH2) rs671 genetic polymorphisms, lifestyles and hypertension risk in Chinese Han people. Sci Rep 7(1):11136. https://doi.org/10.1038/s41598-017-11071-w
45. Mackenzie IS, Maki-Petaja KM, McEniery CM, Bao YP, Wallace SM, Cheriyan J, Monteith S, Brown MJ, Wilkinson IB (2005) Aldehyde dehydrogenase 2 plays a role in the bioactivation of nitroglycerin in humans. Arterioscler Thromb Vasc Biol 25(9):1891–1895. https://doi.org/10.1161/01.ATV.0000179599.71086.89
46. Malhotra A, Begley R, Kang BP, Rana I, Liu J, Yang G, Mochly-Rosen D, Meggs LG (2005) PKC-{epsilon}-dependent survival signals in diabetic hearts. Am J Physiol Heart Circ Physiol 289(4):H1343–H1350. https://doi.org/10.1152/ajpheart.01200.2004
47. Mali VR, Palaniyandi SS (2014) Regulation and therapeutic strategies of 4-hydroxy-2-nonenal metabolism in heart disease. Free Radic Res 48(3):251–263. https://doi.org/10.3109/10715762.2013.864761
48. Mali VR, Ning R, Chen J, Yang XP, Xu J, Palaniyandi SS (2014) Impairment of aldehyde dehydrogenase-2 by 4-hydroxy-2-nonenal adduct formation and cardiomyocyte hypertrophy in mice fed a high-fat diet and injected with low-dose streptozotocin. Exp Biol Med (Maywood) 239(5):610–618. https://doi.org/10.1177/1535370213520109
49. Mali VR, Deshpande M, Pan G, Thandavarayan RA, Palaniyandi SS (2016) Impaired ALDH2 activity decreases the mitochondrial respiration in H9C2 cardiomyocytes. Cell Signal 28(2):1–6. https://doi.org/10.1016/j.cellsig.2015.11.006
50. McAllister SL, Sun K, Gross ER (2016) Developing precision medicine for people of East Asian descent. J Biomed Sci 23(1):80. https://doi.org/10.1186/s12929-016-0299-3
51. Mizuno Y, Harada E, Morita S, Kinoshita K, Hayashida M, Shono M, Morikawa Y, Murohara T, Nakayama M, Yoshimura M, Yasue H (2015) East asian variant of aldehyde dehydrogenase 2 is associated with coronary spastic angina: possible roles of reactive aldehydes and implications of alcohol flushing syndrome. Circulation 131(19):1665–1673. https://doi.org/10.1161/CIRCULATIONAHA.114.013120

52. Mizuno Y, Hokimoto S, Harada E, Kinoshita K, Nakagawa K, Yoshimura M, Ogawa H, Yasue H (2016) Variant aldehyde dehydrogenase 2 (ALDH2*2) is a risk factor for coronary spasm and ST-segment elevation myocardial infarction. J Am Heart Assoc 5(5). https://doi.org/10.1161/JAHA.116.003247
53. Munukutla S, Pan G, Deshpande M, Thandavarayan RA, Krishnamurthy P, Palaniyandi SS (2016) Alcohol toxicity in diabetes and its complications: a double trouble? Alcohol Clin Exp Res 40(4):686–697. https://doi.org/10.1111/acer.13008
54. Musunuru K, Sheikh F, Gupta RM, Houser SR, Maher KO, Milan DJ, Terzic A, Wu JC, American Heart Association Council on Functional G, Translational B, Council on Cardiovascular Disease in the Y, Council on C, Stroke N (2018) Induced Pluripotent Stem Cells for Cardiovascular Disease Modeling and Precision Medicine: A Scientific Statement From the American Heart Association. Circ Genom Precis Med 11(1):e000043. https://doi.org/10.1161/HCG.0000000000000043
55. Nakamura K, Kusano K, Nakamura Y, Kakishita M, Ohta K, Nagase S, Yamamoto M, Miyaji K, Saito H, Morita H, Emori T, Matsubara H, Toyokuni S, Ohe T (2002) Carvedilol decreases elevated oxidative stress in human failing myocardium. Circulation 105(24):2867–2871
56. Nasir S, Aguilar D (2012) Congestive heart failure and diabetes mellitus: balancing glycemic control with heart failure improvement. Am J Cardiol 110(9 Suppl):50B–57B. https://doi.org/10.1016/j.amjcard.2012.08.031
57. Nathan DM, Meigs J, Singer DE (1997) The epidemiology of cardiovascular disease in type 2 diabetes mellitus: how sweet it is … or is it? Lancet 350(Suppl 1):SI4–SI9
58. Ota Y, Kugiyama K, Sugiyama S, Ohgushi M, Matsumura T, Doi H, Ogata N, Oka H, Yasue H (1997) Impairment of endothelium-dependent relaxation of rabbit aortas by cigarette smoke extract--role of free radicals and attenuation by captopril. Atherosclerosis 131(2):195–202
59. Pan G (2016) Abstract 300: Alda-1 attenuates acute ischemia-reperfusion mediated cardiac damage in aldehyde dehydrogenase 2 mutant diabetic mice. Circ Res 119(Suppl 1):A300
60. Pan G, Deshpande M, Thandavarayan RA, Palaniyandi SS (2016) ALDH2 inhibition potentiates high glucose stress-induced injury in cultured cardiomyocytes. J Diabetes Res 2016:1390861. https://doi.org/10.1155/2016/1390861
61. Pan G, Munukutla S, Kar A, Gardinier J, Thandavarayan RA, Palaniyandi SS (2018) Type-2 diabetic aldehyde dehydrogenase 2 mutant mice (ALDH 2*2) exhibiting heart failure with preserved ejection fraction phenotype can be determined by exercise stress echocardiography. PLoS One 13(4):e0195796. https://doi.org/10.1371/journal.pone.0195796
62. Panahpour H, Bohlooli S, Motavallibashi SE (2013) Antioxidant activity-mediated neuroprotective effects of an antagonist of AT1 receptors, candesartan, against cerebral ischemia and edema in rats. Neurophysiology 45(5):441–447. https://doi.org/10.1007/s11062-013-9391-4
63. Peng GS, Chen YC, Wang MF, Lai CL, Yin SJ (2014) ALDH2*2 but not ADH1B*2 is a causative variant gene allele for Asian alcohol flushing after a low-dose challenge: correlation of the pharmacokinetic and pharmacodynamic findings. Pharmacogenet Genomics 24(12):607–617. https://doi.org/10.1097/FPC.0000000000000096
64. Perez-Miller S, Younus H, Vanam R, Chen CH, Mochly-Rosen D, Hurley TD (2010) Alda-1 is an agonist and chemical chaperone for the common human aldehyde dehydrogenase 2 variant. Nat Struct Mol Biol 17(2):159–164. https://doi.org/10.1038/nsmb.1737
65. Pillon N, Soulage C (2012) Lipid peroxidation by-products and the metabolic syndrome. https://doi.org/10.5772/46019
66. Seike N, Noda M, Kadowaki T (2008) Alcohol consumption and risk of type 2 diabetes mellitus in Japanese: a systematic review. Asia Pac J Clin Nutr 17(4):545–551
67. Shen C, Wang C, Fan F, Yang Z, Cao Q, Liu X, Sun X, Zhao X, Wang P, Ma X, Zhu H, Dong Z, Zou Y, Hu K, Sun A, Ge J (2015) Acetaldehyde dehydrogenase 2 (ALDH2) deficiency exacerbates pressure overload-induced cardiac dysfunction by inhibiting Beclin-1 dependent autophagy pathway. Biochim Biophys Acta 1852(2):310–318. https://doi.org/10.1016/j.bbadis.2014.07.014

68. Singh S, Brocker C, Koppaka V, Chen Y, Jackson BC, Matsumoto A, Thompson DC, Vasiliou V (2013) Aldehyde dehydrogenases in cellular responses to oxidative/electrophilic stress. Free Radic Biol Med 56:89–101. https://doi.org/10.1016/j.freeradbiomed.2012.11.010
69. Skinner MD, Lahmek P, Pham H, Aubin HJ (2014) Disulfiram efficacy in the treatment of alcohol dependence: a meta-analysis. PLoS One 9(2):e87366. https://doi.org/10.1371/journal.pone.0087366
70. Sun X, Zhu H, Dong Z, Liu X, Ma X, Han S, Lu F, Wang P, Qian S, Wang C, Shen C, Zhao X, Zou Y, Ge J, Sun A (2017) Mitochondrial aldehyde dehydrogenase-2 deficiency compromises therapeutic effect of ALDH bright cell on peripheral ischemia. Redox Biol 13:196–206. https://doi.org/10.1016/j.redox.2017.05.018
71. Sung YF, Lu CC, Lee JT, Hung YJ, Hu CJ, Jeng JS, Chiou HY, Peng GS (2016) Homozygous ALDH2*2 is an independent risk factor for ischemic stroke in Taiwanese men. Stroke 47(9):2174–2179. https://doi.org/10.1161/STROKEAHA.116.013204
72. Uchida K, Stadtman ER (1993) Covalent attachment of 4-hydroxynonenal to glyceraldehyde-3-phosphate dehydrogenase. A possible involvement of intra- and intermolecular cross-linking reaction. J Biol Chem 268(9):6388–6393
73. Usatyuk PV, Parinandi NL, Natarajan V (2006) Redox regulation of 4-hydroxy-2-nonenal-mediated endothelial barrier dysfunction by focal adhesion, adherens, and tight junction proteins. J Biol Chem 281(46):35554–35566. https://doi.org/10.1074/jbc.M607305200
74. van Dam RM (2018) Alcohol consumption and risk of type 2 diabetes in east Asian populations: do healthy patterns of consumption exist? J Epidemiol 28(3):106–107. https://doi.org/10.2188/jea.JE20170151
75. Wall TLE, Cindy L (1995) Genetic influences affecting alcohol use among Asians. Alcohol Health Res World 19:184–236
76. Wang J, Wang H, Hao P, Xue L, Wei S, Zhang Y, Chen Y (2011) Inhibition of aldehyde dehydrogenase 2 by oxidative stress is associated with cardiac dysfunction in diabetic rats. Mol Med 17(3–4):172–179. https://doi.org/10.2119/molmed.2010.00114
77. Wang HJ, Kang PF, Wu WJ, Tang Y, Pan QQ, Ye HW, Tang B, Li ZH, Gao Q (2013) Changes in cardiac mitochondrial aldehyde dehydrogenase 2 activity in relation to oxidative stress and inflammatory injury in diabetic rats. Mol Med Rep 8(2):686–690. https://doi.org/10.3892/mmr.2013.1524
78. Wang C, Fan F, Cao Q, Shen C, Zhu H, Wang P, Zhao X, Sun X, Dong Z, Ma X, Liu X, Han S, Wu C, Zou Y, Hu K, Ge J, Sun A (2016) Mitochondrial aldehyde dehydrogenase 2 deficiency aggravates energy metabolism disturbance and diastolic dysfunction in diabetic mice. J Mol Med (Berl) 94(11):1229–1240. https://doi.org/10.1007/s00109-016-1449-5
79. Xu F, Chen YG, Geng YJ, Zhang H, Jiang CX, Sun Y, Li RJ, Sagar MB, Xue L, Zhang Y (2007) The polymorphism in acetaldehyde dehydrogenase 2 gene, causing a substitution of Glu > Lys(504), is not associated with coronary atherosclerosis severity in Han Chinese. Tohoku J Exp Med 213(3):215–220
80. Xu F, Chen YG, Xue L, Li RJ, Zhang H, Bian Y, Zhang C, Lv RJ, Feng JB, Zhang Y (2011) Role of aldehyde dehydrogenase 2 Glu504lys polymorphism in acute coronary syndrome. J Cell Mol Med 15(9):1955–1962. https://doi.org/10.1111/j.1582-4934.2010.01181.x
81. Yamagishi S, Nakamura K, Matsui T, Ueda S, Noda Y, Imaizumi T (2008) Inhibitors of advanced glycation end products (AGEs): potential utility for the treatment of cardiovascular disease. Cardiovasc Ther 26(1):50–58. https://doi.org/10.1111/j.1527-3466.2007.00038.x
82. Zambelli VO, Gross ER, Chen CH, Gutierrez VP, Cury Y, Mochly-Rosen D (2014) Aldehyde dehydrogenase-2 regulates nociception in rodent models of acute inflammatory pain. Sci Transl Med 6(251):251ra118. https://doi.org/10.1126/scitranslmed.3009539
83. Zhang Y, Babcock SA, Hu N, Maris JR, Wang H, Ren J (2012) Mitochondrial aldehyde dehydrogenase (ALDH2) protects against streptozotocin-induced diabetic cardiomyopathy: role of GSK3beta and mitochondrial function. BMC Med 10:40. https://doi.org/10.1186/1741-7015-10-40

The Role of ALDH2 in Sepsis and the To-Be-Discovered Mechanisms

Jiaojiao Pang, Yue Zheng, Qi Han, Ying Zhang, Ruru Sun, Jiali Wang, Feng Xu, Yingmei Zhang, Jun Ren, and Yuguo Chen

Abstract Sepsis, defined as life-threatening tissue damage and organ dysfunction caused by a dysregulated host response to infection, is a critical disease which imposes global health burden. Sepsis-induced organ dysfunction, including circulatory and cardiac dysfunction, hepatic dysfunction, renal dysfunction, etc., contributes to high mortality and long-term disability of sepsis patients. Altered inflammatory response, ROS and reactive aldehyde stress, mitochondrial dysfunction, and programmed cell death pathways (necrosis, apoptosis, and autophagy) have been demonstrated to play crucial roles in septic organ dysfunction. Unfortunately, except for infection control and supportive therapies, no specific therapy exists for sepsis. New specific therapeutic targets are highly warranted. Emerging studies suggested a role of potential therapeutic target of ALDH2, a tetrameric enzyme located in mitochondria to detoxify aldehydes, in septic organ

J. Pang · Y. Zheng · Q. Han · Y. Zhang · R. Sun · J. Wang · F. Xu · Y. Chen (✉)
Department of Emergency Medicine and Chest Pain Center, Qilu Hospital of Shandong University, Jinan, China

Clinical Research Center for Emergency and Critical Care Medicine of Shandong Province, Institute of Emergency and Critical Care Medicine of Shandong University, Qilu Hospital of Shandong University, Jinan, China

Key Laboratory of Emergency and Critical Care Medicine of Shandong Province, Key Laboratory of Cardiopulmonary-Cerebral Resuscitation Research of Shandong Province, Qilu Hospital of Shandong University, Jinan, China

The Key Laboratory of Cardiovascular Remodeling and Function Research, Chinese Ministry of Education, Chinese Ministry of Health and Chinese Academy of Medical Sciences, The State and Shandong Province Joint Key Laboratory of Translational Cardiovascular Medicine, Qilu Hospital of Shandong University, Jinan, China
e-mail: chen919085@sdu.edu.cn

Y. Zhang · J. Ren (✉)
Department of Cardiology, Shanghai Institute of Cardiovascular Diseases, Zhongshan Hospital, Fudan University, Shanghai, China

Center for Cardiovascular Research and Alternative Medicine, University of Wyoming College of Health Sciences, Laramie, WY, USA
e-mail: jren@uwyo.edu

J. Ren et al. (eds.), *Aldehyde Dehydrogenases*, Advances in Experimental Medicine and Biology 1193, https://doi.org/10.1007/978-981-13-6260-6_10

dysfunction. In this article, we will review the presentations and pathophysiology of septic organ dysfunction, as well as summarize and discuss the recent insights regarding ALDH2 in sepsis.

Keywords Sepsis · Organ dysfunction · ALDH2 · Mechanisms

1 Introduction

Sepsis, defined as life-threatening tissue damage and organ dysfunction caused by a dysregulated host response to infection (Sepsis-3) [1], is clinically manifested as systemic inflammatory response syndrome (SIRS). Sepsis is a critical medical issue imposing a global health burden. Recently, a WHO resolution has been adopted to call for recognizing sepsis as a global health priority [2]. Based on the available data of hospital-treated sepsis from high-income countries, there are 31.5 million sepsis cases with 5.3 million deaths per year, which may be highly underestimated because of the lack of data from low- and middle-income countries [3]. Owing to an aging population and increasing concomitant factors such as diabetes mellitus, surgical intervention, renal disease, and congestive heart failure, the incidence of sepsis is expected to be rising. Moreover, long-term physical, psychological, and cognitive disabilities are common in sepsis survivors. Infection control and supportive therapies remain main therapeutic strategies for sepsis, including use of antibiotics, elimination of infectious sources, intravenous fluids for resuscitation, and vasopressors for low blood pressure, which are far from sufficient. With the evolution of our knowledge for sepsis, new specific therapeutic targets are highly warranted.

2 Sepsis-Induced Organ Dysfunction

2.1 Circulatory Alterations

Sepsis-induced circulatory alterations are vital for organ dysfunction. Hypovolemia is a main pathological change during the initial period of circulatory dysfunction, which further leads to impaired peripheral perfusion, disturbed microcirculation, and organ injury [4]. Within this period, timely resuscitation is effective to reverse hypovolemia, which is crucial to improve the prognosis of sepsis patients [4, 5]. With progression, more complicated macro- and micro-circulatory disorders appear, for example, vasomotor dysfunction [6, 7], endothelial dysfunction [8, 9], and microthrombus formation [10], making hypovolemia persistent and irreversible. Hypotension, resulting not only from hypovolemia but also vasodilation and cardiac injury, is the most accessible parameter to predict circulatory dysfunction in sepsis, but not always parallel with hypovolemia [11]. qSOFA score (range, 0–3 points, with 1 point each for systolic hypotension [≤100 mm Hg], tachypnea [≥22/min], or altered mentation) [11] is a quick tool to assess organ failure in sepsis, which includes hypotension as one of the three indicators. When sepsis-induced cardiomyopathy and vascular hyporesponsiveness appear, fluid resuscitation and vasopressors become insufficient to reverse hypotension in the "cold shock" period [12].

Microcirculatory system, which is embedded in the organs and tissues, is an important constitution of the circulatory system. Disturbance of microcirculation results in compromised tissue or organ perfusion, low oxygenation, high capillary permeability, and macrohemodynamic alterations [13]. Microvascular reactivity can be assessed by near-infrared spectroscopy (NIRS)-derived techniques in the clinical setting [14, 15]. However, effective treatment of microvascular dysfunction is still lacking, making it a thorny problem in sepsis [13]. Understanding the mechanism of microcirculatory disorder in sepsis is important for identification of potential therapeutic target (Fig. 1).

2.2 *Cardiac Dysfunction*

Cardiac dysfunction commonly develops in patients with severe sepsis and is termed "septic cardiomyopathy" with an incidence rate of approximately 60% [16]. Septic cardiomyopathy is characterized by the presence of left ventricular dilation with normal or low filling pressures and decreased ejection fraction [17–19]. Although it is reversible, efficacious treatment is still lacking. When septic cardiomyopathy presents, the mortality rate of sepsis overtly rises [20]. A number of pathophysiological mechanisms have been proposed for septic cardiomyopathy, including:

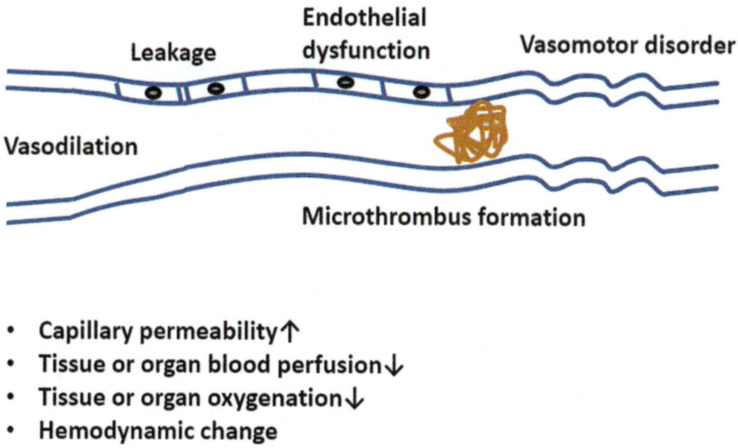

Fig. 1 Microcirculatory disorder in sepsis

1. Myocardial depressive factors such as endotoxins [21], cytokines [21], nitric oxide (NO) [17], extracellular histones [22, 23], and high-mobility group protein B (HMGB) [24]
2. Microcirculatory dysfunction, which leads to vascular leakage, myocardial edema, and disturbed myocardial blood perfusion [25]
3. Mitochondrial dysfunction, which leads to energy depletion and increased production of reactive oxygen and nitrogen species (ROS and RNS) [26–28]
4. Calcium handling disorder, which results from decreased density of calcium L-type channels, blunted myofilament calcium sensitivity, and enhanced calcium leak from the endoplasmic reticulum (ER) [29, 30]
5. Myofibrillar dysfunction, referring to impaired electromechanical coupling at the myofibrillar level [29]
6. Abnormal adrenergic response with elevated catecholamine release and overstimulation of the sympathetic nerve which could have deleterious cardiac effects [31]

However, whether β-adrenergic blockade therapy is beneficial to reduce the mortality of sepsis is still debatable [32–34] (Fig. 2).

2.3 *Hepatic Dysfunction*

The liver plays a crucial role in the regulation of immune response in sepsis. Consisting of hepatocytes, Kupffer cells, and hepatic stellate cells, with abundant blood supply, the liver works as a lymphoid organ to defense against microorganisms invasion by eliminating antigens transported in the blood stream and recruiting

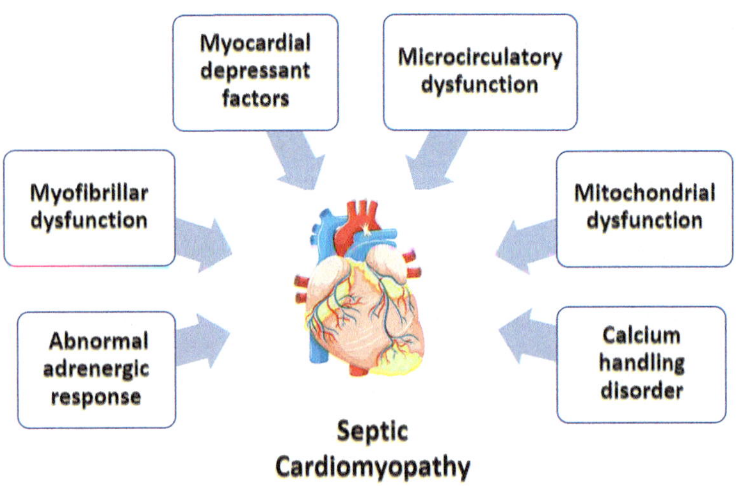

Fig. 2 Mechanisms related to septic cardiomyopathy

mediators to initiate immune response [35–37]. At the same time, anti-inflammatory mediators including IL-10, TGF-β, arginase-1, inducible nitric oxide synthase (iNOS), and regulatory T cells are activated [35]. However, excessive microorganisms may break the immune homeostasis, leading to overwhelming inflammation.

The liver also serves as a main target for sepsis-induced injury, resulting in cholestasis, hepatocellular injury, and increased levels of bilirubin or transaminases [35, 36]. Septic liver injury leads to compromised microorganism clearance and expression of anti-inflammatory mediators [38]. In addition, there is increasing programmed cell death including necroptosis and pyroptosis in the liver during sepsis which promotes inflammation prominently [39].

2.4 Renal Dysfunction

Sepsis-induced acute kidney injury (AKI) presents clinically as oliguria, reduced GFR, and increased serum creatinine [40]. The histologic footprint of AKI is characterized by non-specific, patchy areas of tubular cell vacuolization and remarkable apoptosis or necrosis [41]. Renal ischemia, due to macrohemodynamic changes and heterogeneous distortion of microvascular flow, contributes to renal dysfunction in sepsis [42, 43]. However, with normal or even higher renal blood flow, sepsis-induced AKI still develops, indicating the existence of other potential underlying mechanisms such as inflammation, oxidative stress, microvascular dysfunction, tubular endothelial dysfunction, and mitochondrial injury [42, 43].

3 Pathophysiology of Sepsis

3.1 Inflammatory Response

Specific components within the outer membrane of invasive microorganisms, such as lipopolysaccharide (LPS) in the Gram-negative bacteria and lipoteichoic acid in the Gram-positive bacteria, lead to direct cellular damage and immune response as pathogen-associated molecular patterns (PAMPs) [44]. Meanwhile, the cellular damage causes release of host-derived alarm molecules, termed damage-associated molecular patterns (DAMPs), such as high-mobility group box-1 protein (HMGB-1), heat shock proteins (HSPs), and plasma mitochondrial DNA [45–47]. The PAMPs and DAMPs are then recognized by pattern recognition receptors (PRRs), such as Toll-like receptors (TLRs), Nod-like receptors (NLRs), C-type lectin receptors (CLRs), and RIG-I-like receptors (RLRs), which transduce signals to initiate innate immune responses and prime antigen-specific adaptive immunity [48, 49]. For example, TLRs can act through the myeloid differentiation factor 88 (MyD88)-dependent or MyD88-independent pathways, which then trigger activation of nuclear factor kappa-light-chain-enhancer of activated B cells (NF-κB) and mitogen-associated protein kinase (MAPK) or NF-κB and interferon (IFN) regulatory factors (IRFs), respectively [40]. Appropriate amount of inflammation is essential for pathogen defense. However, uncontrolled inflammation called "cytokine storm" happens in sepsis, as a result of the positive feedback of cytokines and immune cells, which finally results in excessive inflammation [50]. Additional mechanisms, like proinflammatory RIP kinase 1-dependent necrosis, also play important roles in excessive inflammation [51]. Meanwhile, anti-inflammatory responses occur, such as the release of anti-inflammatory cytokines IL-10 and IL-4 [49] (Fig. 3). With the development of sepsis, the normal function and cellular homeostasis of immune cells are disturbed, causing immune suppression [52, 53]. In recent years, "immunoparalysis" in sepsis has been identified, which is more fatal than overwhelming proinflammatory status.

3.2 *Free Radicals and Reactive Aldehydes*

In response to inflammation, increased free radicals are produced, as a result of the accumulation of electrons leaked out from disturbed respiratory electron chain in mitochondria, including reactive oxygen species (ROS), namely, superoxide (O_2^-), hydrogen peroxide (H_2O_2), and hydroxyl radical (·OH), as well as reactive nitrogen

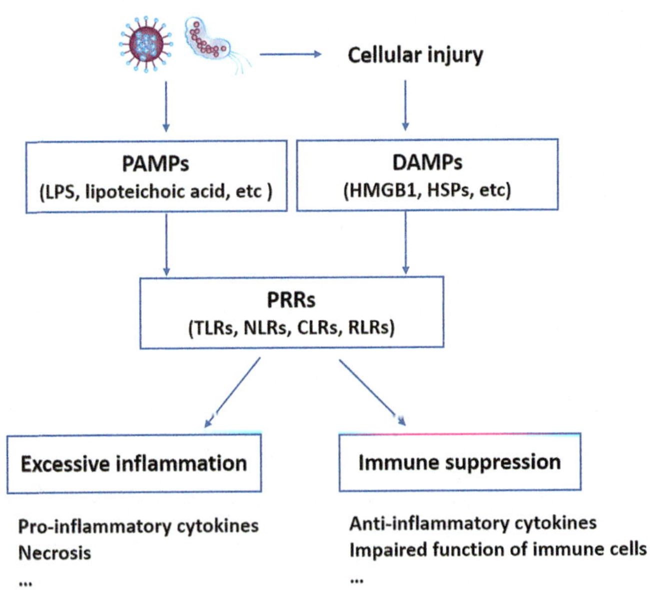

Fig. 3 Initiation of immune response in sepsis. Abbreviations: PAMPs, pathogen-associated molecular patterns; DAMPs, damage-associated molecular patterns; PRRs, pattern recognition receptors; HMGB1, high-mobility group box-1 protein; HSPs, heat shock proteins; TLRs, Toll-like receptors; NLRs, Nod-like receptors; CLRs, C-type lectin receptors; RLRs, RIG-I-like receptors; LPS, lipopolysaccharide

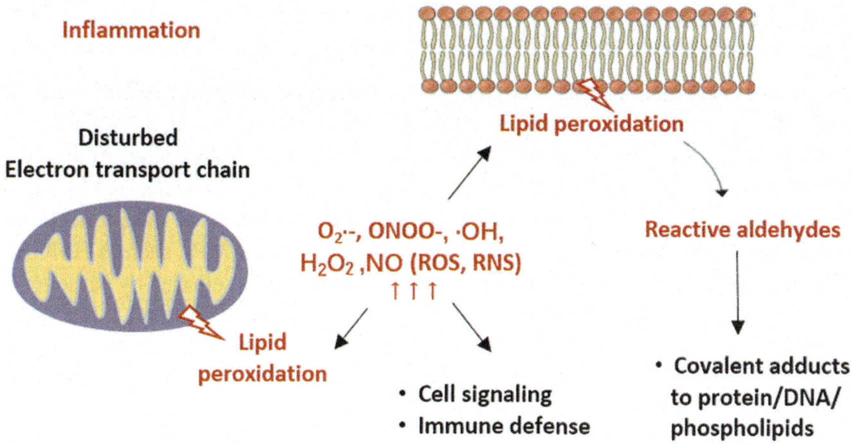

Fig. 4 Generation of ROS, RNS, and reactive aldehydes in sepsis

species (RNS) like nitric oxide (NO) and peroxynitrite ($ONOO^-$) [54, 55]. The highly instable ROS and RNS lead to lipid peroxidation on plasma and mitochondria membranes, and further generate excessive endogenous reactive aldehydes including aldehyde, 4-hydroxy-2-nonenal (4-HNE), malondialdehyde (MDA), and acrolein with various half-life, toxicity, and chemical properties [56] (Fig. 4). On one hand, ROS and RNS play key roles in the initiation of immune defense of macrophagocytes [57–59]. On the other hand, they participate in cell death signaling and inflammatory signaling pathways, for example, apoptotic pathway and NF-kB pathway, causing cell injury and death [57, 58]. Reactive aldehydes may act as second messengers of free radicals, covalently binding with proteins, DNA, and lipids, contributing to the structural and functional changes, and ultimately resulting in cytotoxic effects [60]. Recent studies have demonstrated proinflammatory roles of reactive aldehyde-modified proteins [61–63].

3.3 *Mitochondrial Dysfunction*

During sepsis, mitochondrial injury occurs, including loss of mtDNA integrity, increased production of free radicals, and altered mitochondrial membrane potential (MMP) [64], which further causes organ dysfunction. Mitochondrial structures, including mtDNA, cytochrome c, cardiolipin, ROS, and RNS, are abruptly released in sepsis, which can be recognized as danger-associated molecular patterns (DAMPs) to promote inflammation [64, 65]. It was reported that, accumulation of damaged mitochondria in macrophages caused overactivation of Nod-like receptor family pyrin domain-containing 3 (NLRP3) inflammasome, which initiated inflammatory response [66]. Mitochondria-selective autophagy, termed mitophagy, can prevent hyperinflammation by degradation of damaged mitochondria [67].

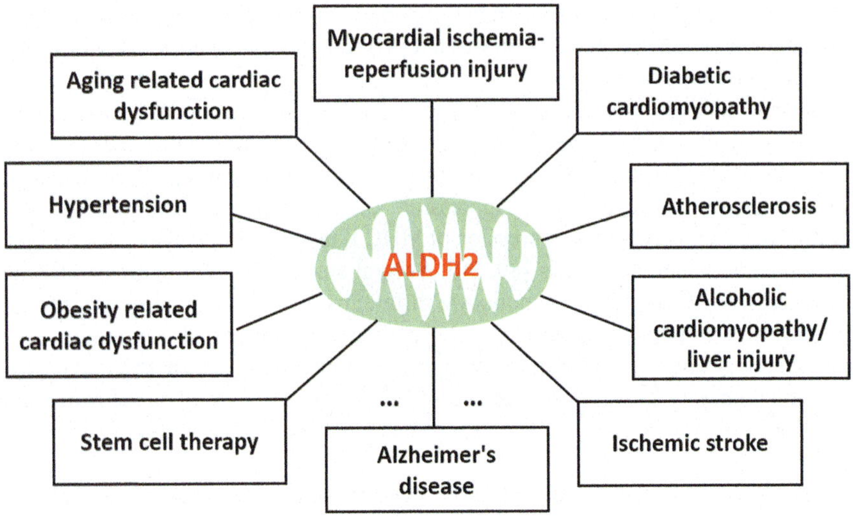

Fig. 5 The role of ALDH2 in various diseases

4.2 *ALDH2 and Inflammatory Response in Sepsis*

Most studies mentioned above didn't fully explore the direct effect of ALDH2 on inflammation. Regarded as an antioxidative agent and an autophagy modulator, we seldom consider the possible role of ALDH2 in inflammation. Interestingly, we and Hu JF [110] et al. detected altered level of inflammatory markers in ALDH2 transgenic mice or ALDH2-selective inhibitor-treated mice compared with wildtype or untreated mice in sepsis. Expanding the spectrum of research fields, ALDH2 was also reported to regulate inflammatory response in atherosclerotic plaque [105], alcoholic liver tissue [111], and myocardial in diabetes [112]. We summarized several possible mechanisms concerning the inflammation regulatory effect of ALDH2 as follows.

Firstly, the crosstalk between ROS and inflammation. In sepsis, ROS can directly cause endothelial cell dysfunction, resulting in increased vascular permeability. Beyond that, ROS was reported to enhance the inflammatory response via the regulation of NF-kB pathway [55], which thus deteriorated sepsis, forming a positive feedback cycle.

Secondly, autophagy and inflammation. As mentioned above, autophagy initiates early after the onset of sepsis and plays essential roles in pathogen defense, monocyte differentiation, and antigen presentation. In 2004, Nakagawa et al. reported that autophagy had pathogen-killing potential, which was supported by a number of reports afterward [113]. Autophagy also presents to be an anti-inflammation machinery, as well as an immunosuppressive effect at the late stage of sepsis. Remarkably, autophagy had strong potential to ameliorate immunosuppressive condition. Numerous studies demonstrated a significant role of ALDH2 on regulating autophagy in different murine models [101, 114–117].

Thirdly, the direct interaction with inflammatory markers. In 2015, Setoh and colleagues performed a genome-wide association study (GWAS) and a two-staged replication study of serum α-1 antitrypsin (AAT) levels in 9359 Japanese healthy individuals to identify genetic loci affecting AAT levels [118]. They found four loci, in which rs671 in *ALDH2* on chromosome 12 showed the strongest association beyond the GWAS significant level (GLM $P = 3.4\Diamond 10^{-11}$). AAT is an acute-phase inflammation marker, controlling inflammation by inhibiting proteases like the neutrophil elastase [119]. AAT deficiency population is prone to develop inflammation in the lung, with an elevated risk of early-onset chronic obstructive pulmonary disease (COPD) [119]. Recent studies have revealed a novel function of AAT including apoptosis [120], autophagy [121], cellular senescence [122], and lipid metabolism [123]. These findings helped to reveal a likely role for ALDH2 in inflammatory response and immunity.

4.3 ALDH2 and Sepsis-Induced Organ Dysfunction

A number of work have noted a role for ALDH2 in sepsis. Chen and coworkers studied liver injury in cecal ligation and puncture (CLP) sepsis model by examining the proteomic alteration of hepatic mitochondria under septic condition [124]. These investigators identified three sites (MP1, MP2, MP3) as potential variants for ALDH2. They suggested that MP1 and MP2 were related to the active form of ALDH2, while MP3 is the inactive form. Although the expression of ALDH2 was not changed obviously, the enzyme activity assay showed that the activity of ALDH2 was decreased slightly during early sepsis, whcrcas decreased significantly in late sepsis, consistent with the change of MP1, MP2, and MP3. Moreover, the disturbed posttranslational phosphorylation of ALDH2 in sepsis also contributed to the decreased enzymatic activity. With heat shock pretreatment, the activity of ALDH2 in sepsis was preserved, especially in the late phase. We know that ALDH2 presents in most tissues, particularly more in the liver. This study showed that in the protein level, ALDH2 was the most significantly disturbed protein in mitochondria under septic condition, revealing a vital role of ALDH2 in sepsis.

In another independent study, Hu and associates studied cardiac injury in sepsis using a lipopolysaccharide (LPS) rat model [125]. They found the beneficial effect of ALDH2 in LPS-induced sepsis model. The expression of ALDH2 in LPS-induced sepsis model was unchanged, although the enzymatic activity was reduced. When treated with Alda-1 to trigger ALDH2 activity, septic cardiac dysfunction was improved. In the sepsis model, opening of MPTP and iNOS activation was observed. Alda-1 treatment preserved the opening of MPTP, and the selective inhibition of iNOS partially reversed the LPS-induced decrease of ALDH2 activity, which suggested possible mechanisms of the protective effect of Alda-1, as well as decreased ALDH2 activity in sepsis. Interested in the same model, our group conducted it in FVB and ALDH2 transgenic mice. We found the same story of the protective effect of ALDH2 in sepsis and further revealed the underlying mechanisms focusing on cellular homeostasis, including ER stress, autophagy, and inflammation. We observed excessive ER stress and autophagy levels in the sepsis model, which were significantly alleviated in the ALDH2 transgenic mice. We also tried to detect the inflammatory factors, finding decreased cytokine levels in ALDH2 transgenic septic mice compared to FVB septic mice[126].

Hu and colleagues recently explored the effect of ALDH2 on the kidney in CLP-induced sepsis rat model [110]. In this model, renal injury was accompanied by the reduction of renal ALDH2 expression. Selective inhibition of ALDH2 aggravated renal injury, associated with the burst of reactive oxygen species and inflammatory reaction.

5 Insights

The beneficial roles of ALDH2 in different septic organ injuries and several important mechanisms have been demonstrated. However, more studies are still needed, especially the studies concerning multi-organ dysfunction, and the underlying mechanisms. Immunoparalysis in the late phase is vital, and whether ALDH2 plays any role is still unknown. Enhanced vascular permeability, leading to microvascular dysfunction, plays an important role to initiate the septic injury. Considering oxidative stress is detrimental for cellular homeostasis, ALDH2 might influence vascular permeability. In conclusion, ALDH2 is a promising therapeutic target for sepsis; however, more studies are urgently needed.

Acknowledgment This study was supported by the National Natural Science Foundation of China (81570401, 81772036, 81571934, 81671952, 81873950, 81873953, 81701952), National Key R&D Program of China (2017YFC0908700, 2017YFC0908703), Taishan Scholar Program of Shandong Province (ts20130911), Taishan Young Scholar Program of Shandong Province (tsqn20161065), Key R&D Program of Shandong Province (2016GSF201235, 2016ZDJS07A14, 2018GSF118003), and the Fundamental Research Funds of Shandong University (2014QLKY04, 2018JC011).

References

1. Singer M, Deutschman CS, Seymour CW, Shankar-Hari M, Annane D, Bauer M, Bellomo R, Bernard GR, Chiche JD, Coopersmith CM, Hotchkiss RS, Levy MM, Marshall JC, Martin GS, Opal SM, Rubenfeld GD, van der Poll T, Vincent JL, Angus DC (2016) The third international consensus definitions for Sepsis and septic shock (Sepsis-3). JAMA 315(8):801–810. https://doi.org/10.1001/jama.2016.0287
2. Reinhart K, Daniels R, Kissoon N, Machado FR, Schachter RD, Finfer S (2017) Recognizing Sepsis as a Global Health priority – a WHO resolution. N Engl J Med 377(5):414–417. https://doi.org/10.1056/NEJMp1707170
3. Fleischmann C, Scherag A, Adhikari NK, Hartog CS, Tsaganos T, Schlattmann P, Angus DC, Reinhart K, International Forum of Acute Care T (2016) Assessment of global incidence and

mortality of hospital-treated Sepsis. Current estimates and limitations. Am J Respir Crit Care Med 193(3):259–272. https://doi.org/10.1164/rccm.201504-0781OC

4. Hernandez G, Bruhn A, Castro R, Regueira T (2012) The holistic view on perfusion monitoring in septic shock. Curr Opin Crit Care 18(3):280–286. https://doi.org/10.1097/MCC.0b013e3283532c08
5. Simpson N, Lamontagne F, Shankar-Hari M (2017) Septic shock resuscitation in the first hour. Curr Opin Crit Care 23(6):561–566. https://doi.org/10.1097/MCC.0000000000000460
6. Terborg C, Schummer W, Albrecht M, Reinhart K, Weiller C, Rother J (2001) Dysfunction of vasomotor reactivity in severe sepsis and septic shock. Intensive Care Med 27(7):1231–1234
7. Horowitz JD, Liu S (2018) ADAM-15 and glycocalyx shedding: a new perspective on sepsis-related vasomotor dysfunction. Cardiovasc Res 114:1694. https://doi.org/10.1093/cvr/cvy199
8. Opal SM, van der Poll T (2015) Endothelial barrier dysfunction in septic shock. J Intern Med 277(3):277–293. https://doi.org/10.1111/joim.12331
9. Vassiliou AG, Mastora Z, Orfanos SE, Jahaj E, Maniatis NA, Koutsoukou A, Armaganidis A, Kotanidou A (2014) Elevated biomarkers of endothelial dysfunction/activation at ICU admission are associated with sepsis development. Cytokine 69(2):240–247. https://doi.org/10.1016/j.cyto.2014.06.010
10. Wang Y, Ouyang Y, Liu B, Ma X, Ding R (2018) Platelet activation and antiplatelet therapy in sepsis: a narrative review. Thromb Res 166:28–36. https://doi.org/10.1016/j.thromres.2018.04.007
11. Seymour CW, Liu VX, Iwashyna TJ, Brunkhorst FM, Rea TD, Scherag A, Rubenfeld G, Kahn JM, Shankar-Hari M, Singer M, Deutschman CS, Escobar GJ, Angus DC (2016) Assessment of clinical criteria for Sepsis: for the third international consensus definitions for Sepsis and septic shock (Sepsis-3). JAMA 315(8):762–774. https://doi.org/10.1001/jama.2016.0288
12. Levy B, Collin S, Sennoun N, Ducrocq N, Kimmoun A, Asfar P, Perez P, Meziani F (2010) Vascular hyporesponsiveness to vasopressors in septic shock: from bench to bedside. Intensive Care Med 36(12):2019–2029. https://doi.org/10.1007/s00134-010-2045-8
13. Potter EK, Hodgson L, Creagh-Brown B, Forni LG (2018) Manipulating the microcirculation in Sepsis - the impact of vasoactive medications on microcirculatory blood flow. A systematic review. Shock:1. https://doi.org/10.1097/SHK.0000000000001239
14. Macdonald SP, Brown SG (2015) Near-infrared spectroscopy in the assessment of suspected sepsis in the emergency department. Emerg Med J 32(5):404–408. https://doi.org/10.1136/emermed-2013-202956
15. Masip J, Mesquida J, Luengo C, Gili G, Goma G, Ferrer R, Teboul JL, Payen D, Artigas A (2013) Near-infrared spectroscopy StO2 monitoring to assess the therapeutic effect of drotrecogin alfa (activated) on microcirculation in patients with severe sepsis or septic shock. Ann Intensive Care 3(1):30. https://doi.org/10.1186/2110-5820-3-30
16. Vieillard-Baron A, Caille V, Charron C, Belliard G, Page B, Jardin F (2008) Actual incidence of global left ventricular hypokinesia in adult septic shock. Crit Care Med 36(6):1701–1706. https://doi.org/10.1097/CCM.0b013e318174db05
17. Flynn A, Chokkalingam Mani B, Mather PJ (2010) Sepsis-induced cardiomyopathy: a review of pathophysiologic mechanisms. Heart Fail Rev 15(6):605–611. https://doi.org/10.1007/s10741-010-9176-4
18. Liu YC, Yu MM, Shou ST, Chai YF (2017) Sepsis-induced cardiomyopathy: mechanisms and treatments. Front Immunol 8:1021. https://doi.org/10.3389/fimmu.2017.01021
19. Ehrman RR, Sullivan AN, Favot MJ, Sherwin RL, Reynolds CA, Abidov A, Levy PD (2018) Pathophysiology, echocardiographic evaluation, biomarker findings, and prognostic implications of septic cardiomyopathy: a review of the literature. Crit Care 22(1):112. https://doi.org/10.1186/s13054-018-2043-8
20. Merx MW, Weber C (2007) Sepsis and the heart. Circulation 116(7):793–802. https://doi.org/10.1161/CIRCULATIONAHA.106.678359

21. Sato R, Nasu M (2015) A review of sepsis-induced cardiomyopathy. J Intensive Care 3:48. https://doi.org/10.1186/s40560-015-0112-5
22. Alhamdi Y, Zi M, Abrams ST, Liu T, Su D, Welters I, Dutt T, Cartwright EJ, Wang G, Toh CH (2016) Circulating histone concentrations differentially affect the predominance of left or right ventricular dysfunction in critical illness. Crit Care Med 44(5):e278–e288. https://doi.org/10.1097/CCM.0000000000001413
23. Kalbitz M, Grailer JJ, Fattahi F, Jajou L, Herron TJ, Campbell KF, Zetoune FS, Bosmann M, Sarma JV, Huber-Lang M, Gebhard F, Loaiza R, Valdivia HH, Jalife J, Russell MW, Ward PA (2015) Role of extracellular histones in the cardiomyopathy of sepsis. FASEB J 29(5):2185–2193. https://doi.org/10.1096/fj.14-268730
24. Ha T, Xia Y, Liu X, Lu C, Liu L, Kelley J, Kalbfleisch J, Kao RL, Williams DL, Li C (2011) Glucan phosphate attenuates myocardial HMGB1 translocation in severe sepsis through inhibiting NF-kappaB activation. Am J Physiol Heart Circ Physiol 301(3):H848–H855. https://doi.org/10.1152/ajpheart.01007.2010
25. Hawiger J (2018) Heartfelt sepsis: microvascular injury due to genomic storm. Kardiol Pol 76(8):1203–1216. https://doi.org/10.5603/KP.a2018.0146
26. Piquereau J, Godin R, Deschenes S, Bessi VL, Mofarrahi M, Hussain SN, Burelle Y (2013) Protective role of PARK2/Parkin in sepsis-induced cardiac contractile and mitochondrial dysfunction. Autophagy 9(11):1837–1851. https://doi.org/10.4161/auto.26502
27. Tsolaki V, Makris D, Mantzarlis K, Zakynthinos E (2017) Sepsis-induced cardiomyopathy: oxidative implications in the initiation and resolution of the damage. Oxidative Med Cell Longev 2017:7393525. https://doi.org/10.1155/2017/7393525
28. Durand A, Duburcq T, Dekeyser T, Neviere R, Howsam M, Favory R, Preau S (2017) Involvement of mitochondrial disorders in septic cardiomyopathy. Oxidative Med Cell Longev 2017:4076348. https://doi.org/10.1155/2017/4076348
29. Wagner S, Schurmann S, Hein S, Schuttler J, Friedrich O (2015) Septic cardiomyopathy in rat LPS-induced endotoxemia: relative contribution of cellular diastolic Ca(2+) removal pathways, myofibrillar biomechanics properties and action of the cardiotonic drug levosimendan. Basic Res Cardiol 110(5):507. https://doi.org/10.1007/s00395-015-0507-4
30. Hobai IA, Edgecomb J, LaBarge K, Colucci WS (2015) Dysregulation of intracellular calcium transporters in animal models of sepsis-induced cardiomyopathy. Shock 43(1):3–15. https://doi.org/10.1097/SHK.0000000000000261
31. de Montmollin E, Aboab J, Mansart A, Annane D (2009) Bench-to-bedside review: beta-adrenergic modulation in sepsis. Crit Care 13(5):230. https://doi.org/10.1186/cc8026
32. Suzuki T, Suzuki Y, Okuda J, Kurazumi T, Suhara T, Ueda T, Nagata H, Morisaki H (2017) Sepsis-induced cardiac dysfunction and beta-adrenergic blockade therapy for sepsis. J Intensive Care 5:22. https://doi.org/10.1186/s40560-017-0215-2
33. Sanfilippo F, Santonocito C, Morelli A, Foex P (2015) Beta-blocker use in severe sepsis and septic shock: a systematic review. Curr Med Res Opin 31(10):1817–1825. https://doi.org/10.1185/03007995.2015.1062357
34. Agac D, Estrada LD, Maples R, Hooper LV, Farrar JD (2018) The beta2-adrenergic receptor controls inflammation by driving rapid IL-10 secretion. Brain Behav Immun 74:176. https://doi.org/10.1016/j.bbi.2018.09.004
35. Yan J, Li S, Li S (2014) The role of the liver in sepsis. Int Rev Immunol 33(6):498–510. https://doi.org/10.3109/08830185.2014.889129
36. Strnad P, Tacke F, Koch A, Trautwein C (2017) Liver - guardian, modifier and target of sepsis. Nat Rev Gastroenterol Hepatol 14(1):55–66. https://doi.org/10.1038/nrgastro.2016.168
37. Marques PE, Oliveira AG, Chang L, Paula-Neto HA, Menezes GB (2015) Understanding liver immunology using intravital microscopy. J Hepatol 63(3):733–742. https://doi.org/10.1016/j.jhep.2015.05.027
38. Crispe IN (2016) Hepatocytes as immunological agents. J Immunol 196(1):17–21. https://doi.org/10.4049/jimmunol.1501668

39. Yang R, Zou X, Tenhunen J, Tonnessen TI (2017) HMGB1 and extracellular histones significantly contribute to systemic inflammation and multiple organ failure in acute liver failure. Mediat Inflamm 2017:5928078. https://doi.org/10.1155/2017/5928078
40. Lelubre C, Vincent JL (2018) Mechanisms and treatment of organ failure in sepsis. Nat Rev Nephrol 14(7):417–427. https://doi.org/10.1038/s41581-018-0005-7
41. Kosaka J, Lankadeva YR, May CN, Bellomo R (2016) Histopathology of septic acute kidney injury: a systematic review of experimental data. Crit Care Med 44(9):e897–e903. https://doi.org/10.1097/CCM.0000000000001735
42. Ma S, Evans RG, Iguchi N, Tare M, Parkington HC, Bellomo R, May CN, Lankadeva YR (2018) Sepsis-induced acute kidney injury: a disease of the microcirculation. Microcirculation:e12483. https://doi.org/10.1111/micc.12483
43. Gomez H, Ince C, De Backer D, Pickkers P, Payen D, Hotchkiss J, Kellum JA (2014) A unified theory of sepsis-induced acute kidney injury: inflammation, microcirculatory dysfunction, bioenergetics, and the tubular cell adaptation to injury. Shock 41(1):3–11. https://doi.org/10.1097/SHK.0000000000000052
44. Raymond SL, Holden DC, Mira JC, Stortz JA, Loftus TJ, Mohr AM, Moldawer LL, Moore FA, Larson SD, Efron PA (2017) Microbial recognition and danger signals in sepsis and trauma. Biochim Biophys Acta Mol basis Dis 1863(10 Pt B):2564–2573. https://doi.org/10.1016/j.bbadis.2017.01.013
45. Di Caro V, Walko TD 3rd, Bola RA, Hong JD, Pang D, Hsue V, Au AK, Halstead ES, Carcillo JA, Clark RS, Aneja RK (2016) Plasma mitochondrial DNA--a novel DAMP in pediatric Sepsis. Shock 45(5):506–511. https://doi.org/10.1097/SHK.0000000000000539
46. Chan JK, Roth J, Oppenheim JJ, Tracey KJ, Vogl T, Feldmann M, Horwood N, Nanchahal J (2012) Alarmins: awaiting a clinical response. J Clin Invest 122(8):2711–2719. https://doi.org/10.1172/JCI62423
47. Yang L, Xie M, Yang M, Yu Y, Zhu S, Hou W, Kang R, Lotze MT, Billiar TR, Wang H, Cao L, Tang D (2014) PKM2 regulates the Warburg effect and promotes HMGB1 release in sepsis. Nat Commun 5:4436. https://doi.org/10.1038/ncomms5436
48. Kawai T, Akira S (2010) The role of pattern-recognition receptors in innate immunity: update on toll-like receptors. Nat Immunol 11(5):373–384. https://doi.org/10.1038/ni.1863
49. Wiersinga WJ, Leopold SJ, Cranendonk DR, van der Poll T (2014) Host innate immune responses to sepsis. Virulence 5(1):36–44. https://doi.org/10.4161/viru.25436
50. Matsuda N, Hattori Y (2006) Systemic inflammatory response syndrome (SIRS): molecular pathophysiology and gene therapy. J Pharmacol Sci 101(3):189–198
51. Zelic M, Roderick JE, O'Donnell JA, Lehman J, Lim SE, Janardhan HP, Trivedi CM, Pasparakis M, Kelliher MA (2018) RIP kinase 1-dependent endothelial necroptosis underlies systemic inflammatory response syndrome. J Clin Invest 128(5):2064–2075. https://doi.org/10.1172/JCI96147
52. Leentjens J, Kox M, van der Hoeven JG, Netea MG, Pickkers P (2013) Immunotherapy for the adjunctive treatment of sepsis: from immunosuppression to immunostimulation. Time for a paradigm change? Am J Respir Crit Care Med 187(12):1287–1293. https://doi.org/10.1164/rccm.201301-0036CP
53. Ho J, Yu J, Wong SH, Zhang L, Liu X, Wong WT, Leung CC, Choi G, Wang MH, Gin T, Chan MT, Wu WK (2016) Autophagy in sepsis: degradation into exhaustion? Autophagy 12(7):1073–1082. https://doi.org/10.1080/15548627.2016.1179410
54. Bar-Or D, Carrick MM, Mains CW, Rael LT, Slone D, Brody EN (2015) Sepsis, oxidative stress, and hypoxia: are there clues to better treatment? Redox Rep 20(5):193–197. https://doi.org/10.1179/1351000215Y.0000000005
55. Prauchner CA (2017) Oxidative stress in sepsis: pathophysiological implications justifying antioxidant co-therapy. Burns 43(3):471–485. https://doi.org/10.1016/j.burns.2016.09.023
56. Sultana R, Perluigi M, Butterfield DA (2013) Lipid peroxidation triggers neurodegeneration: a redox proteomics view into the Alzheimer disease brain. Free Radic Biol Med 62:157–169. https://doi.org/10.1016/j.freeradbiomed.2012.09.027

57. Gostner JM, Becker K, Fuchs D, Sucher R (2013) Redox regulation of the immune response. Redox Rep 18(3):88–94. https://doi.org/10.1179/1351000213Y.0000000044
58. Wink DA, Hines HB, Cheng RY, Switzer CH, Flores-Santana W, Vitek MP, Ridnour LA, Colton CA (2011) Nitric oxide and redox mechanisms in the immune response. J Leukoc Biol 89(6):873–891. https://doi.org/10.1189/jlb.1010550
59. Stoiber W, Obermayer A, Steinbacher P, Krautgartner WD (2015) The role of reactive oxygen species (ROS) in the formation of extracellular traps (ETs) in humans. Biomol Ther 5(2):702–723. https://doi.org/10.3390/biom5020702
60. Forman HJ (2010) Reactive oxygen species and alpha,beta-unsaturated aldehydes as second messengers in signal transduction. Ann N Y Acad Sci 1203:35–44. https://doi.org/10.1111/j.1749-6632.2010.05551.x
61. Antoniak DT, Duryee MJ, Mikuls TR, Thiele GM, Anderson DR (2015) Aldehyde-modified proteins as mediators of early inflammation in atherosclerotic disease. Free Radic Biol Med 89:409–418. https://doi.org/10.1016/j.freeradbiomed.2015.09.003
62. Timucin AC, Basaga H (2017) Pro-apoptotic effects of lipid oxidation products: HNE at the crossroads of NF-kappaB pathway and anti-apoptotic Bcl-2. Free Radic Biol Med 111:209–218. https://doi.org/10.1016/j.freeradbiomed.2016.11.010
63. Frohnert BI, Long EK, Hahn WS, Bernlohr DA (2014) Glutathionylated lipid aldehydes are products of adipocyte oxidative stress and activators of macrophage inflammation. Diabetes 63(1):89–100. https://doi.org/10.2337/db13-0777
64. Lee I, Huttemann M (2014) Energy crisis: the role of oxidative phosphorylation in acute inflammation and sepsis. Biochim Biophys Acta 1842(9):1579–1586. https://doi.org/10.1016/j.bbadis.2014.05.031
65. Harrington JS, Choi AMK, Nakahira K (2017) Mitochondrial DNA in Sepsis. Curr Opin Crit Care 23(4):284–290. https://doi.org/10.1097/MCC.0000000000000427
66. Nakahira K, Haspel JA, Rathinam VA, Lee SJ, Dolinay T, Lam HC, Englert JA, Rabinovitch M, Cernadas M, Kim HP, Fitzgerald KA, Ryter SW, Choi AM (2011) Autophagy proteins regulate innate immune responses by inhibiting the release of mitochondrial DNA mediated by the NALP3 inflammasome. Nat Immunol 12(3):222–230. https://doi.org/10.1038/ni.1980
67. Kim MJ, Yoon JH, Ryu JH (2016) Mitophagy: a balance regulator of NLRP3 inflammasome activation. BMB Rep 49(10):529–535
68. Pasparakis M, Vandenabeele P (2015) Necroptosis and its role in inflammation. Nature 517(7534):311–320. https://doi.org/10.1038/nature14191
69. Huebener P, Pradere JP, Hernandez C, Gwak GY, Caviglia JM, Mu X, Loike JD, Jenkins RE, Antoine DJ, Schwabe RF (2015) The HMGB1/RAGE axis triggers neutrophil-mediated injury amplification following necrosis. J Clin Invest 125(2):539–550. https://doi.org/10.1172/JCI76887
70. Proskuryakov SY, Konoplyannikov AG, Gabai VL (2003) Necrosis: a specific form of programmed cell death? Exp Cell Res 283(1):1–16
71. Rock KL, Kono H (2008) The inflammatory response to cell death. Annu Rev Pathol 3:99–126. https://doi.org/10.1146/annurev.pathmechdis.3.121806.151456
72. Hengartner MO (2000) The biochemistry of apoptosis. Nature 407(6805):770–776. https://doi.org/10.1038/35037710
73. Hsieh YC, Athar M, Chaudry IH (2009) When apoptosis meets autophagy: deciding cell fate after trauma and sepsis. Trends Mol Med 15(3):129–138. https://doi.org/10.1016/j.molmed.2009.01.002
74. Dannappel M, Vlantis K, Kumari S, Polykratis A, Kim C, Wachsmuth L, Eftychi C, Lin J, Corona T, Hermance N, Zelic M, Kirsch P, Basic M, Bleich A, Kelliher M, Pasparakis M (2014) RIPK1 maintains epithelial homeostasis by inhibiting apoptosis and necroptosis. Nature 513(7516):90–94. https://doi.org/10.1038/nature13608
75. Taylor RC, Cullen SP, Martin SJ (2008) Apoptosis: controlled demolition at the cellular level. Nat Rev Mol Cell Biol 9(3):231–241. https://doi.org/10.1038/nrm2312

76. Luan YY, Yao YM, Xiao XZ, Sheng ZY (2015) Insights into the apoptotic death of immune cells in sepsis. J Interf Cytokine Res 35(1):17–22. https://doi.org/10.1089/jir.2014.0069
77. Jorgensen I, Rayamajhi M, Miao EA (2017) Programmed cell death as a defence against infection. Nat Rev Immunol 17(3):151–164. https://doi.org/10.1038/nri.2016.147
78. Han CZ, Ravichandran KS (2011) Metabolic connections during apoptotic cell engulfment. Cell 147(7):1442–1445. https://doi.org/10.1016/j.cell.2011.12.006
79. Lin Z, Guo J, Xue P, Huang L, Deng L, Yang X, Xia Q (2014) Chaiqinchengqi decoction regulates necrosis-apoptosis via regulating the release of mitochondrial cytochrome c and caspase-3 in rats with acute necrotizing pancreatitis. J Tradit Chin Med 34(2):178–183
80. Kim KH, Lee MS (2014) Autophagy--a key player in cellular and body metabolism. Nat Rev Endocrinol 10(6):322–337. https://doi.org/10.1038/nrendo.2014.35
81. Barton LA, Ren J (2018) Pancreatic neoplasms and autophagy. Curr Drug Targets 19(9):1018–1023. https://doi.org/10.2174/1389450117666160622220915
82. Wang S, Ren J (2018) Role of autophagy and regulatory mechanisms in alcoholic cardiomyopathy. Biochim Biophys Acta 1864(6. Pt A):2003–2009. https://doi.org/10.1016/j.bbadis.2018.03.016
83. Zhang Y, Sowers JR, Ren J (2018) Targeting autophagy in obesity: from pathophysiology to management. Nat Rev Endocrinol 14(6):356–376. https://doi.org/10.1038/s41574-018-0009-1
84. Zhang Y, Whaley-Connell AT, Sowers JR, Ren J (2018) Autophagy as an emerging target in cardiorenal metabolic disease: from pathophysiology to management. Pharmacol Ther 191:1. https://doi.org/10.1016/j.pharmthera.2018.06.004
85. Ren J, Sowers JR, Zhang Y (2018) Metabolic stress, autophagy, and cardiovascular aging: from pathophysiology to therapeutics. Trends Endocrinol Metab 29:699. https://doi.org/10.1016/j.tem.2018.08.001
86. Cadwell K (2016) Crosstalk between autophagy and inflammatory signalling pathways: balancing defence and homeostasis. Nat Rev Immunol 16(11):661–675. https://doi.org/10.1038/nri.2016.100
87. Verlhac P, Gregoire IP, Azocar O, Petkova DS, Baguet J, Viret C, Faure M (2015) Autophagy receptor NDP52 regulates pathogen-containing autophagosome maturation. Cell Host Microbe 17(4):515–525. https://doi.org/10.1016/j.chom.2015.02.008
88. Choy A, Roy CR (2013) Autophagy and bacterial infection: an evolving arms race. Trends Microbiol 21(9):451–456. https://doi.org/10.1016/j.tim.2013.06.009
89. Loi M, Gannage M, Munz C (2016) ATGs help MHC class II, but inhibit MHC class I antigen presentation. Autophagy 12(9):1681–1682. https://doi.org/10.1080/15548627.2016.1203488
90. Valecka J, Almeida CR, Su B, Pierre P, Gatti E (2018) Autophagy and MHC-restricted antigen presentation. Mol Immunol 99:163–170. https://doi.org/10.1016/j.molimm.2018.05.009
91. Zhang Y, Morgan MJ, Chen K, Choksi S, Liu ZG (2012) Induction of autophagy is essential for monocyte-macrophage differentiation. Blood 119(12):2895–2905. https://doi.org/10.1182/blood-2011-08-372383
92. Jacquel A, Obba S, Solary E, Auberger P (2012) Proper macrophagic differentiation requires both autophagy and caspase activation. Autophagy 8(7):1141–1143. https://doi.org/10.4161/auto.20367
93. Ren C, Zhang H, Wu TT, Yao YM (2017) Autophagy: a potential therapeutic target for reversing Sepsis-induced immunosuppression. Front Immunol 8:1832. https://doi.org/10.3389/fimmu.2017.01832
94. Zhou R, Yazdi AS, Menu P, Tschopp J (2011) A role for mitochondria in NLRP3 inflammasome activation. Nature 469(7329):221–225. https://doi.org/10.1038/nature09663
95. Ibanez-Cabellos JS, Aguado C, Perez-Cremades D, Garcia-Gimenez JL, Bueno-Beti C, Garcia-Lopez EM, Roma-Mateo C, Novella S, Hermenegildo C, Pallardo FV (2018) Extracellular histones activate autophagy and apoptosis via mTOR signaling in human endothelial cells. Biochim Biophys Acta 1864:3234. https://doi.org/10.1016/j.bbadis.2018.07.010

96. Carchman EH, Rao J, Loughran PA, Rosengart MR, Zuckerbraun BS (2011) Heme oxygenase-1-mediated autophagy protects against hepatocyte cell death and hepatic injury from infection/sepsis in mice. Hepatology 53(6):2053–2062. https://doi.org/10.1002/hep.24324
97. Kuang M, Cen Y, Qin R, Shang S, Zhai Z, Liu C, Pan X, Zhou H (2018) Artesunate attenuates pro-inflammatory cytokine release from macrophages by inhibiting TLR4-mediated Autophagic activation via the TRAF6-Beclin1-PI3KC3 pathway. Cell Physiol Biochem 47(2):475–488. https://doi.org/10.1159/000489982
98. Oami T, Watanabe E, Hatano M, Sunahara S, Fujimura L, Sakamoto A, Ito C, Toshimori K, Oda S (2017) Suppression of T cell autophagy results in decreased viability and function of T cells through accelerated apoptosis in a murine Sepsis model. Crit Care Med 45(1):e77–e85. https://doi.org/10.1097/CCM.0000000000002016
99. Marchitti SA, Brocker C, Stagos D, Vasiliou V (2008) Non-P450 aldehyde oxidizing enzymes: the aldehyde dehydrogenase superfamily. Expert Opin Drug Metab Toxicol 4(6):697–720. https://doi.org/10.1517/17425255.4.6.697
100. Crabb DW, Matsumoto M, Chang D, You M (2004) Overview of the role of alcohol dehydrogenase and aldehyde dehydrogenase and their variants in the genesis of alcohol-related pathology. Proc Nutr Soc 63(1):49–63
101. Zhang Y, Wang C, Zhou J, Sun A, Hueckstaedt LK, Ge J, Ren J (2017) Complex inhibition of autophagy by mitochondrial aldehyde dehydrogenase shortens lifespan and exacerbates cardiac aging. Biochim Biophys Acta 1863(8):1919–1932. https://doi.org/10.1016/j.bbadis.2017.03.016
102. Chen CH, Budas GR, Churchill EN, Disatnik MH, Hurley TD, Mochly-Rosen D (2008) Activation of aldehyde dehydrogenase-2 reduces ischemic damage to the heart. Science 321(5895):1493–1495. https://doi.org/10.1126/science.1158554
103. Pang J, Wang J, Zhang Y, Xu F, Chen Y (2017) Targeting acetaldehyde dehydrogenase 2 (ALDH2) in heart failure-recent insights and perspectives. Biochim Biophys Acta 1863(8):1933–1941. https://doi.org/10.1016/j.bbadis.2016.10.004
104. Liu B, Wang J, Li M, Yuan Q, Xue M, Xu F, Chen Y (2017) Inhibition of ALDH2 by O-GlcNAcylation contributes to the hyperglycemic exacerbation of myocardial ischemia/reperfusion injury. Oncotarget 8(12):19413–19426. https://doi.org/10.18632/oncotarget.14297
105. Pan C, Xing JH, Zhang C, Zhang YM, Zhang LT, Wei SJ, Zhang MX, Wang XP, Yuan QH, Xue L, Wang JL, Cui ZQ, Zhang Y, Xu F, Chen YG (2016) Aldehyde dehydrogenase 2 inhibits inflammatory response and regulates atherosclerotic plaque. Oncotarget 7(24):35562–35576. https://doi.org/10.18632/oncotarget.9384
106. Ma H, Guo R, Yu L, Zhang Y, Ren J (2011) Aldehyde dehydrogenase 2 (ALDH2) rescues myocardial ischaemia/reperfusion injury: role of autophagy paradox and toxic aldehyde. Eur Heart J 32(8):1025–1038. https://doi.org/10.1093/eurheartj/ehq253
107. Ma H, Li J, Gao F, Ren J (2009) Aldehyde dehydrogenase 2 ameliorates acute cardiac toxicity of ethanol: role of protein phosphatase and forkhead transcription factor. J Am Coll Cardiol 54(23):2187–2196. https://doi.org/10.1016/j.jacc.2009.04.100
108. Kato N, Takeuchi F, Tabara Y, Kelly TN, Go MJ, Sim X, Tay WT, Chen CH, Zhang Y, Yamamoto K, Katsuya T, Yokota M, Kim YJ, Ong RT, Nabika T, Gu D, Chang LC, Kokubo Y, Huang W, Ohnaka K, Yamori Y, Nakashima E, Jaquish CE, Lee JY, Seielstad M, Isono M, Hixson JE, Chen YT, Miki T, Zhou X, Sugiyama T, Jeon JP, Liu JJ, Takayanagi R, Kim SS, Aung T, Sung YJ, Zhang X, Wong TY, Han BG, Kobayashi S, Ogihara T, Zhu D, Iwai N, Wu JY, Teo YY, Tai ES, Cho YS, He J (2011) Meta-analysis of genome-wide association studies identifies common variants associated with blood pressure variation in east Asians. Nat Genet 43(6):531–538. https://doi.org/10.1038/ng.834
109. Zhang Y, Ren J (2011) ALDH2 in alcoholic heart diseases: molecular mechanism and clinical implications. Pharmacol Ther 132(1):86–95. https://doi.org/10.1016/j.pharmthera.2011.05.008

110. Hu JF, Wang HX, Li HH, Hu J, Yu Y, Gao Q (2017) Inhibition of ALDH2 expression aggravates renal injury in a rat sepsis syndrome model. Exp Ther Med 14(3):2249–2254. https://doi.org/10.3892/etm.2017.4785
111. Guo R, Xu X, Babcock SA, Zhang Y, Ren J (2015) Aldehyde dedydrogenase-2 plays a beneficial role in ameliorating chronic alcohol-induced hepatic steatosis and inflammation through regulation of autophagy. J Hepatol 62(3):647–656. https://doi.org/10.1016/j.jhep.2014.10.009
112. Wang HJ, Kang PF, Wu WJ, Tang Y, Pan QQ, Ye HW, Tang B, Li ZH, Gao Q (2013) Changes in cardiac mitochondrial aldehyde dehydrogenase 2 activity in relation to oxidative stress and inflammatory injury in diabetic rats. Mol Med Rep 8(2):686–690. https://doi.org/10.3892/mmr.2013.1524
113. Nakagawa I, Amano A, Mizushima N, Yamamoto A, Yamaguchi H, Kamimoto T, Nara A, Funao J, Nakata M, Tsuda K, Hamada S, Yoshimori T (2004) Autophagy defends cells against invading group A Streptococcus. Science 306(5698):1037–1040. https://doi.org/10.1126/science.1103966
114. Zhang B, Zhang Y, La Cour KH, Richmond KL, Wang XM, Ren J (2013) Mitochondrial aldehyde dehydrogenase obliterates endoplasmic reticulum stress-induced cardiac contractile dysfunction via correction of autophagy. Biochim Biophys Acta 1832(4):574–584. https://doi.org/10.1016/j.bbadis.2013.01.013
115. Ge W, Ren J (2012) mTOR-STAT3-notch signalling contributes to ALDH2-induced protection against cardiac contractile dysfunction and autophagy under alcoholism. J Cell Mol Med 16(3):616–626. https://doi.org/10.1111/j.1582-4934.2011.01347.x
116. Sun A, Cheng Y, Zhang Y, Zhang Q, Wang S, Tian S, Zou Y, Hu K, Ren J, Ge J (2014) Aldehyde dehydrogenase 2 ameliorates doxorubicin-induced myocardial dysfunction through detoxification of 4-HNE and suppression of autophagy. J Mol Cell Cardiol 71:92–104. https://doi.org/10.1016/j.yjmcc.2014.01.002
117. Ji W, Wei S, Hao P, Xing J, Yuan Q, Wang J, Xu F, Chen Y (2016) Aldehyde dehydrogenase 2 has Cardioprotective effects on myocardial Ischaemia/reperfusion injury via suppressing mitophagy. Front Pharmacol 7:101. https://doi.org/10.3389/fphar.2016.00101
118. Setoh K, Terao C, Muro S, Kawaguchi T, Tabara Y, Takahashi M, Nakayama T, Kosugi S, Sekine A, Yamada R, Mishima M, Matsuda F (2015) Three missense variants of metabolic syndrome-related genes are associated with alpha-1 antitrypsin levels. Nat Commun 6:7754. https://doi.org/10.1038/ncomms8754
119. Blaurock N, Schmerler D, Hunniger K, Kurzai O, Ludewig K, Baier M, Brunkhorst FM, Imhof D, Kiehntopf M (2016) C-terminal Alpha-1 antitrypsin peptide: a new Sepsis biomarker with immunomodulatory function. Mediat Inflamm 2016:6129437. https://doi.org/10.1155/2016/6129437
120. Hurley K, Lacey N, O'Dwyer CA, Bergin DA, McElvaney OJ, O'Brien ME, OF ME, Reeves EP, McElvaney NG (2014) Alpha-1 antitrypsin augmentation therapy corrects accelerated neutrophil apoptosis in deficient individuals. J Immunol 193(8):3978–3991. https://doi.org/10.4049/jimmunol.1400132
121. Shapira MG, Khalfin B, Lewis EC, Parola AH, Nathan I (2014) Regulation of autophagy by alpha1-antitrypsin: "a foe of a foe is a friend". Mol Med 20:417–426. https://doi.org/10.2119/molmed.2014.00054
122. Saferali A, Lee J, Sin DD, Rouhani FN, Brantly ML, Sandford AJ (2014) Longer telomere length in COPD patients with alpha1-antitrypsin deficiency independent of lung function. PLoS One 9(4):e95600. https://doi.org/10.1371/journal.pone.0095600
123. Frenzel E, Wrenger S, Brugger B, Salipalli S, Immenschuh S, Aggarwal N, Lichtinghagen R, Mahadeva R, Marcondes AM, Dinarello CA, Welte T, Janciauskiene S (2015) alpha1-antitrypsin combines with plasma fatty acids and induces angiopoietin-like protein 4 expression. J Immunol 195(8):3605–3616. https://doi.org/10.4049/jimmunol.1500740

124. Chen HW, Kuo HT, Hwang LC, Kuo MF, Yang RC (2007) Proteomic alteration of mitochondrial aldehyde dehydrogenase 2 in sepsis regulated by heat shock response. Shock 28(6):710–716. https://doi.org/10.1097/shk.0b013e318050c8c2
125. Hu Y, Yan JB, Zheng MZ, Song XH, Wang LL, Shen YL, Chen YY (2015) Mitochondrial aldehyde dehydrogenase activity protects against lipopolysaccharide-induced cardiac dysfunction in rats. Mol Med Rep 11(2):1509–1515. https://doi.org/10.3892/mmr.2014.2803
126. Pang J, Peng H, Wang S, Xu X, Xu F, Wang Q, Chen Y, Barton LA, Chen Y, Zhang Y, Ren J (2019) Mitochondrial ALDH2 protects against lipopolysaccharide-induced myocardial contractile dysfunction by suppression of ER stress and autophagy. Biochim Biophys Acta (BBA) - Mol Basis Dis 1865(6):1627–1641

ALDH2 and Stroke: A Systematic Review of the Evidence

Haixia Xu, Yingmei Zhang, and Jun Ren

Abstract Cerebral stroke is one of the leading causes of mortality and disability worldwide. The prevalence of cerebral stroke is the result of the synergistic effect of genetic susceptibility and numerous vascular risk factors, including hypertension, diabetes, excessive alcohol intake, obesity, and dyslipidemia. Mitochondrial aldehyde dehydrogenase (ALDH2) is a vital enzyme metabolizing various acetaldehyde and toxic aldehydes. The ALDH2 enzymatic activity is severely decreased in the individuals with *ALDH2*2* gene mutation, especially in East Asians. Increasing epidemiological surveys have revealed that ALDH2 genetic polymorphism is closely associated with the increasing incidence of cardiovascular risk factors and cerebral stroke. Evidence from experimental studies has also suggested that ALDH2 facilitates the clearance of reactive aldehydes and reduces the size of cerebral infarct. Therefore, targeting ALDH2 may represent a promising avenue for protection against stroke injury. This review will mainly focus on clinical and epidemiological evidence and the underlying molecular mechanisms involved in the protective effect of ALDH2 in stroke-related injury.

Keywords ALDH2 · Cerebral stroke · Gene polymorphism

1 Introduction

Stroke is an acute cerebrovascular disease characterized by cerebral blood circulatory disorders with prominent functional or structural damage of brain tissues caused by cerebral vascular occlusion [25]. It is reported that over 15 million new stroke patients

H. Xu
Department of Cardiology and Shanghai Institute of Cardiovascular Diseases, Zhongshan Hospital, Fudan University, Shanghai, People's Republic of China

Y. Zhang · J. Ren (✉)
Department of Cardiology, Shanghai Institute of Cardiovascular Diseases, Zhongshan Hospital, Fudan University, Shanghai, China

Center for Cardiovascular Research and Alternative Medicine, University of Wyoming College of Health Sciences, Laramie, WY, USA
e-mail: ren.jun@zs-hospital.sh.cn

J. Ren et al. (eds.), *Aldehyde Dehydrogenases*, Advances in Experimental Medicine and Biology 1193, https://doi.org/10.1007/978-981-13-6260-6_11

are diagnosed per year with five million mortality or permanent disability cases [33]. Cerebral stroke leads to paralysis, aphasia, dementia, and even death, significantly compromising the quality of life for stroke patients, and imposes a heavy burden on families and society. Stroke can generally be divided into two categories, ischemic and hemorrhagic, with ischemic stroke accounting for nearly 87% of all stroke cases [54]. As stroke is a complicate disease where genetic factors interact with environmental factors, exploring its susceptibility genes can help screen high-risk groups and elucidate various causes of the disease. In addition, a better understanding of the susceptible genes should offer clinical value to predict events related to the development of ischemic stroke, so that effective preventive measures may be taken [17, 54, 75].

The aldehyde dehydrogenase 2 (ALDH2) family is an important detoxification enzyme with potent effect to remove acetaldehyde, a toxic product of ethanol oxidation in the liver [93]. Mitochondrial ALDH2 is well-known for its role in alcohol metabolism. However, ALDH2 also removes and detoxifies other toxic reactive aldehydes, such as malondialdehyde (MDA), 4-hydroxy-2-nonenal (4-HNE), and acrolein. A number of independent research groups have examined the underlying mechanisms of ALDH2 in stroke. In this review, we will summarize the current epidemiological evidence and contemporary molecular mechanisms behind the protective function of ALDH2 in stroke-related injury, if any. More importantly, translational potential of ALDH2 in the management of cerebral stroke is reviewed as a new therapeutic strategy.

2 ALDH2

2.1 Structure and Distribution

Aldehyde dehydrogenase (ALDH) is a rather large superfamily, which is responsible for detoxification of a spectrum of biogenic and xenogenic aldehydes through its NAD(P)+-dependent enzymes [32]. These enzymes are commonly involved in mediating various cellular functions, including differentiation, proliferation, growth, and survival. ALDH also exhibits substrate specificity and responds to oxidize a battery of highly reactive aldehydes including 4-HNE and MDA. Human genomics work has revealed that a total of 19 functional ALDH genes expressed in multiple tissues with unique substrate specificity and the prominent functional ALDHs contain ALDH1, ALDH2, ALDH3, and ALDH4. ALDH2 is predominantly located in the mitochondria compared with the other three isozymes (ALDH1, ALDH3, and ALDH4) which are located in cytoplasm. ALDH enzymes play an important role in its dehydrogenase capacity of converting aldehyde into acetic acid with a lower Michaelis constant (Km) for aldehyde [53].

Mitochondrial ALDH2, encoded by a nuclear gene locus on chromosome 12q24.2, consists of 517 amino acids. It plays the role as a homo-tetramer with the molecular mass of 56 kDa [70]. The structure of ALDH2 displays active tetramers with a dimer of two identical subunits with three domains: the NAD-binding, catalytic, and oligomerization domains. Only two catalytic sites on each tetramer complex with an asymmetric structure function to govern the ALDH2 enzyme activity. The protein is transported into the mitochondrial matrix with its N-terminus mitochondrial targeting sequence and cleaved inside mitochondria to complete the folding and maturation of the ALDH2 enzyme [4]. Moreover, ALDH2 also plays the role of reductase activity and esterase activity [40].

ALDH2 is abundantly expressed in the liver, lung, heart, and brain, organs that require high mitochondrial content and high-energy metabolism [65]. The most essential role of ALDH2 is ethanol detoxification, involving two active enzymatic steps. Firstly, ethanol is converted to acetaldehyde by alcohol dehydrogenase (ADH). Acetaldehyde is roughly 10 times for toxic than ethanol, thus making the ethanol metabolism extremely critical for alcoholic organ injury. Secondly, acetaldehyde is diffused across the biological membranes and is circulated in the blood where it is deoxidized into acetic acid by ALDH2 [8]. Moreover, ALDH2 also participates in metabolizing several other aromatic and polycyclic aldehydes and some short-chain aliphatic aldehydes generated from lipid peroxidation, such as MDA and 4-HNE [9]. It is well-known the toxic 4-HNE is associated to mitochondrial membrane integrity and apoptosis, which might promote aging and risk of some neurodegenerative disease. Therefore, clearance of these harmful bio-products by way of ALDH2 is critical for maintaining cellular homeostasis [38, 57].

2.2 ALDH2 Gene Polymorphism

Genetically, ALDH2 gene comprises 13 exons. Exon 12 contains a single nucleotide polymorphism (rs671). As a result of rs671 polymorphism, a guanine (G) substituted into adenine (A) and a codon GAA subsequently changes into AAA [12]. The resultant change alters the 487th amino acid of the protein with lysine (Lys) substituted for glutamic acid (Glu) and causes an allosteric disruption of catalytic sites with an abnormal α-helix structure. Therefore, ALDH2 gene exists in two forms of alleles: the wild-type *ALDH2*1* and mutant *ALDH2*2* allelic variant. A partial variant subunit encoded by the latter allele dramatically decreases the enzyme activity. The wild-type ALDH2, encoded by the *ALDH2*1/*1*, exhibits a tenfold reduced catalytic constant (kcat) and a 200-fold increase in the Km for NAD+ [46]. Compared with wild-type ALDH2 carriers, individuals with heterozygous *ALDH2*1/*2* exhibit overtly reduced activity, whereas individuals with *ALDH2*2/*2* genotype have a ALDH2-deficient phenotype with almost nil enzyme activity [46]. In addition to the rs671 polymorphism, there are other candidate SNPs existing for ALDH2, namely, rs886205, rs2238152, rs441, rs4646778, rs7296651, etc [80].

It is well-known that ALDH2 plays a critical role in the ethanol metabolism. Due to the overtly decreased ALDH2 enzyme activity, individuals carrying the *ALDH2*2/*2* variant are susceptible to the accumulation of aldehyde and other oxidation products in blood circulation after alcohol consumption. This accumulation results in release of vasoactive substances from mast cells, resulting in vasodilation and associated pathophysiological phenotypes, such as facial flushing, palpitation, headache, nausea, sweating, dysphagia, and low blood pressure. Notably, mutation of ALDH2 compromises its protective effect. Moreover, another variation, the His47Arg polymorphism, would also affect an individual's drinking habit independently of Glu504Lys mutation [56].

Previous studies have demonstrated differences in the distribution of ALDH2 alleles in different races [23, 94]. Approximately 560 million populations in the world have the *ALDH2*2* mutation. It is far more prevalent among the Asians (about 30–50%), such as Chinese, Japanese, and South Korean, than the White (<5%), including Caucasoids, Papua New Guineans, Australian Aborigines, and Negroids [43]. Statistically, this might be the most common human enzymopathy, which is identified as "Asian flush" after alcohol consumption. It still remains elusive why the Glu504Lys polymorphism is common in the East Asian population. Certain hypothesis presumed that hepatitis B virus infection probably contributed to the evolution of the ALDH2 mutation in Southeast China, which was postulated to trace back to 2000–3000 years ago [47, 50]. However, the exact causative factors and mechanisms behind the geographical variation of ALDH2 polymorphism still need further investigation.

2.3 Contribution to Diseases

ALDH2*2 variant plays an essential role in limiting excessive alcohol consumption and protecting from alcoholism. This mutant variant has recently been found to be associated with prevalence of various human diseases. Previous studies have revealed that individuals with the *ALDH2*2* allelic variant have a higher prevalence of coronary artery disease, myocardial infarction, hypertension, esophageal cancer, and osteoporosis [8, 15, 30, 68, 91]. Other diseases, such as Parkinson's disease, Alzheimer's disease, Fanconi anemia, diabetes, addiction, and dermatitis, may also be associated to this mutation [6, 21, 86, 89]. Moreover, the *ALDH2*2* variant also causes altered responses to some common drugs, such as nitroglycerin, 5-nitrofuran, and acyclovir [27].

3 ALDH2 and Stroke

The pathogenesis of cerebral stroke is complicated. Epidemiological studies have identified common risk factors of cerebral stroke worldwide, mainly including hypertension, diabetes, smoking, obesity, excessive drinking, dyslipidemia, etc. [3]. However, cerebral stroke occurs even in the absence of these risk factors, which suggests the function of genetic factors in the occurrence of stroke. Therefore, it is believed that the occurrence of cerebral stroke is the result of synergistic effect of genetic susceptibility and environmental factors [54]. Furthermore, *ALDH2*2* variant along with other cardiovascular risk factors collaboratively contributes to the increased prevalence of stroke development. Among various risk factors reported, hypertension, diabetes, and alcohol-related dyslipidemia seem to possess the most profound impact on the risk of stroke development [27].

3.1 ALDH2 Gene Polymorphism and Stroke

Several studies have analyzed the association between *ALDH2*2* allele and ischemic stroke. Nagasawa and associates once revealed that there was no close association between the ALDH2 gene polymorphism and the occurrence of lacunar infarction among Japanese, but the *ALDH2*1/*1* genotype was obviously associated with the development of multiple lacunar infarcts [60]. Moreover, another case-control study conducted in Taiwanese demonstrated that the ALDH2*2/*2 allele might be an independent risk factor for cerebral stroke in male Taiwanese, but not in female Taiwanese. Furthermore, ALDH2 gene polymorphism in male population with cerebral stroke was not associated to other conventional risk factors, vascular stenosis, ischemic stroke subtypes, and outcomes [81]. Another prospective cohort study of Koreans also demonstrated a tight association between ALDH2 polymorphism and cerebral stroke in male population, but not in female patients [78]. While Li et al. recently reported that ALHD2*2 allele was an important risk factor for cerebral ischemia infarction in the Chinese female patients [odds ratio (OR) = 2.207, 95% CI 1.416–3.439], while the influence of ALDH2*2/*2 genotype on cerebral infarction in the male patients was not significant [44].

However, in another study of Han Chinese population, no difference was noted in the distribution of the rs671 genetic variations, but two novel ALDH2 tSNPs-rs886205 and rs7296651 were identified as being related to ischemic stroke susceptibility [80]. Another study also confirmed the genotype of rs10744777 and rs886205 modified the risk of ischemic stroke in Chinese males [11]. In addition, a meta-analysis of genome-wide association study (GWAS) has revealed a close relationship between ALDH2 genetic variants and ischemic stroke among Asian decedents [35]. Most recently, a two-stage GWAS including 16,851 stroke patients showed the 12q24 locus near ALDH2 was associated with all cerebral ischemic stroke, but not specific subtype [63]. An interesting clinical trial was carried out to assess the prognosis in patients with acute cerebral infarction for a short term of 90-day follow-up. It showed *ALDH2* Glu504Lys was an independent risk factor for collateral circulation and negatively predicted the outcomes [69]. Together, ALDH2 has a close association with stroke, which might differ depending on race or ethnicity. Its genetic polymorphism may be an independent risk factor and a negative predictor for cerebral stroke.

3.2 ALDH2 and Hypertension

Hypertension is an extremely important risk factor for cerebral stroke. Previous studies have demonstrated that the *ALDH2*2* allele was associated with the pathophysiology of hypertension. A meta-analysis including 19,608 subjects found that *ALDH2*2* was the most crucial SNP related to the blood pressure variation in East Asians [35]. Another study carried out in East Asians also revealed that individuals with *ALDH2*2* allele were more prone to risk of hypertension compared with those without *ALDH2*2* allele after a follow-up period of over 5 years. Further analysis found that the risk of *ALDH2*2* allele variant is greater for alcohol drinkers than nondrinkers, which provided more evidence for the relationship between *ALDH2*2* mutation and alcohol consumption on the risk of hypertension [7, 55]. Considering that ALDH2 encodes the crucial enzyme involved in ethanol metabolism, its effect on blood pressure is tightly associated with alcohol intake. However, individuals with ALDH2*2 mutation tend to limit or even abstain from alcohol intake due to the adverse effects such as facial flushing and palpitation; therefore this variant may protect these individuals against excessive alcohol-induced cerebral stroke [71]. Several other studies also found that *ALDH2*2* allele was an important risk factor of elevated systolic blood pressure for male drinkers [61, 84]. Another meta-analysis of evaluating the association between the ALDH2 gene polymorphism and hypertension revealed that people with the *ALDH2*1/*1* genotype after average alcohol intake of 25 g/day had significantly higher systolic and diastolic blood pressure than those with the ALDH2*2/*2 genotype [10]. In conclusion, these epidemiologic studies have provided the compelling evidence to highlight the association between ALDH2 genotype and alcohol consumption on the occurrence of hypertension.

Experimental studies have also confirmed the relationship between ALDH2 mutation and hypertension prevalence in the setting of cerebral stroke. The levels of ALDH2 protein is dramatically lower (by approximately 30%) in spontaneously hypertensive rats [28]. In addition, Alda-1 increased the activity of ALDH2 in these rodent models of stroke and decreased the accumulation of numerous reactive aldehydes. Accordingly, pathological indices including cerebral infarct volume, neurological scores, and mortality were significantly improved in these stroke rats [19, 28].

3.3 ALDH2 and Alcohol Intake

The association between alcohol intake and cerebral stroke was extensively evaluated. Multiple epidemiological studies revealed that the association between alcohol and ischemic stroke follows a J-shaped curve, which suggested that the individuals with small to moderate alcohol intake have a lower risk of ischemic stroke compared to those with nonalcohol intake, while people with heavy alcohol consumption have a higher risk of ischemic stroke [41, 55]. Berger and coworkers also found that the risk of ischemic stroke in the male physicians with light to moderate alcohol intake (up to one drink per day) decreased approximately 20% compared with those who had less than one drink per week during a follow-up period of 12.2 years [2]. Furthermore, an interesting case report confirmed two patients with heterozygous ALDH2*2 suffered from sudden cerebral stroke several hours after excessive ethanol consumption. The inactivity of ALDH2 enzyme and amounts of aldehyde accumulation might be the underlying reason [39]. Wang also confirmed the above results and found the overexpression of ALDH2 could show protective effect during the ischemic stroke [88].

On the other hand, there is also some evidence to imply that moderate alcohol intake could protect against the prevalence of ischemic stroke, but is not associated to hemorrhagic stroke [41]. Previous study has revealed that heavy alcohol intake could dramatically increase the risk of hemorrhagic stroke [99]. A large-scale meta-analysis suggested that more than 60 g/d of alcohol intake (four to five drinks) was an essential risk factor of hemorrhagic stroke compared with nondrinkers [odds ratio (OR) = 2.18, 95% CI, 1.48–3.20] [73]. Another meta-analysis found that light to moderate ethanol intake (<150 g/week) is not associated with subarachnoid hemorrhage (SAH), while heavy alcohol consumption significantly increased the risk of SAH [18].

3.4 ALDH2 and Diabetes Mellitus

In addition to the mentioned risk factors of hypertension and heavy alcohol intake, diabetes mellitus is also a considerable risk factor for cerebral stroke. Suzuki and colleagues found the Japanese subjects with a diabetic mother were about 1.5 times higher risk of the development of type 2 diabetes mellitus in the inactive ALDH2 group compared with the individuals in the active ALDH2 group [83]. Another follow-up study revealed that the incidence of having a diabetes mother in those individuals with an *ALDH2*2* mutation was five times than those with wild-type ALDH2 [82]. Interestingly, the ALDH2 polymorphism was found to be associated with increased risk of type 2 diabetes in females with coronary heart disease, but not males [90]. These findings may indicate the *ALDH2*2* gene variation probably leads to the mitochondrial dysfunction and subsequently contribute to diabetic prevalence in the offspring of female diabetic patients. Another independent Japanese study revealed that ALDH2 polymorphism is related to different levels of fasting blood glucose in men with alcohol consumption [92]. In addition, preclinical studies in animal models also revealed the ALDH2 protein participated in the onset and progression of diabetic pathology [87, 97]. Both ALDH2 overexpression in transgenic mice and activation of ALDH2 by Alda-1 alleviated streptozotocin-induced diabetic cardiomyopathy [97].

Similar to hypertension, moderate alcohol intake appears to possess beneficial effect on glucose metabolism, including fasting blood glucose, hemoglobin Alc, and insulin resistance [55]. Population with *ALDH2*2* variant have higher hemoglobin

Alc values compared to those with *ALDH2*1* allele following alcohol consumption [27, 59]. A recent meta-analysis including 15 cohort studies with 12-year follow-up reported a U-shaped relationship between alcohol intake and type 2 diabetes mellitus [36]. All these findings have indicated an association between ALDH2 and diabetes, and more in-depth work is warranted to further explore the underlying mechanism for ALDH2-offered effect in diabetes mellitus.

3.5 ALDH2 and Dyslipidemia

The *ALDH2*2* genotype mutation also associates with other risk factors contributing to the prevalence of stroke, such as dyslipidemia. A meta-analysis reveals the relationship between *ALDH2*2* allele and HDL-C serum concentration in the East Asian population, and the association is stronger in nondrinkers than in drinkers, in male than in female, and in Japanese than in Chinese, the underlying pathogenesis still unknown [31]. Kazuhiko et al. also reported a considerably higher level of hyper-LDL-cholesterolemia in individuals with the ALDH2*2 genotype mutation after adjusting several variables such as drinking status [37]. As reported in the above studies, serum levels of triglycerides were also significantly associated with ALDH2 rs671 polymorphism in individuals with alcohol intake [85].

4 Molecular Mechanisms

Numerous work was carried out to explore the association between ALDH2 and coronary heart disease [24, 51, 67, 79]. Given the similarities existing in molecular mechanisms between myocardial infarction and stroke, the role of ALDH2 in the prevalence of cerebral stroke has also been further explored [8, 58, 68]. In addition to the inflammatory mechanism, oxidative stress is one of the most crucial mechanisms contributing to the ischemic stroke. During the development of cerebral stroke, a majority of aldehydes generated from the lipid peroxidation. Reactive aldehydes can extend the damage caused by reactive oxygen species, including 4-HNE, MDA, 1-palmitoyl-2-oxovaleroyl phosphatidyl choline (POVPC), and so on. ALDH2 protein is capable of detoxifying these reactive aldehydes. It has been demonstrated in the previous studies that expression of 4-HNE protein elevated in the ischemic cerebral cortex in mice within 2 h of stroke induction [64]. In addition, the higher levels of plasma 4-HNE were also found in the genetic stroke-prone rats (stroke-prone spontaneously hypertensive rats) and another stroke rats with middle cerebral artery occlusion (MCAO) [42]. Furthermore, the findings suggested that the cerebral infarct area enlarged, ROS/MPA levels increased, and GSH/GSSG ratio reduced after intravenous inject of 4-HNE, which indicated oxidative stress in cerebral tissue was significantly increased [42]. It is also confirmed that impaired ALDH2 function in the patients with *ALDH2*2* allele caused higher levels of MDA and 4-HNE and induced apoptosis by sustained JNK activation [16]. Furthermore, with the collaborate presence of 4-HNE toxicity in PC12 cells, ALDH2 overexpression could increase neuronal survival [13]. In addition, brain damage area and neuronal cell death were significantly decreased in ALDH2 transgenic mice with 12 weeks of alcohol consumption [72].

The activity of ALDH2 may be regulated bidirectionally. Endogenous protein kinase C-epsilon (PKCε) can enhance the activity of ALDH2 [5], and the exogenous agent, daidzin, can inhibit its activity [49]. Studies have revealed that moderate ethanol administration could alleviate brain tissue injuries in the MCAO rats by the activation of ALDH2 and PKCε, as a molecule upstream of ALDH2, which may be the underline mechanism of cerebral protective effects with moderate ethanol administration [28]. Furthermore, another research showed Alda-1, as an ALDH2 activator, improved outcomes of myocardial infarction, cerebral stroke, and pain in animal models [27]. Accumulating evidence suggests that Alda-1 also creates a protective effect on maintaining enzymatic activity of the ALDH2 *protein in those with ALDH2*2 mutation.* As shown in rodent animal models of cerebral stroke, Alda-1 increased ALDH2 activity, elevated reactive aldehyde clearance, and reduced cerebral infarct volume [19, 28]. Alda-1 administration also could inhibit expression of AQP4 protein, then improve brain edema, and reduce neurological damage through increasing ALDH2 activity [45]. In a nutshell, ALDH2 is a critical mediator in protecting cellular damage from ischemia stroke, and increasing its activity may offer a new perspective on the treatment of cerebral stroke.

ALDH2 confers its protective effects in alcohol detoxification by the regulation of some signaling pathways, such as Akt and AMPK [34]. It was reported that ALDH2 activation or overexpression provides cardiac protection through the balance between Akt and AMPK. Furthermore, the downstream substrates including mTOR, STAT3, Notch1, PP2A, and PP2C related to autophagy and apoptosis were involved in the mentioned mechanism [20, 22]. Other researches also mentioned that in ALDH2-mediated cardiac protection, Akt and AMPK could modulate transcription factor Foxo3 phosphorylation at the sites of Thr32 and Ser413 and subsequently occurred apoptosis and mitochondrial dysfunction [34, 52, 98]. In addition, chronic alcohol consumption induced myocardial hypertrophy and contractile defects in ALDH2 transgenic mice, which is associated with the phosphorylation of ASK-1, GSK-3β, GATA4, and CREB [14]. These signaling pathways mentioned above predominated in cardioprotection, and the role of ALDH2 in stroke needs further exploration.

More recent studies revealed that pharmacologically and genetically improved ALDH2 enzymatic activation improves cardiac contractile function probably via regulation of autophagy [29, 72, 95]. It was also demonstrated that increasing ALDH2 protein activity exerted protective effect in spinal cord ischemia-reperfusion damage by inhibiting apoptosis [48]. While apoptosis has been a better described mechanism in the neuronal damage of ischemic stroke [96], autophagy is also a process worthy of more attention and exploration. Autophagy plays a vital role on turnover of intracellular organelles and focuses on ischemic stroke in recent studies [76]. Moderate autophagy is neuroprotective [74, 77], whereas excessive autophagy is harmful and neurotoxic [1, 26]. Consistently, ALDH2 offers its protective effects in ischemia-reperfusion injury through Akt- and AMPK-mTOR signaling cascades [51]. Up-to-date, much remains unclear for the role of ALDH2 in mitochondrial function and autophagy in ischemia stroke, and further research is needed.

Some findings in recent studies suggested that ALDH2 might play a role by influencing inflammation factors. One vitro study revealed that ALDH2 can inhibit inflammatory molecules of MCP-1 and ICAM-1 and regulate JNK and p38 MAPK and subsequently NF-κB and AP-1 signaling pathways [66]. Another study has depicted that the administration of endothelial progenitor cells (EPCs) with low ALDH2 activity into experimental rats with acute cerebral infarction promoted the migration and accumulation of EPCs into the infarct tissue and induced alleviation of brain infarction. Furthermore, EPCs with low ALDH2 activity were regulated in the accumulation into the infarct area through the stromal cell-derived factor-1 (SDF-1) and CXC chemokine receptor 4 (CXCR4) signaling pathway [62].

5 Conclusion

Mitochondrial ALDH2 enzyme fulfills a crucial role in both alcohol and nonalcoholic metabolism through removing and detoxifying numerous toxic aldehydes and peroxides including acetaldehyde, 4-HNE, and MDA. The unique protective effects of ALDH2 in ischemia stroke possess useful clinical impact, in particular in those individuals with ALDH2 polymorphism. As a beneficial factor for cardiovascular homeostasis, ALDH2 polymorphism is not only independent of the prevalence of cerebral stroke but also associated with other vascular risk factors including hypertension, diabetes, alcohol intake, and dyslipidemia. In addition to the widespread epidemiological evidence, ALDH2 protein also participates in stroke protection through alleviation of oxidative stress, regulation of autophagy, mitochondrial function, and inflammation through various signaling pathways. Nowadays, enhancing ALDH2 enzymatic activity using the specific ALDH2 activator Alda-1 provides new insights for stroke treatment. Moreover, precision medicine development promotes the approach focusing on the population for the *ALDH2*2* genetic variant and will improve the care and health of people.

References

1. Bartlett BJ, Isakson P, Lewerenz J, Sanchez H, Kotzebue RW, Cumming RC, Harris GL, Nezis IP, Schubert DR, Simonsen A, Finley KD (2011) p62, Ref(2)P and ubiquitinated proteins are conserved markers of neuronal aging, aggregate formation and progressive autophagic defects. Autophagy 7(6):572–583
2. Berger K, Ajani UA, Kase CS, Gaziano JM, Buring JE, Glynn RJ, Hennekens CH (1999) Light-to-moderate alcohol consumption and the risk of stroke among U.S. male physicians. N Engl J Med 341(21):1557–1564

3. Boehme AK, Esenwa C, Elkind MS (2017) Stroke risk factors, genetics, and prevention. Circ Res 120(3):472–495. https://doi.org/10.1161/CIRCRESAHA.116.308398
4. Braun T, Bober E, Singh S, Agarwal DP, Goedde HW (1987) Evidence for a signal peptide at the amino-terminal end of human mitochondrial aldehyde dehydrogenase. FEBS Lett 215(2):233–236
5. Budas GR, Churchill EN, Disatnik MH, Sun L, Mochly-Rosen D (2010) Mitochondrial import of PKCepsilon is mediated by HSP90: a role in cardioprotection from ischaemia and reperfusion injury. Cardiovasc Res 88(1):83–92
6. Burke WJ, Bronstein JM, Fitzmaurice A (2014) Aldehyde dehydrogenase variation enhances effect of pesticides associated with Parkinson disease. Neurology 83(20):1879–1880. https://doi.org/10.1212/01.wnl.0000457072.94144.8f
7. Chang YC, Chiu YF, Lee IT, Ho LT, Hung YJ, Hsiung CA, Quertermous T, Donlon T, Lee WJ, Lee PC, Chen CH, Mochly-Rosen D, Chuang LM (2012) Common ALDH2 genetic variants predict development of hypertension in the SAPPHIRe prospective cohort: gene-environmental interaction with alcohol consumption. BMC Cardiovasc Disord 12:58
8. Chen CH, Ferreira JC, Gross ER, Mochly-Rosen D (2014) Targeting aldehyde dehydrogenase 2: new therapeutic opportunities. Physiol Rev 94(1):1–34
9. Chen CH, Sun L, Mochly-Rosen D (2010) Mitochondrial aldehyde dehydrogenase and cardiac diseases. Cardiovasc Res 88(1):51–57
10. Chen L, Smith GD, Harbord RM, Lewis SJ (2008) Alcohol intake and blood pressure: a systematic review implementing a Mendelian randomization approach. PLoS Med 5(3):e52
11. Cheng X, Xu J, Gu M, Wang M, Sun B, Li Z, Ni G, Wang G, Weng Z, Shi Y, Zhang Z, Liu X (2018) Genetic variants in ALDH2 predict risk of ischemic stroke in a Chinese population. Gene 678:49–54. https://doi.org/10.1016/j.gene.2018.08.002
12. Crabb DW, Edenberg HJ, Bosron WF, Li TK (1989) Genotypes for aldehyde dehydrogenase deficiency and alcohol sensitivity. The inactive ALDH2(2) allele is dominant. J Clin Invest 83(1):314–316
13. Deza-Ponzio R, Herrera ML, Bellini MJ, Virgolini MB, Herenu CB (2018) Aldehyde dehydrogenase 2 in the spotlight: the link between mitochondria and neurodegeneration. Neurotoxicology 68:19–24. https://doi.org/10.1016/j.neuro.2018.06.005
14. Doser TA, Turdi S, Thomas DP, Epstein PN, Li SY, Ren J (2009) Transgenic overexpression of aldehyde dehydrogenase-2 rescues chronic alcohol intake-induced myocardial hypertrophy and contractile dysfunction. Circulation 119(14):1941–1949
15. Dura P, Berkers T, van Veen EM, Salomon J, te Morsche RH, Roelofs HM, Kristinsson JO, Wobbes T, Witteman BJ, Tan AC, Drenth JP, Peters WH (2013) Polymorphisms in alcohol-metabolizing enzymes and esophageal carcinoma susceptibility: a Dutch Caucasian case-control study. J Hum Genet 58(11):742–748. https://doi.org/10.1038/jhg.2013.95
16. Ebert AD, Kodo K, Liang P, Wu H, Huber BC, Riegler J, Churko J, Lee J, de Almeida P, Lan F, Diecke S, Burridge PW, Gold JD, Mochly-Rosen D, Wu JC (2014) Characterization of the molecular mechanisms underlying increased ischemic damage in the aldehyde dehydrogenase 2 genetic polymorphism using a human induced pluripotent stem cell model system. Sci Transl Med 6(255):255ra130
17. Falcone GJ, Malik R, Dichgans M, Rosand J (2014) Current concepts and clinical applications of stroke genetics. Lancet Neurol 13(4):405–418
18. Feigin VL, Rinkel GJ, Lawes CM, Algra A, Bennett DA, van Gijn J, Anderson CS (2005) Risk factors for subarachnoid hemorrhage: an updated systematic review of epidemiological studies. Stroke 36(12):2773–2780
19. Fu SH, Zhang HF, Yang ZB, Li TB, Liu B, Lou Z, Ma QL, Luo XJ, Peng J (2014) Alda-1 reduces cerebral ischemia/reperfusion injury in rat through clearance of reactive aldehydes. Naunyn Schmiedeberg's Arch Pharmacol 387(1):87–94
20. Galter D, Buervenich S, Carmine A, Anvret M, Olson L (2003) ALDH1 mRNA: presence in human dopamine neurons and decreases in substantia nigra in Parkinson's disease and in the ventral tegmental area in schizophrenia. Neurobiol Dis 14(3):637–647

21. Garaycoechea JI, Crossan GP, Langevin F, Daly M, Arends MJ, Patel KJ (2012) Genotoxic consequences of endogenous aldehydes on mouse haematopoietic stem cell function. Nature 489(7417):571–575. https://doi.org/10.1038/nature11368
22. Ge W, Guo R, Ren J (2011) AMP-dependent kinase and autophagic flux are involved in aldehyde dehydrogenase-2-induced protection against cardiac toxicity of ethanol. Free Radic Biol Med 51(9):1736–1748
23. Goedde HW, Agarwal DP, Fritze G, Meier-Tackmann D, Singh S, Beckmann G, Bhatia K, Chen LZ, Fang B, Lisker R (1992) Distribution of ADH2 and ALDH2 genotypes in different populations. Hum Genet 88(3):344–346
24. Gong D, Zhang H, Hu S (2013) Mitochondrial aldehyde dehydrogenase 2 activation and cardioprotection. J Mol Cell Cardiol 55:58–63. https://doi.org/10.1016/j.yjmcc.2012.03.017
25. Greenberg SM (2006) Small vessels, big problems. N Engl J Med 354(14):1451–1453
26. Grishchuk Y, Ginet V, Truttmann AC, Clarke PG, Puyal J (2011) Beclin 1-independent autophagy contributes to apoptosis in cortical neurons. Autophagy 7(10):1115–1131
27. Gross ER, Zambelli VO, Small BA, Ferreira JC, Chen CH, Mochly-Rosen D (2015) A personalized medicine approach for Asian Americans with the aldehyde dehydrogenase 2*2 variant. Annu Rev Pharmacol Toxicol 55:107–127
28. Guo JM, Liu AJ, Zang P, Dong WZ, Ying L, Wang W, Xu P, Song XR, Cai J, Zhang SQ, Duan JL, Mehta JL, Su DF (2013) ALDH2 protects against stroke by clearing 4-HNE. Cell Res 23(7):915–930
29. Guo R, Zhong L, Ren J (2009) Overexpression of aldehyde dehydrogenase-2 attenuates chronic alcohol exposure-induced apoptosis, change in Akt and Pim signalling in liver. Clin Exp Pharmacol Physiol 36(5–6):463–468
30. Han H, Wang H, Yin Z, Jiang H, Fang M, Han J (2013) Association of genetic polymorphisms in ADH and ALDH2 with risk of coronary artery disease and myocardial infarction: a meta-analysis. Gene 526(2):134–141. https://doi.org/10.1016/j.gene.2013.05.002
31. Hao PP, Xue L, Wang XL, Chen YG, Wang JL, Ji WQ, Xu F, Wei SJ, Zhang Y (2010) Association between aldehyde dehydrogenase 2 genetic polymorphism and serum lipids or lipoproteins: a meta-analysis of seven East Asian populations. Atherosclerosis 212(1):213–216
32. Jackson B, Brocker C, Thompson DC, Black W, Vasiliou K, Nebert DW, Vasiliou V (2011) Update on the aldehyde dehydrogenase gene (ALDH) superfamily. Hum Genomics 5(4):283–303
33. Johnston SC, Mendis S, Mathers CD (2009) Global variation in stroke burden and mortality: estimates from monitoring, surveillance, and modelling. Lancet Neurol 8(4):345–354
34. Karliner JS (2009) Lessons from the besotted heart. J Am Coll Cardiol 54(23):2197–2198
35. Kato N, Takeuchi F, Tabara Y, Kelly TN, Go MJ, Sim X, Tay WT, Chen CH, Zhang Y, Yamamoto K, Katsuya T, Yokota M, Kim YJ, Ong RT, Nabika T, Gu D, Chang LC, Kokubo Y, Huang W, Ohnaka K, Yamori Y, Nakashima E, Jaquish CE, Lee JY, Seielstad M, Isono M, Hixson JE, Chen YT, Miki T, Zhou X, Sugiyama T, Jeon JP, Liu JJ, Takayanagi R, Kim SS, Aung T, Sung YJ, Zhang X, Wong TY, Han BG, Kobayashi S, Ogihara T, Zhu D, Iwai N, Wu JY, Teo YY, Tai ES, Cho YS, He J (2011) Meta-analysis of genome-wide association studies identifies common variants associated with blood pressure variation in east Asians. Nat Genet 43(6):531–538
36. Koppes LL, Dekker JM, Hendriks HF, Bouter LM, Heine RJ (2005) Moderate alcohol consumption lowers the risk of type 2 diabetes: a meta-analysis of prospective observational studies. Diabetes Care 28(3):719–725
37. Kotani K, Sakane N, Yamada T (2012) Association of an aldehyde dehydrogenase 2 (ALDH2) gene polymorphism with hyper-low-density lipoprotein cholesterolemia in a Japanese population. Ethn Dis 22(3):324–328
38. Kruman I, Bruce-Keller AJ, Bredesen D, Waeg G, Mattson MP (1997) Evidence that 4-hydroxynonenal mediates oxidative stress-induced neuronal apoptosis. J Neurosci Off J Soc Neurosci 17(13):5089–5100

39. Lai CL, Liu MT, Yin SJ, Lee JT, Lu CC, Peng GS (2012) Heavy binge drinking may increase risk of stroke in nonalcoholic hypertensives carrying variant ALDH2*2 gene allele. Acta Neurol Taiwanica 21(1):39–43
40. Larson HN, Zhou J, Chen Z, Stamler JS, Weiner H, Hurley TD (2007) Structural and functional consequences of coenzyme binding to the inactive Asian variant of mitochondrial aldehyde dehydrogenase: roles of residues 475 and 487. J Biol Chem 282(17):12940–12950
41. Larsson SC, Wallin A, Wolk A, Markus HS (2016) Differing association of alcohol consumption with different stroke types: a systematic review and meta-analysis. BMC Med 14(1):178. https://doi.org/10.1186/s12916-016-0721-4
42. Lee WC, Wong HY, Chai YY, Shi CW, Amino N, Kikuchi S, Huang SH (2012) Lipid peroxidation dysregulation in ischemic stroke: plasma 4-HNE as a potential biomarker? Biochem Biophys Res Commun 425(4):842–847
43. Li H, Borinskaya S, Yoshimura K, Kal'ina N, Marusin A, Stepanov VA, Qin Z, Khaliq S, Lee MY, Yang Y, Mohyuddin A, Gurwitz D, Mehdi SQ, Rogaev E, Jin L, Yankovsky NK, Kidd JR, Kidd KK (2009) Refined geographic distribution of the oriental ALDH2*504Lys (nee 487Lys) variant. Ann Hum Genet 73(Pt 3):335–345
44. Li QY, Zhao NM, Ma JJ, Duan HF, Ma YC, Zhang W, Zhao HW, Qin YH (2015) ALDH2*2 allele is a negative risk factor for cerebral infarction in Chinese women. Biochem Genet 53(9–10):260–267
45. Li Y, Liu SL, Qi SH (2018) ALDH2 protects against ischemic stroke in rats by facilitating 4-HNE clearance and AQP4 down-regulation. Neurochem Res. https://doi.org/10.1007/s11064-018-2549-0
46. Li Y, Zhang D, Jin W, Shao C, Yan P, Xu C, Sheng H, Liu Y, Yu J, Xie Y, Zhao Y, Lu D, Nebert DW, Harrison DC, Huang W, Jin L (2006) Mitochondrial aldehyde dehydrogenase-2 (ALDH2) Glu504Lys polymorphism contributes to the variation in efficacy of sublingual nitroglycerin. J Clin Invest 116(2):506–511
47. Lin YP, Cheng TJ (2002) Why can't Chinese Han drink alcohol? Hepatitis B virus infection and the evolution of acetaldehyde dehydrogenase deficiency. Med Hypotheses 59(2):204–207
48. Liu XZ, Sun X, Shen KP, Jin WJ, Fu ZY, Tao HR, Xu ZX (2017) Aldehyde dehydrogenase 2 overexpression inhibits neuronal apoptosis after spinal cord ischemia/reperfusion injury. Neural Regen Res 12(7):1166–1171. https://doi.org/10.4103/1673-5374.211198
49. Lowe ED, Gao GY, Johnson LN, Keung WM (2008) Structure of daidzin, a naturally occurring anti-alcohol-addiction agent, in complex with human mitochondrial aldehyde dehydrogenase. J Med Chem 51(15):4482–4487
50. Luo HR, Wu GS, Pakstis AJ, Tong L, Oota H, Kidd KK, Zhang YP (2009) Origin and dispersal of atypical aldehyde dehydrogenase ALDH2487Lys. Gene 435(1–2):96–103
51. Ma H, Guo R, Yu L, Zhang Y, Ren J (2011) Aldehyde dehydrogenase 2 (ALDH2) rescues myocardial ischaemia/reperfusion injury: role of autophagy paradox and toxic aldehyde. Eur Heart J 32(8):1025–1038
52. Ma H, Li J, Gao F, Ren J (2009) Aldehyde dehydrogenase 2 ameliorates acute cardiac toxicity of ethanol: role of protein phosphatase and forkhead transcription factor. J Am Coll Cardiol 54(23):2187–2196
53. Marchitti SA, Brocker C, Stagos D, Vasiliou V (2008) Non-P450 aldehyde oxidizing enzymes: the aldehyde dehydrogenase superfamily. Expert Opin Drug Metab Toxicol 4(6):697–720
54. Marsh JD, Keyrouz SG (2010) Stroke prevention and treatment. J Am Coll Cardiol 56(9):683–691
55. Matsumoto C, Miedema MD, Ofman P, Gaziano JM, Sesso HD (2014) An expanding knowledge of the mechanisms and effects of alcohol consumption on cardiovascular disease. J Cardiopulm Rehabil Prev 34(3):159–171
56. Matsuo K, Wakai K, Hirose K, Ito H, Saito T, Tajima K (2006) Alcohol dehydrogenase 2 His47Arg polymorphism influences drinking habit independently of aldehyde dehydrogenase 2 Glu487Lys polymorphism: analysis of 2,299 Japanese subjects. Cancer epidemiology,

biomarkers & prevention: a publication of the American Association for Cancer Research, cosponsored by the American Society of Preventive. Oncology 15(5):1009–1013
57. Milkovic L, Cipak Gasparovic A, Zarkovic N (2015) Overview on major lipid peroxidation bioactive factor 4-hydroxynonenal as pluripotent growth-regulating factor. Free Radic Res 49(7):850–860
58. Munzel T, Daiber A (2018) The potential of aldehyde dehydrogenase 2 as a therapeutic target in cardiovascular disease. Expert Opin Ther Targets 22(3):217–231. https://doi.org/10.1080/14728222.2018.1439922
59. Murata C, Suzuki Y, Muramatsu T, Taniyama M, Atsumi Y, Matsuoka K, Watanabe T, Okazaki I (2000) Inactive aldehyde dehydrogenase 2 worsens glycemic control in patients with type 2 diabetes mellitus who drink low to moderate amounts of alcohol. Alcohol Clin Exp Res 24(4 Suppl):5S–11S
60. Nagasawa H, Wada M, Arawaka S, Kawanami T, Kurita K, Daimon M, Adachi M, Hosoya T, Emi M, Muramatsu M, Kato T (2007) A polymorphism of the aldehyde dehydrogenase 2 gene is a risk factor for multiple lacunar infarcts in Japanese men: the Takahata Study. Eur J Neurol 14(4):428–434
61. Nakagawa T, Kajiwara A, Saruwatari J, Hamamoto A, Kaku W, Oniki K, Mihara S, Ogata Y, Nakagawa K (2013) The combination of mitochondrial low enzyme-activity aldehyde dehydrogenase 2 allele and superoxide dismutase 2 genotypes increases the risk of hypertension in relation to alcohol consumption. Pharmacogenet Genomics 23(1):34–37
62. Nakamura K, Tsurushima H, Marushima A, Nagano M, Yamashita T, Suzuki K, Ohneda O, Matsumura A (2012) A subpopulation of endothelial progenitor cells with low aldehyde dehydrogenase activity attenuates acute ischemic brain injury in rats. Biochem Biophys Res Commun 418(1):87–92
63. Network NSG, International Stroke Genetics C (2016) Loci associated with ischaemic stroke and its subtypes (SiGN): a genome-wide association study. Lancet Neurol 15(2):174–184. https://doi.org/10.1016/S1474-4422(15)00338-5
64. Okun E, Dikshtein Y, Carmely A, Saida H, Frei G, Sela BA, Varshavsky L, Ofir A, Levy E, Albeck M, Sredni B (2007) The organotellurium compound ammonium trichloro(dioxoethylene-o,o′) tellurate reacts with homocysteine to form homocystine and decreases homocysteine levels in hyperhomocysteinemic mice. FEBS J 274(12):3159–3170
65. Oyama T, Isse T, Kagawa N, Kinaga T, Kim YD, Morita M, Sugio K, Weiner H, Yasumoto K, Kawamoto T (2005) Tissue-distribution of aldehyde dehydrogenase 2 and effects of the ALDH2 gene-disruption on the expression of enzymes involved in alcohol metabolism. Front Biosci 10:951–960
66. Pan C, Xing JH, Zhang C, Zhang YM, Zhang LT, Wei SJ, Zhang MX, Wang XP, Yuan QH, Xue L, Wang JL, Cui ZQ, Zhang Y, Xu F, Chen YG (2016) Aldehyde dehydrogenase 2 inhibits inflammatory response and regulates atherosclerotic plaque. Oncotarget 7(24):35562–35576
67. Pang JJ, Barton LA, Chen YG, Ren J (2015) Mitochondrial aldehyde dehydrogenase in myocardial ischemia-reperfusion injury: from bench to bedside. Sheng li xue bao: [Acta Physiol Sinica] 67(6):535–544
68. Panisello-Rosello A, Lopez A, Folch-Puy E, Carbonell T, Rolo A, Palmeira C, Adam R, Net M, Rosello-Catafau J (2018) Role of aldehyde dehydrogenase 2 in ischemia reperfusion injury: an update. World J Gastroenterol 24(27):2984–2994. https://doi.org/10.3748/wjg.v24.i27.2984
69. Qu Y, Zhang H, Li H, Yu L, Sun Y, Chen Y (2017) Aldehyde dehydrogenase 2 (ALDH2) Glu504Lys polymorphism affects collateral circulation and short-term prognosis of acute cerebral infarction patients. Med Sci Monit Int Med J Exp Clin Res 23:4559–4566
70. Raghunathan L, Hsu LC, Klisak I, Sparkes RS, Yoshida A, Mohandas T (1988) Regional localization of the human genes for aldehyde dehydrogenase-1 and aldehyde dehydrogenase-2. Genomics 2(3):267–269
71. Rantakömi SH, Laukkanen JA, Sivenius J, Kauhanen J, Kurl S (2013) Hangover and the risk of stroke in middle-aged men. Acta Neurol Scand 127(3):186–191

72. Ren J, Babcock SA, Li Q, Huff AF, Li SY, Doser TA (2009) Aldehyde dehydrogenase-2 transgene ameliorates chronic alcohol ingestion-induced apoptosis in cerebral cortex. Toxicol Lett 187(3):149–156
73. Reynolds K, Lewis B, Nolen JD, Kinney GL, Sathya B, He J, Lewis BL (2003) Alcohol consumption and risk of stroke: a meta-analysis. JAMA 289(5):579–588
74. Rodríguez-Muela N, Boya P (2012) Axonal damage, autophagy and neuronal survival. Autophagy 8(2):286–288
75. Seshadri S, Wolf PA (2007) Lifetime risk of stroke and dementia: current concepts, and estimates from the Framingham Study. Lancet Neurol 6(12):1106–1114
76. Shen HM, Codogno P (2011) Autophagic cell death: Loch Ness monster or endangered species. Autophagy 7(5):457–465
77. Sheng R, Zhang LS, Han R, Liu XQ, Gao B, Qin ZH (2010) Autophagy activation is associated with neuroprotection in a rat model of focal cerebral ischemic preconditioning. Autophagy 6(4):482–494
78. Shin C, Kwack K, Cho NH, Kim SH, Baik I (2015) Sex-specific differences in the association of a common aldehyde dehydrogenase 2 gene polymorphism and alcohol consumption with stroke risk in a Korean population: a prospective cohort study. Nutr Res Pract 9(1):79–86. https://doi.org/10.4162/nrp.2015.9.1.79
79. Sun A, Zou Y, Wang P, Xu D, Gong H, Wang S, Qin Y, Zhang P, Chen Y, Harada M, Isse T, Kawamoto T, Fan H, Yang P, Akazawa H, Nagai T, Takano H, Ping P, Komuro I, Ge J (2014) Mitochondrial aldehyde dehydrogenase 2 plays protective roles in heart failure after myocardial infarction via suppression of the cytosolic JNK/p53 pathway in mice. J Am Heart Assoc 3(5):e000779. https://doi.org/10.1161/JAHA.113.000779
80. Sun S, He J, Zhang Y, Xiao R, Yan M, Ren Y, Zhu Y, Jin T, Xia Y (2017) Genetic polymorphisms in the ALDH2 gene and the risk of ischemic stroke in a Chinese han population. Oncotarget 8(60):101936–101943
81. Sung YF, Lu CC, Lee JT, Hung YJ, Hu CJ, Jeng JS, Chiou HY, Peng GS (2016) Homozygous ALDH2*2 is an independent risk factor for ischemic stroke in Taiwanese men. Stroke 47(9):2174–2179
82. Suzuki Y, Kuriyama S, Atsumi Y, Murata C, Matsuoka K, Suzuki Y, Taniyama M, Muramatsu T, Suzuki Y, Ohta S (2003) Maternal inheritance of diabetes is associated with inactive ALDH2 genotype in diabetics with renal failure in Japanese. Diabetes Res Clin Pract 60(2):143–145
83. Suzuki Y, Muramatsu T, Taniyama M, Atsumi Y, Kawaguchi R, Higuchi S, Hosokawa K, Asahina T, Murata C, Matsuoka K (1996) Association of aldehyde dehydrogenase with inheritance of NIDDM. Diabetologia 39(9):1115–1118
84. Takagi S, Baba S, Iwai N, Fukuda M, Katsuya T, Higaki J, Mannami T, Ogata J, Goto Y, Ogihara T (2001) The aldehyde dehydrogenase 2 gene is a risk factor for hypertension in Japanese but does not alter the sensitivity to pressor effects of alcohol: the Suita study. Hypertens Res 24(4):365–370
85. Tan A, Sun J, Xia N, Qin X, Hu Y, Zhang S, Tao S, Gao Y, Yang X, Zhang H, Kim ST, Peng T, Lin X, Li L, Mo L, Liang Z, Shi D, Huang Z, Huang X, Liu M, Ding Q, Trent JM, Zheng SL, Mo Z, Xu J (2012) A genome-wide association and gene-environment interaction study for serum triglycerides levels in a healthy Chinese male population. Hum Mol Genet 21(7):1658–1664. https://doi.org/10.1093/hmg/ddr587
86. Wang B, Wang J, Zhou S, Tan S, He X, Yang Z, Xie YC, Li S, Zheng C, Ma X (2008) The association of mitochondrial aldehyde dehydrogenase gene (ALDH2) polymorphism with susceptibility to late-onset Alzheimer's disease in Chinese. J Neurol Sci 268(1–2):172–175. https://doi.org/10.1016/j.jns.2007.12.006
87. Wang J, Wang H, Hao P, Xue L, Wei S, Zhang Y, Chen Y (2011) Inhibition of aldehyde dehydrogenase 2 by oxidative stress is associated with cardiac dysfunction in diabetic rats. Mol Med (Cambridge, Mass) 17(3–4):172–179

88. Wang W, Lin LL, Guo JM, Cheng YQ, Qian J, Mehta JL, Su DF, Luan P, Liu AJ (2015) Heavy ethanol consumption aggravates the ischemic cerebral injury by inhibiting ALDH2. Int J Stroke 10(8):1261–1269
89. Wong JW, Harris K, Powell D (2011) Alcohol urticaria syndrome. Dermatitis 22(6):350–354. https://doi.org/10.2310/6620.2011.11067
90. Xu F, Chen Y, Lv R, Zhang H, Tian H, Bian Y, Feng J, Sun Y, Li R, Wang R, Zhang Y (2010) ALDH2 genetic polymorphism and the risk of type II diabetes mellitus in CAD patients. Hypertens Res 33(1):49–55
91. Yamaguchi J, Hasegawa Y, Kawasaki M, Masui T, Kanoh T, Ishiguro N, Hamajima N (2006) ALDH2 polymorphisms and bone mineral density in an elderly Japanese population. Osteoporos Int 17(6):908–913. https://doi.org/10.1007/s00198-006-0077-2
92. Yin G, Naito M, Wakai K, Morita E, Kawai S, Hamajima N, Suzuki S, Kita Y, Takezaki T, Tanaka K, Morita M, Uemura H, Ozaki E, Hosono S, Mikami H, Kubo M, Tanaka H, Japan Multi-institutional Collaborative Cohort Study Group (2016) ALDH2 polymorphism is associated with fasting blood glucose through alcohol consumption in Japanese men. Nagoya J Med Sci 78(2):183–193
93. Yin SJ (1994) Alcohol dehydrogenase: enzymology and metabolism. Alcohol Alcohol (Oxford, Oxfordshire) Suppl 2:113–119
94. Yoshida A, Huang IY, Ikawa M (1984) Molecular abnormality of an inactive aldehyde dehydrogenase variant commonly found in Orientals. Proc Natl Acad Sci U S A 81(1):258–261
95. Zhang B, Zhang Y, La Cour KH, Richmond KL, Wang XM, Ren J (2013) Mitochondrial aldehyde dehydrogenase obliterates endoplasmic reticulum stress-induced cardiac contractile dysfunction via correction of autophagy. Biochim Biophys Acta 1832(4):574–584
96. Zhang F, Yin W, Chen J (2004) Apoptosis in cerebral ischemia: executional and regulatory signaling mechanisms. Neurol Res 26(8):835–845
97. Zhang Y, Babcock SA, Hu N, Maris JR, Wang H, Ren J (2012) Mitochondrial aldehyde dehydrogenase (ALDH2) protects against streptozotocin-induced diabetic cardiomyopathy: role of GSK3β and mitochondrial function. BMC Med 10:40
98. Zhang Y, Ren J (2011) ALDH2 in alcoholic heart diseases: molecular mechanism and clinical implications. Pharmacol Ther 132(1):86–95
99. Zhang Y, Tuomilehto J, Jousilahti P, Wang Y, Antikainen R, Hu G (2011) Lifestyle factors on the risks of ischemic and hemorrhagic stroke. Arch Intern Med 171(20):1811–1818. https://doi.org/10.1001/archinternmed.2011.443

Targeting ALDH2 in Atherosclerosis: Molecular Mechanisms and Therapeutic Opportunities

Sai Ma and Feng Cao

Abstract Aldehyde dehydrogenase 2 (ALDH2) is an important member of the functional aldehyde dehydrogenases (ALDHs) family in human beings, playing a fundamental role in the detoxification of acetaldehyde and other aldehydes. In recent years, a number of researches have given attention to the association between ALDH2 and atherosclerosis, which provided insights on targeting ALDH2 for therapeutic intervention of atherosclerosis. In this review, these inspiring studies will be discussed, and the clinical implications and concerns will be expounded.

Keywords ALDH2 (aldehyde dehydrogenase 2) · Atherosclerosis · Cardioprotection · Molecular mechanisms · Clinical implications

Abbreviations

4-HNE	4-hydroxy-2-nonenal
ALDH2	Aldehyde dehydrogenase 2
ALDHs	Aldehyde dehydrogenases
$apoE^{-/-}$	Apolipoprotein E knockout
ASCVD	Atherosclerotic cardiovascular disease
CABG	Coronary artery bypass grafting
CVD	Cardiovascular disease
EMP	Empagliflozin

S. Ma
Department of Cardiology, & National Clinical Research Center of Geriatrics Disease, Chinese PLA General Hospital, Beijing, China

Jinling Hospital Department of Cardiology, Nanjing University, School of Medicine, Nanjing, China

F. Cao (✉)
Department of Cardiology & National Clinical Research Center of Geriatrics Disease, Chinese PLA General Hospital, Beijing, China

J. Ren et al. (eds.), *Aldehyde Dehydrogenases*, Advances in Experimental Medicine and Biology 1193, https://doi.org/10.1007/978-981-13-6260-6_12

ER	Endoplasmic reticulum
HDL-C	High-density lipoprotein cholesterol
HUVECs	Human umbilical vein endothelial cells
I/R injury	Ischemia-reperfusion injury
LDL-C	Low-density lipoprotein cholesterol
MDA	Malondialdehyde
MI	Myocardial infarction

1 Introduction

Throughout life span, living beings are exposed to numerous damaging agents, both endogenously and ectogenicly. In consequence, enzymes responsible for detoxifying toxic agents are indispensable for metabolism. Among the enzymes, the aldehyde dehydrogenases (ALDHs) family has attracted great interest in recent years, due to its fundamental role in detoxifying aldehydes. Among the 19 ALDHs genes, aldehyde dehydrogenase 2 (ALDH2) is considered as the most important isoenzyme member in human pathophysiological situations and complications. Recently, the role of ALDH2 in combating cardiovascular diseases and the process of aging has been elucidated, making ALDH2 a promising target for cardiovascular and aging-related diseases.

Atherosclerosis is the main pathophysiological situation which eventually leads to cardiovascular disease (CVD) [1, 2]. The pathogenesis of atherosclerosis is characterized with the process that fatty streaks gradually develop into atheroma and characteristic atherosclerotic plaques and finally evolve into stenosis or even occlusion inside the coronary arteries, leading to the major or fatal cardiovascular events [3]. To this day, atherosclerotic cardiovascular disease (ASCVD) is still a global health and economical burden [4, 5], triggering off the unfailing search for new therapeutic targets [6]. Recently, growing evidence supports a possible association between the function of ALDH2 and the progression of atherosclerosis. In the following mini-review, we mainly focus on the latest studies about the regulatory functions of ALDH2 in atherosclerosis.

2 ALDH2: An Enzyme for the Detoxification of Biogenic and Xenogenic Aldehydes

2.1 ALDHs Family

Over the years, the ALDHs gene superfamily is well known for its role in the detoxification of aldehydes, preventing the accumulation of detrimental aldehydes derived from both endogenous production and exogenous exposures [7]. Having an ancient origin, the ALDHs could be found from bacteria and yeasts to plants and animals [8]. Although the major function of ALDHs is aldehyde oxidation, recent literatures uncovered the multiple properties of ALDHs, including catalytic, binding, antioxidant, structural, and regulatory functions [9]. There are totally 19 functional ALDHs genes, with different tissue expressions and substrate specificities. Here, we mainly focus on the latest studies on ALDH2.

2.2 Cellular Location and Biological Functions of ALDH2

The human ALDH2 gene is located on chromosome 12q24, coding a 517-amino acid polypeptide. The mature ALDH2 protein is located in the mitochondria, serving as the most efficient ALDH isozyme with the lowest K_m for acetaldehyde [10, 11]. ALDH2 is one of the total 19 functional ALDH genes in humans, playing a fundamental role in the detoxification of acetaldehyde (mainly derived from ethanol metabolism) in various tissues and cell types. In addition to acetaldehyde, ALDH2 is also involved in the detoxification of other aldehydes including malondialdehyde (MDA) and 4-hydroxy-2-nonenal (4-HNE) [12]. Although ALDH2 protein is best known for its critical function in alcohol metabolism which mainly occurs in the liver, its greater role in other organs and pathologies have been revealed in recent years [13, 14], such as diabetes mellitus, cancer, neurodegenerative diseases, and cardiovascular diseases [15–19].

2.3 Evidence of ALDH2 in Cardioprotection

It is of interest to note that ALDH2 activity is associated with cardioprotection. The best evidence of ALDH2 in cardioprotection comes from the epidemiological data that ALDH2*2 mutation could be considered as an independent risk factor for acute coronary syndrome. A large population of East Asians carry an ALDH2 variant, ALDH2*2, resulting in a unique loss of function in ALDH2 enzyme, exhibiting significantly lowered dehydrogenase, esterase, and nitroglycerin enzymatic activities [20, 21]. In a clinical trial based on 202 Japanese patients, it was manifested that patients with polymorphism of ALDH2*2 had higher frequencies of coronary spasm and more severe myocardial injury [22]. Moreover, another piece of clinical study revealed that patients with the ALDH2*2 genotype suffered from higher post-operative oxidative stress levels and poorer clinical prognosis after coronary artery bypass grafting (CABG) [23].

For animal models, the beneficial role of ALDH2 in cardioprotection is best illustrated in alcohol-related myocardial injuries, as ALDH2 serving as an enzyme for detoxifying ethanol metabolite. A large body of studies have demonstrated the beneficial effects of ALDH2 in alleviating alcoholic cardiomyopathy [24–26].

Furthermore, the cardioprotection function of ALDH2 in myocardial ischemia-reperfusion injury (I/R injury) was revealed and gained considerable attention in recent years [16, 27, 28]. As the popularity of ALDH2 in the field of cardiovascular diseases arises, its role in the regulation of atherosclerosis gradually gains increasing attention.

3 Targeting ALDH2 in Atherosclerosis

Atherosclerosis is an aging and age-related disease. Recent researches from both animal and human evidence have implicated the potential function of ALDH2 in regulating the process of aging, suggesting that the elevated ALDH2 activity may depict a new picture for atherosclerosis prevention and treatment. In the following review, we mainly focus on the new findings of the potential benefits of ALDH2 regulation in atherosclerosis, which might shed light on new therapeutic targets of atherosclerosis treatment under clinical settings.

3.1 Molecular Mechanisms and Pathways of ALDH2 in Regulating Atherosclerosis

In recent years, the anti-atherosclerotic effects of ALDH2 activation have been reported by several groups based on the evidence from both animal study and human data. In the work by Stachowicz A and his colleagues, they applied the ALDH2 activator Alda-1 in apolipoprotein E knockout ($apoE^{-/-}$) mice. They observed the decrease of atherosclerotic lesions (using both *en face* and cross-sectional methods). However, no changes in plasma lipid profile, inflammatory markers, and plaque composition within the lesions were observed, as well as markers of anti-oxidative defense, apoptosis, mitogenesis, and autophagy. Their work reported the anti-atherosclerotic functions of ALDH2 activation without uncovering the underlying mechanisms [29]. Similarly, Lapenna D et al. revealed that the degree of patients' atherosclerosis severity was correlated inversely with plaque ALDH2 enzymatic activities, indicating the latent association between ALDH2 activity and atherosclerosis progression [30]. However, these reports lacked mechanism analysis. The research work by Yang MY et al. went further. With the clinical data of 248 patients with CVD, they found out that ALDH2 gene polymorphism is strongly related with the severity of atherosclerosis and stenosis. Moreover, using an in vitro model of atherosclerosis, they demonstrated that ALDH2 activation by Alda-1 markedly attenuated that endoplasmic reticulum (ER) stress and apoptosis in vascular smooth muscle cells, which might contribute to the inhibition of atherosclerotic progression [31]. Furthermore, using an in vitro human umbilical vein endothelial cells (HUVECs) model, Pan C et al. documented that decreased ALDH2 activity resulted in elevated inflammatory responses, while increased ALDH2 activity triggered contrary effects, which was mediated by NF-κB signaling pathways [32]. These findings collectively suggested that ALDH2 could influence the atherosclerotic lesion vulnerability and development *via* directly regulating plaque cell function, metabolism, and inflammatory responses.

In addition to this direct function of ALDH2 in atherosclerotic plaques, indirect mechanisms were also demonstrated. To our knowledge, serum lipid profile, including the low-density lipoprotein cholesterol (LDL-C) and high-density lipoprotein cholesterol (HDL-C) levels, plays a vital role in the progression of atherosclerosis [6, 33–35]. Results of the meta-analysis by Hao P et al. indicated an association of ALDH2*2 allele and low HDL-C level among East Asians, suggesting that ALDH2 might influence atherosclerotic progression *via* regulating serum lipid profile [36].

Furthermore, plentiful clinical studies have documented the association of elevated blood pressure in the progression of atherosclerosis, indicating that blood pressure control might be a rational approach to the regulation of atherosclerotic process [37, 38]. High blood pressure itself adversely contributes to the pathogenesis of atherosclerosis, as well as to the morbidity and mortality of ASCVD. In view of these fundamental findings, it is interesting to notice the role of ALDH2 in blood pressure development, which might in consequence affect the progression of atherosclerosis. A Japanese cohort study revealed the association of ALDH2 genotype with the presence of hypertension, confirming the influence of the ALDH2 enzyme on blood pressure [39]. This association was further confirmed in a prospective Chinese cohort [40]. In addition to these clinical trials, experimental animal studies further provided evidence and insights into how ALDH2 regulates blood pressure. By definition, this regulation is multifaceted. On one hand, ALDH2 played a key role in determining the level of circulating acetaldehyde, which has been documented to alter blood pressure in several animal studies [41–43]. On the other hand, ALDH2 has also been shown to influence the bio-activation of nitroglycerin, which markedly affected vascular smooth muscle cell relaxation and blood pressure level [44–46]. These findings indicated the role of ALDH2 activity in blood pressure regulation, which in turn influenced the pathogenesis and clinical prognosis of atherosclerosis.

Atherosclerosis is a complicated pathologic situation with complex mechanisms, including elevated circulated levels of lipoprotein cholesterol, chronic inflammatory responses, and increased oxidative stress and mitochondrial dysfunctions in arterial wall cells. Up to now, the association between ALDH2 and these pathologic mechanisms is far from well elucidated. These unsolved questions, doubtlessly, need further exploration.

3.2 Clinical Implications and Concerns

For clinical purposes, the elevation of ALDH2 activity could be realized by the application of different pharmacological agents, depending on the ALDH2 genotype. For instance, Alda-1, a classical ALDH2 agonist, has been demonstrated to augment ALDH2 enzymatic activity in several animal models [47–49]. Interestingly, precision medicine approach was also developed specially for ALDH2*2 mutant patients. Using a ALDH2*2 mutant mouse model which could mimic the ALDH2*2 phenotype carriers, Pan G et al. demonstrated that empagliflozin (EMP) ameliorated diabetic cardiomyopathy by decreasing 4-HNE protein adducts, indicating EMP might serve as an substitutive ALDH2 activator for ALDH2*2 patients [50]. These inspiring findings implied the feasibility of ALDH2 activation for therapeutic purposes.

Even though the aforementioned findings based on both animal studies and clinical data shed light on the therapeutic potential of ALDH2 manipulation in atherosclerosis prevention and treatment, controversies still exist. Firstly, the anti-atherosclerotic effects of ALDH2 are still being questioned. For instance, Xu F et al. revealed that ALDH2 gene mutation was not related to the atherosclerosis severity among Han Chinese, which was contradictory with previous clinical findings that ALDH2 genotypes of 1*2 and 2*2 were independent risk factors for myocardial infarction (MI) [51]. Secondly, the signaling pathway of ALDH2 is still unclear and needs further investigation. In the recent work by Jun R et al., who is one of the leading researchers in ALDH2 field, they revealed that ALDH2 enzyme accentuated the myocardial remodeling and dysfunction in aging, mainly *via* suppressing myocardial autophagy the JNK-Bcl-2 and IKKβ-AMPK-dependent signaling pathway in an experimental animal model [52]. In another piece of their work, they proposed that the AMPK/Sirt1-mediated mitochondria-associated mechanism was also involved in the regulatory function of ALDH2 in aging-related myocardial remodeling and contractile dysfunction [53]. Similarly, Dassanayaka S et al. recently reported that ALDH2 overexpression augments pressure overload-induced cardiac dysfunction, suggesting that decreased ALDH2 activity may be an adaptive response to certain cardiac pathologies [54]. In contrary, Guo Y et al. revealed that ALDH2 significantly attenuated myocardial dysfunction possibly through an AMPK-dependent regulation of autophagy, implying the beneficial effects of ALDH2 upregulation [55]. Other signaling pathways were also revealed, as Li C et al. reported that ALDH2 may effectively attenuate myocardial remodeling and contractile dysfunction through regulating the JNK/AP-1 and IRS-1/Akt signaling pathways, protecting cardiac function from lipotoxic cardiomyopathy [17]. These important experimental findings, together with the works of other research groups, indicated the complexity of ALDH2 function, which hinders the application of ALDH2 activation in atherosclerosis under clinical settings.

4 Concluding Remarks

Herein, we reviewed the recent findings regarding the association between ALDH2 and atherosclerosis, which provides insights on the ALDH2 as a novel target for therapeutic intervention of atherosclerosis. The listed results are inspiring, especially for the recent work, which provides both clinical and laboratory evidence that

ALDH2 functions in the inhibition of atherosclerotic progression [31]. However, for eventual clinical application, concerns and controversies still exist. On one hand, more strong evidence is needed to prove the crucial benefits of ALDH2 activation in atherosclerosis prevention and treatment. On the other hand, the mechanism analysis needs further exploration and elucidation. Collectively, the manipulation of ALDH2 activity may shed light on new therapeutic targets of atherosclerosis treatment.

References

1. Kucharska-Newton A, Griswold M, Yao ZH, Foraker R, Rose K, Rosamond W et al (2017) Cardiovascular disease and patterns of change in functional status over 15 years: findings from the Atherosclerosis Risk in Communities (ARIC) Study. J Am Heart Assoc 6(3)
2. Herrington W, Lacey B, Sherliker P, Armitage J, Lewington S (2016) Epidemiology of atherosclerosis and the potential to reduce the global burden of atherothrombotic disease. Circ Res 118(4):535–546
3. Barquera S, Pedroza-Tobias A, Medina C, Hernandez-Barrera L, Bibbins-Domingo K, Lozano R et al (2015) Global overview of the epidemiology of atherosclerotic cardiovascular disease. Arch Med Res 46(5):328–338
4. Levenson JW, Skerrett PJ, Gaziano JM (2002) Reducing the global burden of cardiovascular disease: the role of risk factors. Prev Cardiol 5(4):188–199
5. Gaziano TA, Bitton A, Anand S, Abrahams-Gessel S, Murphy A (2010) Growing epidemic of coronary heart disease in low- and middle-income countries. Curr Probl Cardiol 35(2):72–115
6. Solanki A, Bhatt L, Johnston T (2018) Evolving targets for the treatment of atherosclerosis. Pharmacol Ther 187:1–12
7. Chiang CP, Jao SW, Lee SP, Chen PC, Chung CC, Lee SL et al (2012) Expression pattern, ethanol-metabolizing activities, and cellular localization of alcohol and aldehyde dehydrogenases in human large bowel: association of the functional polymorphisms of ADH and ALDH genes with hemorrhoids and colorectal cancer. Alcohol 46(1):37–49
8. Vasiliou V (1997) Aldehyde dehydrogenase genes. Adv Exp Med Biol 414:595–600
9. Vasiliou V, Nebert DW (2005) Analysis and update of the human aldehyde dehydrogenase (ALDH) gene family. Hum Genomics 2(2):138–143
10. Parrilla R, Okawa K, Lindros KO, Zimmerman UJ, Kobayashi K, Williamson JR (1974) Functional compartmentation of acetaldehyde oxidation in rat liver. J Biol Chem 249(15):4926–4933
11. Eriksson CJ, Marselos M, Koivula T (1975) Role of cytosolic rat liver aldehyde dehydrogenase in the oxidation of acetaldehyde during ethanol metabolism in vivo. Biochem J 152(3):709–712
12. Zhu Q, He G, Wang J, Wang Y, Chen W (2017) Pretreatment with the ALDH2 agonist Alda-1 reduces intestinal injury induced by ischaemia and reperfusion in mice. Clin Sci 131(11):1123–1136
13. Chen C, Ferreira J, Gross E, Mochly-Rosen D (2014) Targeting aldehyde dehydrogenase 2: new therapeutic opportunities. Physiol Rev 94(1):1–34
14. Stickel F, Moreno C, Hampe J, Morgan M (2017) The genetics of alcohol dependence and alcohol-related liver disease. J Hepatol 66(1):195–211
15. Baig M, Adil M, Khan R, Dhadi S, Ahmad K, Rabbani G et al (2017) Enzyme targeting strategies for prevention and treatment of cancer: implications for cancer therapy. Semin Cancer Biol

16. Panisello-Roselló A, Lopez A, Folch-Puy E, Carbonell T, Rolo A, Palmeira C et al (2018) Role of aldehyde dehydrogenase 2 in ischemia reperfusion injury: an update. World J Gastroenterol 24(27):2984–2994
17. Li C, Li X, Chang Y, Zhao L, Liu B, Wei S et al (2018) Aldehyde dehydrogenase-2 attenuates myocardial remodeling and contractile dysfunction induced by a high-fat diet. Cell Physiol Biochem 48(5):1843–1853
18. Sung Y, Lu C, Lee J, Hung Y, Hu C, Jeng J et al (2016) Homozygous ALDH2*2 is an independent risk factor for ischemic stroke in Taiwanese men. Stroke 47(9):2174–2179
19. Zhang X, Ye Y, Wang Y, Liu F, Liu X, Hu B et al (2015) Aldehyde dehydrogenase 2 genetic variations may increase susceptibility to Parkinson's disease in Han Chinese population. Neurobiol Aging 36(9):2660. e9–13
20. Beretta M, Gorren AC, Wenzl MV, Weis R, Russwurm M, Koesling D et al (2010) Characterization of the East Asian variant of aldehyde dehydrogenase-2: bioactivation of nitroglycerin and effects of Alda-1. J Biol Chem 285(2):943–952
21. Gross E, Zambelli V, Small B, Ferreira J, Chen C, Mochly-Rosen D (2015) A personalized medicine approach for Asian Americans with the aldehyde dehydrogenase 2*2 variant. Annu Rev Pharmacol Toxicol 55:107–127
22. Mizuno Y, Harada E, Morita S, Kinoshita K, Hayashida M, Shono M et al (2015) East Asian variant of aldehyde dehydrogenase 2 is associated with coronary spastic angina: possible roles of reactive aldehydes and implications of alcohol flushing syndrome. Circulation 131(19):1665–1673
23. Gong D, Zhang L, Zhang Y, Wang F, Zhou X, Sun H (2019) East Asian variant of aldehyde dehydrogenase 2 is related to worse cardioprotective results after coronary artery bypass grafting. Interact Cardiovasc Thorac Surg 28(1):79–84
24. Zhang Y, Ren J (2011) ALDH2 in alcoholic heart diseases: molecular mechanism and clinical implications. Pharmacol Ther 132(1):86–95
25. Doser T, Turdi S, Thomas D, Epstein P, Li S, Ren J (2009) Transgenic overexpression of aldehyde dehydrogenase-2 rescues chronic alcohol intake-induced myocardial hypertrophy and contractile dysfunction. Circulation 119(14):1941–1949
26. Liu B, Zhang R, Wei S, Yuan Q, Xue M, Hao P et al (2018) ALDH2 protects against alcoholic cardiomyopathy through a mechanism involving the p38 MAPK/CREB pathway and local renin-angiotensin system inhibition in cardiomyocytes. Int J Cardiol 257:150–159
27. Ueta C, Campos J, Albuquerque R, Lima V, Disatnik M, Sanchez A et al (2018) Cardioprotection induced by a brief exposure to acetaldehyde: role of aldehyde dehydrogenase 2. Cardiovasc Res 114(7):1006–1015
28. Kang P, Wu W, Tang Y, Xuan L, Guan S, Tang B et al (2016) Activation of ALDH2 with low concentration of ethanol attenuates myocardial ischemia/reperfusion injury in diabetes rat model. Oxidative Med Cell Longev 2016:6190504
29. Stachowicz A, Olszanecki R, Suski M, Wiśniewska A, Totoń-Żurańska J, Madej J et al (2014) Mitochondrial aldehyde dehydrogenase activation by Alda-1 inhibits atherosclerosis and attenuates hepatic steatosis in apolipoprotein E-knockout mice. J Am Heart Assoc 3(6):e001329
30. Lapenna D, Ciofani G, Ucchino S, Giamberardino M, Di Ilio C, Cuccurullo F (2015) Reactive aldehyde-scavenging enzyme activities in atherosclerotic plaques of cigarette smokers and nonsmokers. Atherosclerosis 238(2):190–194
31. Yang MY, Wang YB, Han B, Yang B, Qiang YW, Zhang Y et al (2018) Activation of aldehyde dehydrogenase 2 slows down the progression of atherosclerosis via attenuation of ER stress and apoptosis in smooth muscle cells. Acta Pharmacol Sin 39(1):48–58
32. Pan C, Xing JH, Zhang C, Zhang YM, Zhang LT, Wei SJ et al (2016) Aldehyde dehydrogenase 2 inhibits inflammatory response and regulates atherosclerotic plaque. Oncotarget 7(24):35562–35576
33. Tsujita K, Sugiyama S, Sumida H, Shimomura H, Yamashita T, Yamanaga K et al (2015) Impact of dual lipid-lowering strategy with ezetimibe and atorvastatin on coronary plaque regression in patients with percutaneous coronary intervention: the multicenter randomized controlled PRECISE-IVUS trial. J Am Coll Cardiol 66(5):495–507

34. Patel K, Strong A, Tohyama J, Jin X, Morales C, Billheimer J et al (2015) Macrophage sortilin promotes LDL uptake, foam cell formation, and atherosclerosis. Circ Res 116(5):789–796
35. Rader D, Hovingh G (2014) HDL and cardiovascular disease. Lancet 384(9943):618–625
36. Hao P, Xue L, Wang X, Chen Y, Wang J, Ji W et al (2010) Association between aldehyde dehydrogenase 2 genetic polymorphism and serum lipids or lipoproteins: a meta-analysis of seven East Asian populations. Atherosclerosis 212(1):213–216
37. Newman WP, Freedman DS, Voors AW, Gard PD, Srinivasan SR, Cresanta JL et al (1986) Relation of serum lipoprotein levels and systolic blood pressure to early atherosclerosis. The Bogalusa Heart Study. N Engl J Med 314(3):138–144
38. Kaplan H, Thompson R, Trumble B, Wann L, Allam A, Beheim B et al (2017) Coronary atherosclerosis in indigenous South American Tsimane: a cross-sectional cohort study. Lancet 389(10080):1730–1739
39. Takagi S, Baba S, Iwai N, Fukuda M, Katsuya T, Higaki J et al (2001) The aldehyde dehydrogenase 2 gene is a risk factor for hypertension in Japanese but does not alter the sensitivity to pressor effects of alcohol: the Suita study. Hypertens Res 24(4):365
40. Chang YC, Yen-Feng C, I-Te L, Low-Tone H, Yi-Jen H, Hsiung CA et al (2012) Common ALDH2 genetic variants predict development of hypertension in the SAPPHIRe prospective cohort: gene-environmental interaction with alcohol consumption. BMC Cardiovasc Disord 12(1):58
41. Brandt M, Wenzel P (2018) Alcohol puts the heart under pressure: acetaldehyde activates a localized renin angiotensin aldosterone system within the myocardium in alcoholic cardiomyopathy. Int J Cardiol 257:220–221
42. Hellström E, Tottmar O (1982) Acute effects of ethanol and acetaldehyde on blood pressure and heart rate in disulfiram-treated and control rats. Pharmacol Biochem Behav 17(6):1103–1109
43. Satoh Y, Ide Y, Sugano T, Koda K, Momose Y, Tagami M (2008) Hypotensive and hypertensive effects of acetaldehyde on blood pressure in rats. Nihon Arukoru Yakubutsu Igakkai Zasshi 43(3):188–193
44. Chen Z, Stamler JS (2006) Bioactivation of nitroglycerin by the mitochondrial aldehyde dehydrogenase. Trends Cardiovasc Med 16(8):259–265
45. Chen Z, Foster MW, Zhang J, Mao L, Rockman HA, Kawamoto T et al (2005) An essential role for mitochondrial aldehyde dehydrogenase in nitroglycerin bioactivation. Proc Natl Acad Sci U S A 102(34):12159–12164
46. Kollau A, Hofer A, Russwurm M, Koesling D, Keung WM, Schmidt K et al (2005) Contribution of aldehyde dehydrogenase to mitochondrial bioactivation of nitroglycerin: evidence for the activation of purified soluble guanylate cyclase through direct formation of nitric oxide. Biochem J 385(Pt 3):769–777
47. Zhong W, Zhang W, Li Q, Xie G, Sun Q, Sun X et al (2015) Pharmacological activation of aldehyde dehydrogenase 2 by Alda-1 reverses alcohol-induced hepatic steatosis and cell death in mice. J Hepatol 62(6):1375–1381
48. Hua Y, Chen H, Zhao X, Liu M, Jin W, Yan W et al (2018) Alda-1, an aldehyde dehydrogenase-2 agonist, improves long-term survival in rats with chronic heart failure following myocardial infarction. Mol Med Rep 18(3):3159–3166
49. Chen C, Cruz L, Mochly-Rosen D (2015) Pharmacological recruitment of aldehyde dehydrogenase 3A1 (ALDH3A1) to assist ALDH2 in acetaldehyde and ethanol metabolism in vivo. Proc Natl Acad Sci U S A 112(10):3074–3079
50. Pan G, Deshpande M, Pang H, Palaniyandi S (2018) Precision medicine approach: Empagliflozin for diabetic cardiomyopathy in mice with aldehyde dehydrogenase (ALDH) 2*2 mutation, a specific genetic mutation in millions of East Asians. Eur J Pharmacol 839:76–81
51. Xu F, Chen YG, Geng YJ, Zhang H, Jiang CX, Sun Y et al (2007) The polymorphism in acetaldehyde dehydrogenase 2 gene, causing a substitution of Glu > Lys(504), is not associated with coronary atherosclerosis severity in Han Chinese. Tohoku J Exp Med 213(3):215–220

52. Zhang Y, Wang C, Zhou J, Sun A, Hueckstaedt LK, Ge J et al (2017) Complex inhibition of autophagy by mitochondrial aldehyde dehydrogenase shortens lifespan and exacerbates cardiac aging. Biochim Biophys Acta Mol basis Dis 1863(8):1919–1932
53. Zhang Y, Mi SL, Hu N, Doser TA, Sun A, Ge J et al (2014) Mitochondrial aldehyde dehydrogenase 2 accentuates aging-induced cardiac remodeling and contractile dysfunction: role of AMPK, Sirt1, and mitochondrial function. Free Radic Biol Med 71:208–220
54. Dassanayaka S, Zheng Y, Gibb A, Cummins T, McNally L, Brittian K et al (2018) Cardiac-specific overexpression of aldehyde dehydrogenase 2 exacerbates cardiac remodeling in response to pressure overload. Redox Biol 17:440–449
55. Guo Y, Yu W, Sun D, Wang J, Li C, Zhang R et al (2015) A novel protective mechanism for mitochondrial aldehyde dehydrogenase (ALDH2) in type i diabetes-induced cardiac dysfunction: role of AMPK-regulated autophagy. Biochim Biophys Acta 1852(2):319–331

ALDH2 and Cancer Therapy

Li-Shun Wang and Zhao-Xia Wu

Abstract Aldehyde dehydrogenase 2 (ALDH2) is a member of ALDH family. ALDH1 has been widely recognized for its roles in carcinogenesis and cancer therapy; however, investigation for ALDH2 in cancer is seldom mentioned. The ALDH2 point mutation ALDH2*2 is the most frequent human gene variant, and it is present in approximately 560 million East Asians. ALDH2*2 demonstrates its effect on alcohol consumption limiting and alcoholism developing protection, and this variant is recently found to have an important impact on human health. This chapter focuses on its potential effect on cancer therapy, especially for chemotherapeutics with anthracyclines.

1 Introduction

Aldehyde dehydrogenase 2 (ALDH2) belongs to the aldehyde dehydrogenase family of enzymes. ALDH1 has been recognized for its roles in carcinogenesis and cancer therapy [1]; however, ALDH2 is well known for its effect on alcohol consumption limiting and alcoholism developing protection. Intriguingly, this gene has recently been revealed to have an important impact on pathogenesis and therapeutics [2]. Notably, the point mutation in ALDH2 identified as ALDH2*2, which lost 90% its enzyme activity, is the most frequent human gene variant, and it is found in 8% of the world's population, and approximately 560 million East Asians present this variant [3].

ALDH2 has a major contribution in oxidizing endogenous aldehydic products arising from lipid peroxidation under oxidative stress, and these accumulations, such as 4-hydroxy-2-nonenal (4-HNE) and malondialdehyde, are thought to contribute to many diseases [4]. For example, ALDH2 has demonstrated to be a crucial

L.-S. Wang (✉) · Z.-X. Wu
Institute of Fudan-Minhang Academic Health System, Minhang Branch, Zhongshan Hospital, Fudan University, Shanghai, China
e-mail: lishunwang@fudan.edu.cn

J. Ren et al. (eds.), *Aldehyde Dehydrogenases*, Advances in Experimental Medicine and Biology 1193, https://doi.org/10.1007/978-981-13-6260-6_13

enzyme in protecting the heart from oxidative stress via depletion of 4-HNE [5]. ALDH2 dysfunction may be involved in a broad spectrum of human diseases such as cardiovascular diseases, diabetes, Fanconi anemia, pain, osteoporosis, as well as cancer [6].

A close connection between the variant ALDH2*2 and many kinds of cancer had been revealed [7]. A prognostic analysis showed that patients with aldehyde dehydrogenase 2 (ALDH2) variants had a shorter time to first recurrence in bladder cancer patients [8]. On the other hand, the metabolism of ROS and 4-HNE is thought to be deeply involved in cell death of cancer cells, and thus ALDH2 is proposed to have effect on the cancer therapy. In this chapter, we have collected these interesting clinical and translational studies on ALDH2 variant and cancer therapy (Table 1).

2 ALDH2 and Melphalan

High-dose chemotherapy in combination with stem cell transplantation has been a major breakthrough in multiple myeloma treatment, and high-dose melphalan presents a central component in the treatment of patients under the age of 65. Dumontet et al. showed the overall survival time of high-dose melphalan-treated multiple myeloma individuals in ALDH2 rs886205 vs other genotype as 3628 and 2572 days ($P = 0.01$), respectively [9]. These results indicate that the detection of mutation of this gene will contribute to the treatment and prognosis of this disease. Notably, this melphalan-related SNP is rs886205 instead of re671 the most common one in ALDH2 gene.

3 ALDH2 and Cisplatin

ALDH2 plays a major role in detoxifying acetaldehydes and ROS derived from ethanol metabolism [4]. Cisplatin, which can induce cytotoxicity partly by invoking reactive oxygen species (ROS) and DNA damaging, is a first-line chemotherapeutic agents for many solid tumors and [10]. Mice with ALDH2*1/*2 mutation demonstrate even higher ROS production under cisplatin treatment, which indicate that cisplatin usage in the cancer patients should be adjusted based on their ALDH2 variant to reduce serious side effects [11].

4 ALDH2 and Microtubule Inhibitors

Microtubule inhibitors comprise two categories: the microtubule-stabilizing drugs, such as Taxol and docetaxel, and the microtubule-destabilizing drugs, such as vincristine and vinorelbine. Microtubule inhibitors are one of the most important

Table 1 ALDH2 and cancer therapy

Genotype	Subjects	Cancer	Treatment	Phenotype	Reference(s)
ALDH2 RS886205	Patients	Multiple myeloma	High-dose melphalan	Increases the survival time in patients receiving HDM	[9]
Knockdown of ALDH2	Cell lines	Human lung cancer	Taxol and vincristine	Enhances the sensitivity to taxol and vincristine	[15]
Overexpression of ALDH2	Cell lines	Human leukemia and lung cancer	Doxorubicin	Increase the drug resistance to doxorubicin	[19]
Knockdown of ALDH2	Cell lines	Human clear cell renal cell carcinoma	Doxorubicin	Enhances the sensitivity to doxorubicin	[15]
ADH1B*2, ALDH2*1/*2	Patients	Esophageal cancer	Surgery and/or chemoradiation	Increases susceptibility to Macrocytic anemia and leukocytopenia	[26]
ALDH2*1/*2	Knock-in mice		Cisplatin	High sensitivity and cytotoxicity	[11]
ALDH2 knockout	Mice		Doxorubicin	Accentuates cardiac contractile and autophagic anomalies	[6](d)
ALDH2 overexpression	Mice			Improve cardiac function and suppress autophagy	
Agonist Alda-1	Mice		Doxorubicin	Attenuates cardiotoxicity by inhibiting oxidative stress, NOX2 expression and reducing myocardial apoptosis	[22]
Antagonist Daidzin	Mice		Doxorubicin	Aggravate cardiotoxicity	
				Oxidative stress, NOX2 expression and myocardial apoptosis	
ALDH2 transgenic mice	Mice		Doxorubicin	Rescue against doxorubicin cardiac toxicity through a TRPV1-mediated protection of mitochondrial integrity	[23]

chemotherapeutic agents for various cancers. However, drug resistance is always the barrier of these drugs to treat cancers effectively, even that some of the key mechanisms of resistance to microtubule inhibitors have been identified [12–14]. Intriguingly, Wang NN et al. [15] have found that microtubule inhibitor resistance in cancer cells can be reversed by disulfiram/copper via suppressing ALDH2 expression, which reveals a novel target to overcome the drug resistance of microtubule inhibitor.

5 ALDH2 and Doxorubicin

Anthracyclines, mainly consisting of daunorubicin (DAU), doxorubicin (DOX), epirubicin (EPI), and idarubicin (4-demethoxydaunorubicin, IDA), have been the most effective chemotherapeutics ever introduced to clinic. They still remain an essential component of modern chemotherapy regimens for a broad spectrum of human cancers, including leukemia, lymphoma, breast, bladder, and thyroid carcinoma, despite the advent of targeted therapy [16]. However, the potential of anthracyclines is severely limited by their severe cardiotoxicity of a dose-dependent cardiomyopathy, which irreversibly develops to congestive heart failure [17]. The molecular mechanisms by which anthracyclines cause cell death or cardiotoxicity are largely unknown and remain to be deeply investigated [18].

5.1 ALDH2 and the Cytotoxicity of Doxorubicin in Cancer

ALDH2 has been found to regulate the cytotoxicity of doxorubicin in cancer cells. Overexpression of ALDH2 in leukemia cell and lung cancer cell has demonstrated to promote cell proliferation and clone-forming ability and increase drug resistance to doxorubicin [19]. Wang LS et al. [20] found that von Hippel-Lindau (VHL)-deficient ccRCC cells show higher sensitivity to anthracyclines than VHL-proficient cells in vivo or in vitro. Intriguingly, hypoxia-inducible factors (HIF-αs), as the major substrates of VHL, don't contribute to the increased anthracycline cytotoxicity. Next, by comparative proteomics and following with RNAi or overexpression verification in ccRCC cells, ALDH2 is found to be transcriptionally regulated by VHL, which is independent on its E3 ubiquitin ligase activity. The VHL deficiency and thus ALDH2 suppression contribute to enhanced anthracycline cytotoxicity. As reported, hepatocyte nuclear factor 4α (HNF-4α) is a transcription factor for ALDH2 [21], and by Chip and luciferase activity analysis, the data indicated that VHL regulates ALDH2 expression through directly binding the site of −130 bp to −160 bp on the promoter to activate the transcription of HNF-4α. Furthermore, HNF-4α mediates the anthracycline cytotoxicity in ccRCC via transcriptional regulation of ALDH2. These results indicate ALDH2 regulates the sensitivity to anthracyclines (Fig. 1). In consideration that the inactive variant of ALDH2 presents in 40% of East Asian populations, it is of great interest to explore the effect of ALDH2 on anthracyclines in other cancers.

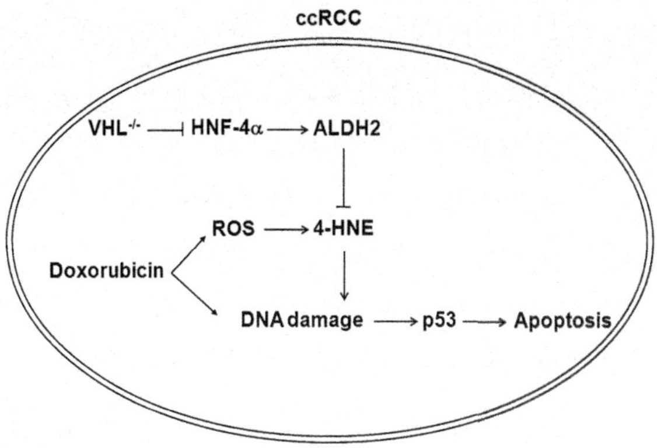

Fig. 1 Schematic diagram depicting proposed mechanism for higher sensitivity to doxorubicin in VHL-deficient ccRCC cells. It is proposed that VHL deficiency inactivates transcription of HNF-4 alpha, which decreases the expression of ALDH2. As a result, the HNE accumulates and induces the DNA damage to apoptosis. *VHL*, von Hippel-Lindau; *HNF-4*, hepatocyte nuclear factor 4 alpha; *DOX*, doxorubicin; *ROS*, reactive oxygen species; *ccRCC*, clear cell renal cell carcinoma

5.2 ALDH2 and the Cardiotoxicity of Doxorubicin in Cancer Chemotherapy

Cardiotoxicity is the major life-threatening side effect encountered in clinical cancer chemotherapy with anthracyclines. Recently, the effect ALDH2 on cytotoxicity of anthracyclines in cardiac cells has been documented by several groups. ALDH2 knockout mice have demonstrated further aggravated doxorubicin-induced myocardial cellular toxicity, while ALDH2-transfected mice have demonstrated ameliorated toxicity through detoxification of 4-HNE[6d]. Consistently, the doxorubicin plus ALDH2 inhibitor daidzin is also found to have aggravated cardiotoxicity, while doxorubicin plus ALDH2 agonist Alda-1 would partially or completely reduce the cardiotoxicity [22]. Furthermore, Ge et al. have found that doxorubicin cardiotoxicity could be reduced by ALDH2 through a transient receptor potential channel vanilloid 1-mediated mechanism in ALDH2 transgenic mice [23]. These data indicate that the DOX-induced cardiomyocyte dysfunction is regulated by ALDH2.

6 ALDH2 and Dermatitis in Cancer Radiation Therapy

Radiation therapy is one of major therapeutics for cancer. Unfortunately it has to face a debilitating clinical problem of radiation-induced dermatitis, which is a possible outcome of too high levels of radiation exposure. Intriguingly, delayed onset of radiation dermatitis and substantially reduced symptoms have been found in a clinically relevant model of radiation-induced dermatitis when the allosteric agonist, Alda-1, is used to activate ALDH2 enzyme and significantly reduce

4-hydroxynonenal adduct accumulation. However, Alda-1 has no radioprotection for tumors in mice; instead, Alda-1 increases the sensitivity to radiation therapy [24]. These results indicate that ALDH2 is a potential a novel target for the treatment of radiation dermatitis and increase the sensitivity of radiotherapy to cancer cells.

7 ALDH2 and Alcohol Consumption in Cancer Therapy

Chemotherapy always induces nausea and vomiting, which significantly affect the life quality and treatment compliance of cancer patients. Alcohol consumption demonstrates an inverse correlation with chemotherapy-induced nausea and vomiting. And thus alcohol consumption is frequently seen in cancer patients to relieve their suffering from cancer. Recently, the correlations of alcohol consumption, ALDH2 variant, and chemotherapy-induced nausea and vomiting have been investigated in 81 Japanese breast cancer individuals under adjuvant chemotherapy containing high-emetic drugs. The ALDH2*1/*2 patients who habitually consumed alcohol had less vomiting after chemotherapy even without anti-emetics, which indicated the important role of ALDH2 variant [25].

However, the inactivation of ALDH2 is also thought to increase susceptibility to macrocytic anemia and leukocytopenia in alcoholics. Another study in Japan esophageal cancer patients indicated that mean values of hemoglobin and hematocrit were the lowest, and those of mean corpuscular volume were markedly the highest in the group with combination of alcohol dehydrogenase-1B (ADH1B*2 allele) and inactive heterozygous aldehyde dehydrogenase-2 (ALDH2*1/*2) when they drink alcohol. And they return to normal, remaining the significant intergroup differences, after abstinence. Under alcohol admission, this group has lowest mean leukocyte count. This group has highest frequencies of MCV, hemoglobin levels, hemoglobin levels, and leukocytopenia [26].

8 Conclusion and Perspective

The differences of human ALDH2 genotype, such as the ALDH2 rs671, present in East Asian people widely, which suggest a personalized medicine based on an ALDH2 genotype may shed light on the individual health by evaluation of potential disease risk and medication efficacy. A seamless clinical and translational effort needs clinicians and scientists with common interest to join together, which could maximize the health and wellness of the population.

On the other hand, ALDH2, even in high abundance, stands behind the other genes and is occupied by nothing but alcohol metabolism at most times. In fact, the biological roles of ALDH2 are largely underestimated previously. Very recently, ALDH2 began to demonstrate its great biological roles under the given certain condition. For example, ALDH2 is found to play important roles in aplastic anemia and leukemia when it is knock out in combination with Fanconi gene [27]. In addition, animal models with specific variants, such as the transgenic knock-in mouse for ALDH2*2, have been created [4], which can verify the observations from the clinic.

Additionally, Alda-1, the small molecule restoring enzymatic activity to the ALDH2*2 enzyme, has the potential to reverse some pathophysiology in the ALDH2*2 population [5]. Such agonist could be used to prevent or modify ALDH2-related diseases. In addition, CRISPR-Cas9, as a useful gene editing method, will be potential to apply to correct ALDH2 mutation, restore the normal functions of ALDH2, and thus lower the cancer risk [7].

In summary, more and more attentions have been attracted to the biological, medical, and chemical research of ALDH2. We believe that ALDH2 has the potential to play important roles in cancer screening, diagnosis, therapeutics, and prognosis in the near future.

References

1. (a) Shao C, Sullivan JP, Girard L, Augustyn A, Yenerall P, Rodriguez-Canales J, Liu H, Behrens C, Shay JW, Wistuba II, Minna JD (2014) Clin Cancer Res 20:4154–4166; (b) Liu X, Wang L, Cui W, Yuan X, Lin L, Cao Q, Wang N, Li Y, Guo W, Zhang X, Wu C, Yang J (2016) Oncotarget 7:58516–58530
2. Gross ER, Zambelli VO, Small BA, Ferreira JC, Chen CH, Mochly-Rosen D (2015) Annu Rev Pharmacol Toxicol 55:107–127
3. Li H, Borinskaya S, Yoshimura K, Kal'ina N, Marusin A, Stepanov VA, Qin Z, Khaliq S, Lee MY, Yang Y, Mohyuddin A, Gurwitz D, Mehdi SQ, Rogaev E, Jin L, Yankovsky NK, Kidd JR, Kidd KK (2009) Ann Hum Genet 73:335–345
4. Chen CH, Ferreira JC, Gross ER, Mochly-Rosen D (2014) Physiol Rev 94:1–34
5. Chen CH, Budas GR, Churchill EN, Disatnik MH, Hurley TD, Mochly-Rosen D (2008) Science 321:1493–1495
6. (a) Burke WJ, Li SW, Williams EA, Nonneman R, Zahm DS (2003) Brain Res 989:205–213; (b) Chan JC, Malik V, Jia W, Kadowaki T, Yajnik CS, Yoon KH, Hu FB (2009) JAMA 301:2129–2140; (c) Wenzel P, Muller J, Zurmeyer S, Schuhmacher S, Schulz E, Oelze M, Pautz A, Kawamoto T, Wojnowski L, Kleinert H, Munzel T, Daiber A (2008) Biochem Biophys Res Commun 367:137–143; (d) Sun A, Cheng Y, Zhang Y, Zhang Q, Wang S, Tian S, Zou Y, Hu K, Ren J, Ge J (2014) J Mol Cell Cardiol 71:92–104
7. Li R, Zhao Z, Sun M, Luo J, Xiao Y (2016) Life Sci 147:59–66
8. Andrew AS, Gui J, Hu T, Wyszynski A, Marsit CJ, Kelsey KT, Schned AR, Tanyos SA, Pendleton EM, Ekstrom RM, Li Z, Zens MS, Borsuk M, Moore JH, Karagas MR (2015) BJU Int 115:238–247
9. Dumontet C, Landi S, Reiman T, Perry T, Plesa A, Bellini I, Barale R, Pilarski LM, Troncy J, Tavtigian S, Gemignani F (2010) Bone Marrow Transplant 45:1316–1324
10. Bar-Ad V, Palmer J, Yang H, Cognetti D, Curry J, Luginbuhl A, Tuluc M, Campling B, Axelrod R (2014) Semin Oncol 41:798–806
11. Kim J, Chen CH, Yang J, Mochly-Rosen D (2017) J Biomed Sci 24:33
12. (a) Yabuki N, Sakata K, Yamasaki T, Terashima H, Mio T, Miyazaki Y, Fujii T, Kitada K (2007) Cancer Genet Cytogenet 173:1–9; (b) Sato H, Fukumoto K, Hada S, Hagiwara H, Fujimoto E, Negishi E, Ueno K, Yano T (2007) Cancer Chemother Pharmacol 60:449–457

13. (a) Kavallaris M, Burkhart CA, Horwitz SB (1999) Br J Cancer 80:1020–1025; (b) Seve P, Isaac S, Tredan O, Souquet PJ, Pacheco Y, Perol M, Lafanechere L, Penet A, Peiller EL, Dumontet C (2005) Clin Cancer Res 11:5481–5486
14. Dean M, Fojo T, Bates S (2005) Nat Rev Cancer 5:275–284
15. Wang NN, Wang LH, Li Y, Fu SY, Xue X, Jia LN, Yuan XZ, Wang YT, Tang X, Yang JY, Wu CF (2018) Exp Cell Res 362:72–82
16. (a) Minotti G, Menna P, Salvatorelli E, Cairo G, Gianni L (2004) Pharmacol Rev 56:185–229; (b) Ewer MS, Ewer SM (2010) Nat Rev Cardiol 7:564–575
17. (a) Gianni L, Herman EH, Lipshultz SE, Minotti G, Sarvazyan N, Sawyer DB (2008) J Clin Oncol 26:3777–3784; (b) Sawyer DB (2013) N Engl J Med 368:1154–1156
18. (a) Mordente A, Meucci E, Silvestrini A, Martorana GE, Giardina B (2009) Curr Med Chem 16:1656–1672; (b) Ichikawa Y, Ghanefar M, Bayeva M, Wu R, Khechaduri A, Naga Prasad SV, Mutharasan RK, Naik TJ, Ardehali H (2014) J Clin Invest 124:617–630
19. Moreb JS, Ucar D, Han S, Amory JK, Goldstein AS, Ostmark B, Chang LJ (2012) Chem Biol Interact 195:52–60
20. Gao YH, Wu ZX, Xie LQ, Li CX, Mao YQ, Duan YT, Han B, Han SF, Yu Y, Lu HJ, Yang PY, Xu TR, Xia JL, Chen GQ, Wang LS (2017) Nat Commun 8:15337
21. (a) Stewart MJ, Dipple KM, Estonius M, Nakshatri H, Everett LM, Crabb DW (1998) Biochim Biophys Acta 1399:81–186; (b) You M, Fischer M, Cho WK, Crabb D (2002) Arch Biochem Biophys 398:79–86
22. Gao Y, Xu Y, Hua S, Zhou S, Wang K (2015) Int J Clin Exp Med 8:6794–6803
23. Ge W, Yuan M, Ceylan AF, Wang X, Ren J (2016) Biochim Biophys Acta 1862:622–634
24. Ning S, Budas GR, Churchill EN, Chen CH, Knox SJ, Mochly-Rosen D (2012) Radiat Res 178:69–74
25. Uomori T, Horimoto Y, Mogushi K, Matsuoka J, Saito M (2017) Breast Cancer 24:702–707
26. Yokoyama A, Brooks PJ, Yokoyama T, Mizukami T, Shiba S, Nakamoto N, Maruyama K (2017) Jpn J Clin Oncol 47:306–312
27. Langevin F, Crossan GP, Rosado IV, Arends MJ, Patel KJ (2011) Nature 475:53–58

ALDH2 Polymorphism and Ethanol Consumption: A Genetic-Environmental Interaction in Carcinogenesis

Mingjie Yang, Yingmei Zhang, and Jun Ren

Abstract Aldehyde dehydrogenase 2 (ALDH2) is a key enzyme in the detoxification of toxic aldehydes, especially acetaldehyde, which is commonly considered as a carcinogen. ALDH2 mutation and impaired enzymatic activity will cause acetaldehyde accumulation and thus participate in the development of cancers. It deserves more attention since around 40% of East Asian population carry the inactive ALDH2 allele. Moreover, the risk for cancers will be even higher when ALDH2 mutation combined with heavy alcohol consumption, suggesting a genetic-environmental interaction in carcinogenesis. This may provide us with a potential target for cancer prevention and treatment.

Keywords ALDH2 mutation · Alcohol consumption · Cancer risk · Neoplasm

1 ALDH2 Polymorphism and Enzymatic Activity

Aldehyde dehydrogenase 2 (ALDH2) is one of 19 members of aldehyde dehydrogenase family in human, which is encoded by ALDH2 gene on chromosome 12 and located in mitochondrial matrix. Structurally, it is a tetrameric enzyme composed of 517 amino acids. The most common genetic mutation is G1510A, which means substituting glutamate with lysine at position 504 (known as Glu504Lys, E504K, rs671, or Glu487Lys, E487K before). The wild-type allele 504Glu is noted as ALDH2*1 and the mutated allele 504Lys as ALDH2*2.

M. Yang
Department of Cardiology and Shanghai Institute of Cardiovascular Diseases, Zhongshan Hospital, Fudan University, Shanghai, People's Republic of China

Y. Zhang · J. Ren (✉)
Department of Cardiology, Shanghai Institute of Cardiovascular Diseases, Zhongshan Hospital, Fudan University, Shanghai, China

Center for Cardiovascular Research and Alternative Medicine, University of Wyoming College of Health Sciences, Laramie, WY, USA
e-mail: jren@uwyo.edu

J. Ren et al. (eds.), *Aldehyde Dehydrogenases*, Advances in Experimental Medicine and Biology 1193, https://doi.org/10.1007/978-981-13-6260-6_14

ALDH2 and alcohol dehydrogenase (ADH) are two major enzymes involved in ethanol metabolism in the liver. Ethanol was first catalyzed to acetaldehyde by ADH and then oxidized into acetic acid by ALDH2. Aside from acetaldehyde, ALDH2 also participates in the oxidation of other aldehydes including 4-hydroxy-2-nonenal (4-HNE). Besides, it is also involved in the conversion of nitroglycerin, which means turning glyceryl trinitrate into 1,2-glyceryl dinitrate [26].

Due to its important role in ethanol metabolism, ALDH2 and ADH polymorphism inevitably affect its enzymatic activity and can be either pathogenic or protective to human health, which is depicted in Table 1. Epidemiological studies show that up to 40% of East Asian populations are carrying heterozygous ALDH2*1/*2 genotype, which only have 30–40% enzymatic capacity as compared to Caucasians with ALDH2*1/*1 genotype [18]. Therefore, they are more prone to alcoholic com-

Table 1 Part of single nucleotide variants of ALDH2 and ADH1B genes

Variant ID	Location	Gene	Molecular consequences	Clinical significance	1000G MAF
rs886205	111,766,623	ALDH2	2KB upstream variant		A = 0.4912
rs4767939	111,769,091	ALDH2	Intron variant		G = 0.4197
rs4767944	111,771,537	ALDH2	Intron variant		T = 0.1619
rs2238151	111,774,029	ALDH2	Intron variant		T = 0.3013
rs2238152	111,776,655	ALDH2	Intron variant		T = 0.2017
rs7312055	111,782,373	ALDH2	Intron variant		A = 0.0891
rs4648328	111,784,984	ALDH2	Intron variant		T = 0.2003
rs7311852	111,787,500	ALDH2	Intron variant		G = 0.0202
rs10849970	111,788,629	ALDH2	Intron variant		A = 0.1625
rs440	111,790,910	ALDH2	Intron variant		C = 0.1987
rs441	111,791,045	ALDH2	Intron variant		C = 0.2021
rs4646776	111,792,215	ALDH2	Intron variant		C = 0.0353
rs4646777	111,792,232	ALDH2	Intron variant		A = 0.1977
rs10744777	111,795,214	ALDH2	Intron variant		T = 0.2969
rs968529	111,796,564	ALDH2	Intron variant		T = 0.0302
rs671	111,803,962	ALDH2	Missense variant	Pathogenic [19]	A = 0.0357
rs2158029	111,804,401	ALDH2	Intron variant		A = 0.1841
rs16941667	111,806,609	ALDH2	Intron variant		T = 0.0713
rs11066028	111,807,366	ALDH2	Intron variant		A = 0.3728
rs16941669	111,807,833	ALDH2	Intron variant		G = 0.1250
rs7296651	111,809,150	ALDH2	Intron variant		C = 0.4978
rs2066702	99,307,860	ADH1B	Missense variant	Protective [23]	A = 0.0531
rs7673353	99,309,919	ADH1B	Intron variant		T = 0.0533
rs10033960	99,310,185	ADH1B	Intron variant		A = 0.3650
rs17028834	99,311,443	ADH1B	Intron variant		C = 0.0423
rs1229985	99,311,721	ADH1B	Intron variant		G = 0.0333
rs6850217	99,312,463	ADH1B	Intron variant		T = 0.1739
rs1789882	99,313,896	ADH1B	Synonymous variant		A = 0.1703

(continued)

Table 1 (continued)

Variant ID	Location	Gene	Molecular consequences	Clinical significance	1000G MAF
rs2018417	99,313,983	ADH1B	Synonymous variant		A = 0.0190
rs28626993	99,314,037	ADH1B	Synonymous variant		A = 0.0220
rs1789883	99,315,218	ADH1B	Intron variant		A = 0.0236
rs1693457	99,315,605	ADH1B	Intron variant		C = 0.1733
rs2066701	99,317,256	ADH1B	Intron variant		A = 0.3672
rs2075633	99,317,841	ADH1B	Intron variant		C = 0.3674
rs4147536	99,317,955	ADH1B	Intron variant		A = 0.2594
rs1229984	99,318,162	ADH1B	Synonymous variant, missense variant	Protective [7]	T = 0.1585
rs1229983	99,318,845	ADH1B	Synonymous variant, 5 prime UTR variant		C = 0.0343
rs1235416	99,319,764	ADH1B	Intron variant		T = 0.0236
rs1353621	99,320,418	ADH1B	Intron variant		C = 0.1589
rs698	99,339,632	ADH1C	Noncoding transcript variant, missense variant	Protective [13]	C = 0.2143
rs1693482	99,342,808	ADH1C	Intron variant, missense variant	Protective	T = 0.2143
rs283413	99,347,033	ADH1C	Synonymous variant, noncoding transcript variant, nonsense	Risk factor	A = 0.0072
rs932546861	99,347,033	ADH1C	Noncoding transcript variant, nonsense, frameshift variant	Risk factor	

According to https://www.ncbi.nlm.nih.gov/variation/view/

plications such as facial flushing, headaches, nausea, and cardiac palpitations after drinking alcohol, because of high acetaldehyde concentration in blood.

Studies have shown ALDH2 dysfunction is correlated with different kinds of diseases, including cardiovascular and cerebrovascular diseases, diabetes mellitus, neurodegenerative diseases, and cancers [4]. As chronic alcohol consumption is considered a risk factor for neoplasms, ALDH2 deficiency will inevitably affect the development of cancers, especially those in the esophagus, stomach, colorectum, and liver. A meta-analysis which included 16,774 patients and 32,060 controls showed people with ALDH2 rs671 mutation were at 20% higher risk for developing cancers, compared to those with wild-type alleles (Odds ratio/OR = 1.20, 95% Confidence interval/CI: 1.03-1.39) [3].

2 ALDH2 Polymorphism and Upper Aerodigestive Track Cancers

A study showed that ALDH2 rs671 mutation increases 39% risk for carcinogenesis in upper aerodigestive track (UADT), including oral cavity, oropharynx, hypopharynx, larynx, and esophagus (95% CI, 1.11–1.73) [3]. A case-control study including

585 patients and 1170 matched controls in Japan also confirmed that ALDH2 504Lys allele increased the risk of UADT cancers, especially among those heavy drinkers (drink ≥46 g/d and ≥5 days/week). The ORs for heavy drinkers with or without 504Lys allele were 14.41 and 1.59 compared to non-drinkers [17]. This may suggest that ALDH2 polymorphism affects the development of UADT cancers due to its role in ethanol metabolism.

Esophageal cancer ranks first in the morbidity and mortality among upper aerodigestive cancers [2], which deserves special attention. ALDH2 rs671 mutation carriers were at 7–12 times higher risk for esophageal cancer, whether they were drinkers or not [12]. Another case-control study that followed 102 patients and 241 controls from Aichi Cancer Center suggested that the OR for esophageal cancer in ALDH2 504Lys allele carriers was 3.43 compared to wild-type controls (95% CI, 1.74–6.75) after adjusting the age, sex, smoking, and drinking conditions [14].

Further studies show the susceptibility for esophageal cancer significantly increases if these mutation carriers drink heavily. In the Aichi study, the OR for heavy drinkers with ALDH2 504Lys was 6.84 (95% CI, 2.39–19.6), much higher than the average risk [14]. In a similar case-control study with 858 patients and 1081 controls in China, those moderate or heavy drinkers with ALDH2 rs671 mutation were at 64% higher risk for esophageal cancer (95% CI, 1.12–2.40), while not the same case in light or non-drinkers [22]. Another Chinese cohort study also reached the same conclusion. Impaired ALDH2 enzymatic activity was positively associated with esophageal cancer risk among those daily drinkers, but not in light drinkers [25]. All of these studies show a strong interaction between ALDH2 polymorphism and ethanol consumption in developing esophageal cancer.

Apart from alcohol, cigarette is also another important source of acetaldehyde. Smoking together with heavy drinking made things even worse for those ALDH2 504Lys carriers, with OR 50.1 for esophageal squamous cell carcinoma relative to those wild-type non-smokers and non-drinkers [16]. A Japanese genome-wide study including 1070 patients and 2836 controls revealed that ALDH2 504Lys mutation, drinking, and smoking were all independent risk factors for esophageal cell carcinoma, with ORs 1.66, 1.92, and 1.79, respectively. However, the risk was 190-folds higher if these risk factors combined [6]. It suggests that genetic and environmental risk factors are closely interacted during carcinogenesis.

3 ALDH2 Polymorphism and Gastric Cancer

The incidence and mortality of stomach cancer ranks high among different types of cancers worldwide, especially in Eastern Asia like Japan, Korea, and China [2, 21]. Therefore, studies have been carried out in these countries on the relationship of alcohol consumption and ALDH2 polymorphism in developing gastric cancer.

Chinese researchers found ALDH2*2/*2 mutation significantly raised the risk of stomach cancer in comparison with wild-type controls [5]. A Japanese case-control study including 697 cases and 1372 controls also showed the ORs for Glu/Lys and

Lys/Lys genotypes are 1.40 (95% CI, 1.11–1.76) and 1.73 (95% CI, 1.12–2.68), relative to Glu/Glu carriers. The risk was even higher among heavy drinkers with mutation (OR = 3.93; 95% CI, 1.99–5.79) compared to non-drinkers without mutation [15]. Other studies in Japan also suggested ALDH2 504Lys mutation elevated the risk of gastric cancer under heavy ethanol ingestion [8, 10]. A Korean study including 445 older patients and 370 controls showed that heavy drinkers with ALDH2*1/*2 genotype were 4 times easier to develop gastric cancer, relative to those without rs671 mutation (OR = 4.26; 95% CI, 1.10–16.47), while the risk didn't increase in those drank less [20].

These studies suggest that the amount of ethanol consumption must be considered when considering the relationship between ALDH2 polymorphism and susceptibility of stomach cancer.

4 ALDH2 Polymorphism and Colorectal Cancer

In a case-control study of 440 cases and 800 controls in Southwest China, the OR of developing colorectal cancer was 1.86 (95% CI, 1.12–3.09) for those ALDH2 504Lys/Lys carriers compared to Glu/Glu genotypes. If the amount of alcohol consumption is taken into consideration, the OR of heavy drinkers turned out to be 7.17 (95% CI, 2.01–25.53) [24].

However, the conclusion is absolutely different according to several meta-analysis. In a study comprising 2909 patients and 4903 controls, the cancer risk decreased in those with ALDH2 504Lys mutation compared with wild-type controls (OR = 0.81, 95% CI = 0.68–0.98) [27]. Another meta-analysis including 1664 patients and 2777 controls showed a similar conclusion, and the ORs for ALDH2*1/*2 and ALDH2*2/*2 are 0.85 and 0.95 relative to wild-type alleles [28]. Further expanded study is required to clarify the relationship of ALDH2 rs671 polymorphism and colorectal cancer risk.

5 ALDH2 Polymorphism and Hepatocellular Carcinoma

Heavy and persistent alcohol consumption contributes to liver cirrhosis, even without hepatitis B virus (HBV) or hepatitis C virus (HCV) infection. A Japanese study comprising 67 patients and matched controls suggested ALDH2 rs671 mutation was an independent risk factor for carcinogenesis in the liver [1].

A research group from China also detected decreased level of ALDH2 in human hepatocellular carcinoma (HCC) tissues, with a poorer prognosis. Defective ALDH2 enzyme led to accumulation of acetaldehyde and activated AMP-activated protein kinase (AMPK) pathway through a redox-dependent mechanism [9]. ALDH2 E504K-mutated mice showed more severe DNA damage in hepatocytes and were more prone to develop HCC if treated with a chemical carcinogen [11]. Since the liver is the most important organ where acetaldehyde is detoxicated by ALDH2, the enzyme is considered protective in the development and migration of hepatocellular carcinoma.

6 Conclusion

Most human studies have suggested ALDH2 504Lys allele disturbs ethanol metabolism and thus increases the risk of carcinogenesis, particularly in the heavy drinkers. Obviously, the genetic risk factor ALDH2 mutation and environmental risk factor alcohol consumption are closely interacted in this pathological process. The linkage is acetaldehyde, a carcinogen causing DNA damage. Studies showed ALDH2 level increased in the wild-type mice to eliminate acetaldehyde upon drinking. However, in ALDH2 knockout mice with alcohol feeding, acetaldehyde-derived DNA damage deteriorated as represented by the elevation of N(2)-ethyl-2′-deoxyguanosine in the esophagus, stomach, and liver [4].

Therefore, ALDH2 is regarded not only as a novel biological marker for predicting and assessing cancer risks but a potential therapeutic target for cancers especially in the esophagus and stomach. In the East Asian population with impaired ALDH2 enzyme, ALDH2 activators may help reduce the toxicity of acetaldehyde after alcohol drinking, thus reducing the risk for developing neoplasms.

However, most researches concentrate on the role of ALDH2 enzyme in gastrointestinal cancers and hepatic carcinoma, and little attention has been paid to other cancers. Lung cancer is commonly considered as smoking-related, and cigarette is another important source of acetaldehyde. A Japanese case-control study showed impaired ALDH2 enzyme increased cumulative smoking dose and the incidence of lung cancer [25]. Another case-control study revealed that ALDH2 504Glu/Lys raised the bladder cancer risk compared to wild-type alleles (OR = 2.03; 95% CI, 1.14–3.62), but for drinkers only [26].

Due to the extremely high mutation rate in East Asian population, it is of great significance to clarify the relationship between ADLH2 polymorphism and carcinogenesis, and far more efforts are needed from bench to bedside.

References

1. Abe H, Aida Y, Seki N, Sugita T, Tomita Y, Nagano T, Itagaki M, Sutoh S, Nagatsuma K, Itoh K, Matsuura T, Aizawa Y (2015) Aldehyde dehydrogenase 2 polymorphism for development to hepatocellular carcinoma in East Asian alcoholic liver cirrhosis. J Gastroenterol Hepatol 30(9):1376–1383. https://doi.org/10.1111/jgh.12948
2. Bray F, Jemal A, Grey N, Ferlay J, Forman D (2012) Global cancer transitions according to the Human Development Index (2008–2030): a population-based study. Lancet Oncol 13(8):790–801. https://doi.org/10.1016/s1470-2045(12)70211-5

3. Cai Q, Wu J, Cai Q, Chen EZ, Jiang ZY (2015) Association between Glu504Lys polymorphism of ALDH2 gene and cancer risk: a meta-analysis. PLoS One 10(2):e0117173. https://doi.org/10.1371/journal.pone.0117173
4. Chen CH, Ferreira JC, Gross ER, Mochly-Rosen D (2014) Targeting aldehyde dehydrogenase 2: new therapeutic opportunities. Physiol Rev 94(1):1–34. https://doi.org/10.1152/physrev.00017.2013
5. Chen ZH, Xian JF, Luo LP (2016) Analysis of ADH1B Arg47His, ALDH2 Glu487Lys, and CYP4502E1 polymorphisms in gastric cancer risk and interaction with environmental factors. Genet Mol Res 15(4). https://doi.org/10.4238/gmr15048904
6. Cui R, Kamatani Y, Takahashi A, Usami M, Hosono N, Kawaguchi T, Tsunoda T, Kamatani N, Kubo M, Nakamura Y, Matsuda K (2009) Functional variants in ADH1B and ALDH2 coupled with alcohol and smoking synergistically enhance esophageal cancer risk. Gastroenterology 137(5):1768–1775
7. Galinsky KJ, Bhatia G, Loh PR, Georgiev S, Mukherjee S, Patterson NJ, Price AL (2016) Fast principal-component analysis reveals convergent evolution of ADH1B in Europe and East Asia. Am J Hum Genet 98(3):456–472. https://doi.org/10.1016/j.ajhg.2015.12.022
8. Hidaka A, Sasazuki S, Matsuo K, Ito H, Sawada N, Shimazu T, Yamaji T, Iwasaki M, Inoue M, Tsugane S (2015) Genetic polymorphisms of ADH1B, ADH1C and ALDH2, alcohol consumption, and the risk of gastric cancer: the Japan Public Health Center-based prospective study. Carcinogenesis 36(2):223–231. https://doi.org/10.1093/carcin/bgu244
9. Hou G, Chen L, Liu G, Li L, Yang Y, Yan H, Zhang H, Tang J, Yang Y, Lin X, Chen X, Luo G, Zhu Y, Tang S, Zhang J, Liu H, Gu Q, Zhao L, Li Y, Liu L, Zhou W, Wang H (2017) Aldehyde dehydrogenase-2 (ALDH2) opposes hepatocellular carcinoma progression by regulating AMP-activated protein kinase signaling in mice. Hepatology 65(5):1628–1644
10. Ishioka K, Masaoka H, Ito H, Oze I, Ito S, Tajika M, Shimizu Y, Niwa Y, Nakamura S, Matsuo K (2018) Association between ALDH2 and ADH1B polymorphisms, alcohol drinking and gastric cancer: a replication and mediation analysis. Gastric Cancer 21:936–945
11. Jin S, Chen J, Chen L, Histen G, Lin Z, Gross S, Hixon J, Chen Y, Kung C, Chen Y, Fu Y, Lu Y, Lin H, Cai X, Yang H, Cairns RA, Dorsch M, Su SM, Biller S, Mak TW, Cang Y (2015) ALDH2(E487K) mutation increases protein turnover and promotes murine hepatocarcinogenesis. Proc Natl Acad Sci U S A 112(29):9088–9093. https://doi.org/10.1073/pnas.1510757112
12. Kutler DI, Auerbach AD, Satagopan J, Giampietro PF, Batish SD, Huvos AG, Goberdhan A, Shah JP, Singh B (2003) High incidence of head and neck squamous cell carcinoma in patients with Fanconi anemia. Arch Otolaryngol Head Neck Surg 129(1):106–112
13. Lawlor DA, Benn M, Zuccolo L, De Silva NM, Tybjaerg-Hansen A, Smith GD, Nordestgaard BG (2014) ADH1B and ADH1C genotype, alcohol consumption and biomarkers of liver function: findings from a Mendelian randomization study in 58,313 European origin Danes. PLoS One 9(12):e114294. https://doi.org/10.1371/journal.pone.0114294
14. Matsuo K, Hamajima N, Shinoda M, Hatooka S, Inoue M, Takezaki T, Tajima K (2001) Gene-environment interaction between an aldehyde dehydrogenase-2 (ALDH2) polymorphism and alcohol consumption for the risk of esophageal cancer. Carcinogenesis 22(6):913–916
15. Matsuo K, Oze I, Hosono S, Ito H, Watanabe M, Ishioka K, Ito S, Tajika M, Yatabe Y, Niwa Y, Yamao K, Nakamura S, Tajima K, Tanaka H (2013) The aldehyde dehydrogenase 2 (ALDH2) Glu504Lys polymorphism interacts with alcohol drinking in the risk of stomach cancer. Carcinogenesis 34(7):1510–1515. https://doi.org/10.1093/carcin/bgt080
16. Morita M, Kumashiro R, Kubo N, Nakashima Y, Yoshida R, Yoshinaga K, Saeki H, Emi Y, Kakeji Y, Sakaguchi Y, Toh Y, Maehara Y (2010) Alcohol drinking, cigarette smoking, and the development of squamous cell carcinoma of the esophagus: epidemiology, clinical findings, and prevention. Int J Clin Oncol 15(2):126–134. https://doi.org/10.1007/s10147-010-0056-7
17. Oze I, Matsuo K, Hosono S, Ito H, Kawase T, Watanabe M, Suzuki T, Hatooka S, Yatabe Y, Hasegawa Y, Shinoda M, Tajima K, Tanaka H (2010) Comparison between self-reported facial flushing after alcohol consumption and ALDH2 Glu504Lys polymorphism for risk of upper aerodigestive tract cancer in a Japanese population. Cancer Sci 101(8):1875–1880. https://doi.org/10.1111/j.1349-7006.2010.01599.x

18. Pang JJ, Barton LA, Chen YG, Ren J (2015) Mitochondrial aldehyde dehydrogenase in myocardial ischemia-reperfusion injury: from bench to bedside. Sheng Li Xue Bao: [Acta Physiologica Sinica] 67(6):535–544
19. Quillen EE, Chen XD, Almasy L, Yang F, He H, Li X, Wang XY, Liu TQ, Hao W, Deng HW, Kranzler HR, Gelernter J (2014) ALDH2 is associated to alcohol dependence and is the major genetic determinant of "daily maximum drinks" in a GWAS study of an isolated rural Chinese sample. Am J Med Genet B Neuropsychiatr Genet 165b(2):103–110. https://doi.org/10.1002/ajmg.b.32213
20. Shin CM, Kim N, Cho SI, Kim JS, Jung HC, Song IS (2011) Association between alcohol intake and risk for gastric cancer with regard to ALDH2 genotype in the Korean population. Int J Epidemiol 40(4):1047–1055. https://doi.org/10.1093/ije/dyr067
21. Torre LA, Bray F, Siegel RL, Ferlay J, Lortet-Tieulent J, Jemal A (2015) Global cancer statistics, 2012. CA Cancer J Clin 65(2):87–108. https://doi.org/10.3322/caac.21262
22. Wu M, Chang SC, Kampman E, Yang J, Wang XS, Gu XP, Han RQ, Liu AM, Wallar G, Zhou JY, Kok FJ, Zhao JK, Zhang ZF (2013) Single nucleotide polymorphisms of ADH1B, ADH1C and ALDH2 genes and esophageal cancer: a population-based case-control study in China. Int J Cancer 132(8):1868–1877. https://doi.org/10.1002/ijc.27803
23. Xu K, Kranzler HR, Sherva R, Sartor CE, Almasy L, Koesterer R, Zhao H, Farrer LA, Gelernter J (2015) Genomewide association study for maximum number of alcoholic drinks in European Americans and African Americans. Alcohol Clin Exp Res 39(7):1137–1147. https://doi.org/10.1111/acer.12751
24. Yang H, Zhou Y, Zhou Z, Liu J, Yuan X, Matsuo K, Takezaki T, Tajima K, Cao J (2009) A novel polymorphism rs1329149 of CYP2E1 and a known polymorphism rs671 of ALDH2 of alcohol metabolizing enzymes are associated with colorectal cancer in a southwestern Chinese population. Cancer Epidemiol Biomarkers Prev 18(9):2522–2527. https://doi.org/10.1158/1055-9965.epi-09-0398
25. Yu C, Guo Y, Bian Z, Yang L, Millwood I, Walters R, Chen Y, Chen Y, Zhang X, Lei Y, Chen J, Chen Z, Lv J, Li L (2018) Association of low-activity ALDH2 and alcohol consumption with risk of esophageal cancer in Chinese adults: a population-based cohort study. Int J Cancer 143:1652–1661
26. Zhang Y, Ren J (2011) ALDH2 in alcoholic heart diseases: molecular mechanism and clinical implications. Pharmacol Ther 132(1):86–95. https://doi.org/10.1016/j.pharmthera.2011.05.008
27. Zhao H, Liu KJ, Lei ZD, Lei SL, Tian YQ (2014) Meta-analysis of the aldehyde dehydrogenases-2 (ALDH2) Glu487Lys polymorphism and colorectal cancer risk. PLoS One 9(2):e88656. https://doi.org/10.1371/journal.pone.0088656
28. Zhao S, Du XM, Ma SS, Wang LM (2016) Association between aldehyde dehydrogenase 2 (ALDH2) Glu504Lys polymorphism and susceptibility to colorectal cancer: a meta-analysis. Genet Mol Res 15(3). https://doi.org/10.4238/gmr.15037872

Aldehyde Dehydrogenase 2 (ALDH2) and Aging: Is There a Sensible Link?

Ne Natalie Wu and Jun Ren

Abstract Aging is a complex irreversible biological process associated with increased prevalence of chronic disease and high healthcare burden. Several theories have been proposed for the biology of aging including free radical accumulation, DNA damage, apoptosis, telomere shortening, autophagy failure, and disturbed autonomic response. Aging is also closely associated with progressive deterioration of cardiovascular and neurological functions. Linkage, genome-wide association (GWAS), and next-generation sequencing analysis have confirmed a number of susceptibility loci for aging, in particular, Alzheimer's disease. Recent evidence from our group and others also revealed a tie between genetic mutation of mitochondrial aldehyde dehydrogenase (ALDH2) and life span as well as cardiovascular aging. ALDH2 represents the single most gene with the greatest number of human genetic polymorphism and is deemed an important enzyme for detoxification of reactive aldehydes. Here, we will briefly review the tie between ALDH2 and cardiovascular aging process. While recent work on ALDH2 research has broadened the pathogenic mechanisms of ALDH2 mutation or deficiency, therapeutic potential targeting ALDH2 in the elderly still remains debatable.

Keywords ALDH2 · Aging · Mitochondria · Oxidative stress · Autophagy

N. N. Wu
Department of Cardiology, Shanghai Institute of Cardiovascular Diseases, Zhongshan Hospital, Fudan University, Shanghai, China

J. Ren (✉)
Department of Cardiology, Shanghai Institute of Cardiovascular Diseases, Zhongshan Hospital, Fudan University, Shanghai, China

Center for Cardiovascular Research and Alternative Medicine, University of Wyoming College of Health Sciences, Laramie, WY, USA
e-mail: jren@uwyo.edu

J. Ren et al. (eds.), *Aldehyde Dehydrogenases*, Advances in Experimental Medicine and Biology 1193, https://doi.org/10.1007/978-981-13-6260-6_15

1 Introduction

Aging is an inevitable biological process, characterized by age-dependent progressive deterioration of cellular and organismal function, onset of multiple diseases, and ultimately increased mortality [1]. With the prolonged human life expectancy over the past century, the elderly population (>65 years of age) is steadily rising and is estimated to be at ~22% by 2050 as opposed to 10% in 2000 [2]. Searching for ways to improve the health span or healthy aging in particular management of aging-associated diseases (e.g., cancer, diabetes mellitus, heart failure, vascular stiffness and dysfunction) has become a burning issue worldwide to reduce healthcare cost with the aging population [3–5]. Accumulating evidence has depicted a close association between aging and genetic or epigenetic alterations [6, 7]. A number of (>20) susceptibility loci have been identified using genome-wide association (GWAS) analysis for the increased prevalence of aging-associated comorbidities such as Alzheimer's disease [8]. A number of theories have been proposed for the biology of aging including telomere shortening, cellular senescence, mitochondrial dysfunction, defective autophagy, and the most classical one being the oxidative stress theory [9]. Aging is a process determined by a combined effect of the complex cellular and molecular signaling networks interacting with environmental factors. Modulation of oxidative stress is considered as a critical mechanism in maintaining cellular homeostasis and the biology of aging process [10]. Declines in the capacity of resistance against exogenous harm and recycling of cellular components are hallmarks in the aging process. With the "wear and tear" aspects of aging on individual's well-being, a better understanding of the mechanism behind the aging process should provide promising targets for therapeutics against aging-related diseases and moreover life span extension. Recent evidence from our group and others has suggested a unique role for mitochondrial acetaldehyde dehydrogenase 2 (ALDH2), a detoxification enzyme in ethanol metabolism, in health and disease prevalence [11–15]. Moreover, ALDH2 rs671 mutation (ALDH2*2) represents the single most common point mutation in human among all human genomes. In recent years, the linkage between this single point mutation and risk for aging-related diseases was indicated from epidemiological studies along with experimental findings, attracting much interest in the mechanism behind ALDH2 and aging or aging complications. While some reports revealed a protective role for ALDH2 through detoxifying oxidative stress-induced lipid-derived aldehyde products [16], we and others found surprising unfavorable effects of ALDH2 in life span and cardiac aging. Data from our own lab found ALDH2 overexpression accentuated myocardial dysfunction through the JNK-Bcl-2 and mTOR signaling cascades in aging [15]. These discrepancies have raised an issue of whether ALDH2 genetic mutation may impact human aging process.

2 ALDH2

2.1 *Characteristics of ALDH2*

Aldehyde dehydrogenase (ALDH) is a detoxifying enzyme superfamily known as a catalyst for oxidation of endogenous and exogenous aldehydes and aldehyde derivatives. Up to date, 19 members of isozymes have been identified, among which ALDH2 is the most active one consisting of 517 amino acid to form 4 56-kDa subunits, each containing 3 domains [17]. Encoded by a nuclear gene on chromosome 12q24, ALDH2 is located in mitochondria and is widely distributed in the liver, kidney, heart, lung, brain, and other tissues. ALDH2 is responsible for catalytic oxidation of aldehydes and lipid peroxidation-produced alkenols, such as 4-hydroxynonenal (4-HNE) and malondialdehyde (MDA) [11]. A total of 84 single nucleotide polymorphism loci have been found on human ALDH2 gene, most are rs671 G > A mutations which present in exon 12 and result in amino acid substitutions of Glu504Lys [18]. The ALDH2 genotypes include wild homozygous GG (ALDH2*1/*1), owning normal catalytic activity of the enzyme, mutant heterozygous GA (ALDH2*1/*2), expected to possess approximately 10%–45% enzymatic activity of wild-type homozygous. Mutant homozygotes AA (ALDH2 *2/*2), on the other hand, display only 1–5% the wild-type enzyme catalytic activity [19]. This deactivation mutant is present in nearly 8% of world populations and approximately 40% of the East Asian population. Further scrutiny in ALDH2 and its mutants have revealed a broader scope of pathologies implicated in human health and aging [17, 20]. Earlier findings from our group and others have demonstrated a protective role of ALDH2 in coronary artery diseases [21, 22], cancer [23, 24], metabolic disorder [25], and neurodegeneration diseases [26] in both clinical and experimental settings. Paradoxically, recent finding from our lab revealed that ALDH2 overexpression accelerated aging-associated myocardial dysfunction [27], which provoked further research into the intrinsic mechanisms.

2.2 *Bioengineered Animal Models of ALDH2*

A number of technological advances have offered promising leads toward addressing the specific role and mechanism of ALDH2 in the overall human health and prevalence of diseases. Mammalian models such as mice are most genetic proximity to humans with feasible access to gene knockout/in and could be utilized to recapitulate the diverse phenotypes [17]. Up to date, murine models of ALDH2 exhibit different ALDH2 genotype and express different enzyme activity, such as ALDH2 knockout mice, ALDH2*2 or ALDH2-overexpressing (transgenic) mice controlled by EF1α and chicken β-actin promoters, and ALDH2 mutation knock-in mice. Available pharmacological tools that activate or inhibit ALDH2 enzyme such as Alda-1 and Benomyl make it possible to decipher the rationale of ALDH2 manipulation to be harnessed therapeutically [23].

3 Aging

During the aging process, almost all biochemical processes that underpin intricate cellular functions remarkably decline with time [28]. Excessive fluctuations in physiological factors perturb homeostasis and increase vulnerability to major human pathologies. More than 70% of people over 65 have at least two or more forms of concurrent chronic diseases [29]. About half of human deaths are attributed to chronic aging-associated diseases (AADs) [30], most prominently cardiovascular diseases, metabolic disorder, chronic obstructive pulmonary disease (COPD), neurodegeneration diseases, stroke, and cancer [31]. Aging is defined as a slowly progressive decline of fitness and adaptation, which is determined by the balance between pro-aging and antiaging systems [32]. Both parallel forces (such as DNA reparation systems, quality control of proteins [33], cellular senescence [34]), and opposite forces (such as mutations, oxidation stress [35], hypoxia, mitochondrial dysfunction, nutrition deficiency [36]) were defined in existing theories of aging. Over the past several decades, these theories have received validations from various emerging experimental evidences [37]. Given the multifactorial nature of aging, how each influencing factor acts in concert and how they interact to contribute to the aging process still remain somewhat elusive.

3.1 Oxidative Stress

Among various theories and hypotheses for aging, the "oxidative stress/free radical theory" represents perhaps the most classical and well-perceived theory that highlights the toxic role of accumulation of reactive oxygen species (ROS) in aging process. ROS can be generated in oxidative phosphorylation (OXPHOS) in mitochondrial and arise dramatically under pathological state such as ionizing radiation and aging. Reduction-oxidation (redox) disequilibrium occurs when excrescent ROS accumulate beyond the capacity of scavenging. The term "oxidative stress" was first coined half a century ago to elucidate the potential detrimental impact of oxidization reactivity on cellular functions [38]. Traditional hypothesis persists, while recent findings invite nuanced explication that appreciate ROS as signaling molecules when produced at a controlled level to alleviate the cellular stress [39].

Experimental findings from our own laboratory and others have depicted increased oxidative stress in aged animals using sensitive biomarkers of peroxidation and oxidative stress [27, 40]. Oxidative stress is known to independently contribute to increased all-cause mortality according to population-based cohort studies [41]. Conversely, various molecules with antioxidant capacity have shown beneficial effects in maintaining cellular functions and longevity through reducing oxidative stress [42]. There is a considerable evidence that large amounts of ROS promote aging through oxidative damage on vital intracellular content [43], including chronic impairment on mitochondrial DNA and clonal extension of the mutations which lead to fatal energy deficiency [44].

3.2 *Mitochondrial Dysfunction in Aging*

A drastic decline in mitochondrial function and presence of abundant mitochondrial DNA (mtDNA) mutations or deficiency has been documented in aged mammalian species including aged tissue from human [45]. Given that mitochondria lie at the crossroad of the global metabolic homeostasis, dysfunctional mitochondria may impinge whole organismal function [46] and may predispose individuals to aging-related comorbidities [47]. According to the oxidative stress theory, enhanced ROS stress as a result of deficient mitochondrial oxidative phosphorylation (OXPHOS) contributes to aging. In turn, amplified oxidative stress could accelerate decay of mitochondria [48]. However, the initial trigger for declined mitochondrial integrity and function remains elusive.

Over the past decades, much insights have emerged gaining the intrinsic mechanisms behind aging-associated mitochondrial deterioration. Sirtuins, which belong to a family of conserved nicotinamide adenine dinucleotide (NAD^+)-dependent histone deacetylates, play pivotal roles in longevity and survival [49]. Peroxisome proliferator-activated receptor γ coactivator 1-α (PGC1-α) is a key regulator of mitochondrial biogenesis depending on nuclear-mitochondrial communication. SIRT1 counters starvation by deacetylating PGC1-α to induce gluconeogenesis [50]. Age-dependent decline of nuclear NAD^+ reduces SIRT1 deacetylate activity, resulting in the attenuated PGC1-α activity [51]. Further studies found that reduced SIRT1 function is causally associated with the mitochondrial decline through down-regulating the expression of a nucleus gene which encodes mitochondrial transcription factor A (TFAM). This process is mediated by the accumulation of hypoxia-inducible factor 1a (HIF1α) and cMyc [52]. AMP-activated protein kinase (AMPK) may promote the SIRT1 activity through increasing NAD^+ in response to elevated AMP/ATP or ADP/ATP ratios in the face of nutrient restriction. AMPK activation extends life span in animal models [53], and a dampened AMPK activity was noticed in the aging process [54, 55].

3.3 Autophagy and Aging

Autophagy is a catabolism process that targeted cellular constituents are sequestered by a structure called autophagosome before being delivered to the lysosomes for degradation to maintain cellular homeostasis [56]. Constitutive autophagy happens at basal level, while adaptive autophagy is simulated under stress conditions [57]. A bidirectional linkage between autophagy and aging-related pathologies has been revealed [58]. Autophagy alters with age; meanwhile, deficient autophagy contributes to diverse aging-related diseases, including cardiovascular diseases [59], metabolic disorder [58], cancer [60], neurodegeneration [61], and so on. In addition, long-lived *Caenorhabditis elegans* mutants exhibit an increased level of autophagy [62]. Autophagy may have complex, dual effects on the pathological aging process such as ischemia/reperfusion damage. Too much or too little autophagy may produce an unfavorable biological effect. Moderate autophagy induced by a temporary ischemia and mild oxidative stress serves as a protective process against adversity, while severe I/R may elicit excessive autophagy, which accelerates cell death and impairs organs [16].

A set of gene, known as autophagy-related genes (ATGs), drive autophagy [63]. It is believed that autophagy virtually communicates with signaling networks for metabolism, cellular survival, and cell cycle regulation [64]. A number of metabolism pathways converge to a nexus point termed mammalian target of rapamycin (mTOR), a serine/threonine-protein kinase. mTOR suppresses autophagy in starvation through inhibiting unc-51-like autophagy activating kinase-1 (ULK1) and the Beclin-1–VPS34 complexes [46]. Rapamycin inhibits mTOR, therefore triggering autophagy initiation [65]. Besides, endoplasmic reticulum (ER) stress modulates autophagy through a stress stimuli-sensitive molecular, c-Jun N-terminal kinase (JNK) [63]. In addition, SIRT1 may activate nuclear LC3 (microtubule-associated protein 1 light chain 3) and FoxO1 through deacetylation, which were both necessary and sufficient for restoring autophagy flux [13].

4 The Janus Faces of ALDH2 in Cardiac Aging?

By the year of 2035, more than 130 million (45.1%) adults in the USA are anticipated to have some forms of cardiovascular diseases (CVD), and the economic burden due to CVD is expected to reach $1.1 trillion in 2035 [66]. As mentioned earlier, aging is an independent risk factor for CVD and weighs most commonly in employed risk scores [67]. Cardiovascular morbidity increases yearly as societies age [68]. CVD still ranks first in the causes of death in America. More than half of the mortality are related to CVD striking individuals over 65 [69]. Unraveling the molecular pathways of aging-related myocardial dysfunction and the promising therapeutic targets attracted much attention.

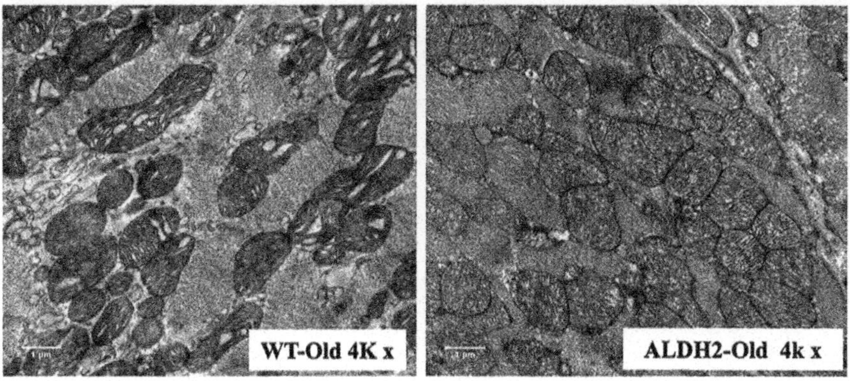

Fig. 1 Myocardial ultrastructure in 24-month-old WT and ALDH2 transgenic mice using transmission electron microscopy (x 4000). Scale bar = 1 μm. Myofilament alignment is distorted in both groups with mitochondrial volume likely enlarged in ALDH2-old group (possibly due to decreased mitochondrial autophagy)

Previously, a number of studies have demonstrated the distinct impacts of ALDH2 on the pathogenesis in cardiovascular diseases, while few focused on the aging heart. To this end, our laboratory conducted a series of studies. Young (4–5-month-old) and old (26–28-month-old) wild-type and ALDH2-overexpressing transgene mice were examined. We found that ALDH2 augmented aging-related pathological cardiac hypertrophy but not fibrosis. Increased contractile dysfunction, reactive oxygen species (ROS), apoptosis, mitochondrial injury, and intracellular Ca^{2+} handling in the aged mice were exaggerated by ALDH2, while no such effect was observed in the young group [27]. Figure 1 displays representative ultrastructural images of myocardium from old WT and ALDH2 transgenic mice, suggesting lack of protection for ALDH2 against cardiac aging and perhaps a rise in mitochondrial volume (which may be due to decreased mitophagy). In addition, ALDH2 overexpression shortened life span by 7.7% in mice without notable variation in aging-related changes of plasma profiles.

However and paradoxically, Wu and associates reported that ALDH2 knockout mice exhibited an accelerated age-related heart dysfunction and a shortened life span along with retarded autophagy flux, which can be abrogated by sustained treatment of Alda-1. ALDH2 activity was decreased in aged (22-month-old) mice, leading to a heavier aldehydic load as characterized by 4-HNE accumulation. ALDH2 activation in aged mice (22-month-old) by Alda-1-ameliorated autophagy flux through activating SIRT1 via reducing carbonylation [13]. SIRT1 is found to be definitely necessary in restoring autophagy flux. Besides, experiments carried out by Wenzel and colleagues showed a similar damage to vascular function in middle-aged (6-month-old) $ALDH2^{-/-}$ mice as old (12-month-old) $ALDH2^{-/-}$ mice, indicating a likely protective role of ALDH2 in vascular aging [70]. Mitochondrial ROS production and oxidative mtDNA damage observed in aged mice were parallel to endothelial dysfunction, suggesting that this protection may be related to the antioxidant effects of ALDH2.

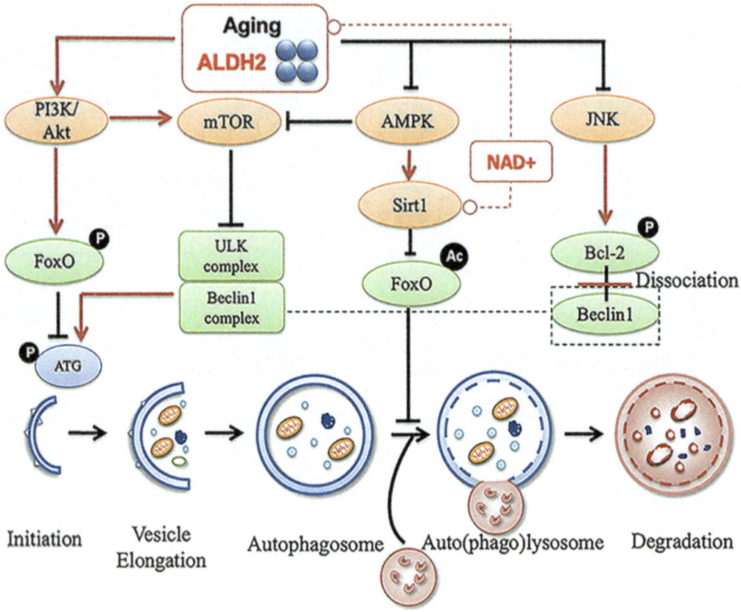

Fig. 2 Signaling mechanism in ALDH2-accentuated suppression of autophagy. (1) ALDH2 promotes aging-related activation of Akt, which inhibits the initiation of autophagy through phosphorylation of FoxO and activation of mTOR. (2) ALDH2 promotes suppression of AMPK in aging, leading to reduced activity of SIRT1 and activated mTOR; (3) Diminished SIRT1 leads to a reduced deacetylation of FoxO, thus inhibiting autophagolysosome formation. (4) Upregulated mTOR suppresses autophagy through negatively regulating ULK complex. (5) ALDH2 and aging reduce phosphorylation of JNK and Bcl-2, which suppress autophagy through dampened dissociation of Bcl-2 and Beclin1. (6) Both ALDH2 and SIRT1 depend on NAD^+, and thus ALDH2 may compete with SIRT1 for NAD+

These seeming discrepancies invited deeper insights. Our data further revealed that aging significantly suppresses AMPK phosphorylation and downregulates the levels of SIRT1 and several pivotal mitochondrial proteins (PGC-1α and UCP-2). These alternations are exaggerated by ALDH2. Besides, autophagy promotes survival under harmful conditions and plays a key role in maintaining cardiac homeostasis. Compromised autophagy in the aging heart [71] showed an association with heart failure and aging [56]. We have clarified that ALDH2 and aging jointly suppressed autophagy and autophagy flux permissively through both JNK-Bcl-2- and AMPK-mTOR-dependent pathway [15]. A scheme is provided summarizing ALDH2-offered regulation of cell signaling pathways (Fig. 2).

Moreover, we analyzed the genetic polymorphism in those who have a family history of longevity (with a life span over 90 years). Interestingly, it seems that longevity favors the ALDH2*1/*2 (G/A) mutation most (32.8% vs 20.9%) and the frequency of G/G is much lower in the longevity group (65.7%), compared with the control group (77.6%). We next examined the genotype distribution in 411 individuals with normal life span (<90 years). The frequency of G/A and A/A added up to 32.85%, while the G/G genotype was 67.15% in the normal age group [27].

Examination on left ventricle systolic function suggests a much better cardiac performance in aged ALDH2 mutants. Moreover, a subdued cardiac remodeling was noticed in the GA/AA group. Another candidate gene research in Koreans confirmed that the rs671 (A) allele of the ALDH2 gene was associated with longevity only in men ($P = 0.008$) [72]. Although Mizuno's group revealed a linkage between ALDH2*2 mutation and coronary spastic angina [73, 74], coronary spasm is caused by a combination of multiple triggers, thus whether this result causally associated with aging seems unclear.

Although the precise mechanism of action behind the discrepant findings in ALDH2 and aging still remain unclear, some explanations may be considered in addition to disparity in experimental condition and animal models. Logically, both overexpressed and insufficient ALDH2 may be harmful in some pathological conditions. Emerging evidence supported the supposition. As mentioned above, loss of ALDH2 may exaggerate aging. Dassanayaka and colleagues found that following treatment of 12 weeks of transverse aortic constriction (TAC), ALDH2 overexpression mice exhibited larger hearts and lower capillary density than the wild-type [75]. In line with the favor of natural selection for the G/A genotype which represents less active but not inactive ALDH2, these studies collectively suggested that a "moderate" level of ALDH2 is needed to induce favorable effects. Furthermore, both ALDH2 and SIRT1 depend on NAD^+ as a cofactor. This creates a natural "competition" for NAD^+ between the two and more crucial molecules. As a result, ALDH2 may display unfavorable effects on other processes dependent on the NAD^+ availability. Thus the specific effect of ALDH2 in aging may be determined by the level of NAD^+, which remains to be confirmed by more evidence. According to the findings from our lab and others, the ALDH2-induced restoration of autophagy flux required activated SIRT1. When NAD^+ is insufficient, ALDH2 overexpression will grab scanty NAD^+, which is also desperately needed for SIRT1. In addition, we found that ALDH2 markedly enhanced the suppression of AMPK in aging, which in turn reduced the activity the SIRT1. Diminished SIRT1 leads to a reduced deacetylation of FoxO, which turns out to hamper autophagy. Consequently, the potential protective effect of ALDH2 may fail to work or even be reversed due to the damage on other NAD^+-dependent metabolism pathways in the shortage of NAD^+. Last but not least, ALDH2 enhanced the aging-related activation of Akt, a serine/threonine-protein kinase which inhibits autophagy through phosphorylation of FoxO.

5 Role of ALDH2 in Other Aging-Related Diseases

5.1 *Neurodegeneration*

Although systemic diseases take up a large proportion of health, the brain is the arbiter of death preceded by a progressive deterioration of cognitive and motor function which occurs during aging [76]. Pervasive mutations caused by pathological conditions along with inherent risk-associated genotypes work together in the development of multiple neurodegenerative disorders [77]. Alzheimer's disease and Parkinson disease are typical neurodegenerative diseases raising escalating concern in aging society.

D'Souza and colleagues noted an endothelial dysfunction and brain atrophy along with an age-related increase of amyloid-β, p-tau, and caspases in ALDH2$^{-/-}$ mice. ALDH2$^{-/-}$ mice also exhibit notable age-related memory deficits. Taken together, ALDH2 deficiency accelerated age-dependent neurodegeneration [78]. Ohsawa found that ALDH2 deficiency deteriorated cognitive and memory impairment and even caused a loss of life span. The median and maximal life span of mice with ALDH2*2-expressing neurons were significantly shorter than that of non-transgenic control mice. At 1 year, 1.5 year after birth, 20%, 77.8% of ALDH2*2 mice showed notable neurodegeneration, whereas non-transgenic control mice averted this deterioration [79].

However, epidemiology studies provided equivocal evidence for these benefits. Early statistics from a Japanese sample showed the possible linkage between ALDH2*2 mutation and the risk of Alzheimer's disease (AD). They found a higher frequency of the ALDH2∗2 in AD patients (27.5–28.0%) than in controls (19.9–22.7%). A study of 690 Koreans aged 65 and over showed no significant association between ALDH2*2 and cognitive outcomes [80]. Several considerations for this conflict should be proposed: (1) possibly different clinical diagnostic criteria and accuracy; (2) the lower frequency of ALDH2*2 in the Korean study (15.7%) compared to that of Japanese (23.6–34.7%), which may obscure the potential association; and (3) sample representativeness and different demographic characteristics, such as education level. Thus prospective studies are warranted to clarify the issue.

In addition, another research carried out on 584 Parkinson disease (PD) patients and 582 controls in Han Chinese population indicated ALDH2 rs4767944 (T < C) ($p = 0.002$), but not rs441 and rs671, were associated with Parkinson disease [26]. PD patients presented higher frequency of CC and CT (15.8%, 54.1%) than the controls (10.7%, 51.7%). Further analysis found that the frequency of C allele is much higher in the PD patients (36.5%) compared to the controls (42.8%), which provided a novel ALDH2 variation associated with PD.

5.2 Cancer

Aging acts as an independent risk factor for cancer. Cancer is formed as a result of the accumulation of genetic mutations that either inhibit the activity of tumor suppressor genes or activate oncogenes [30]. Accumulating evidence has indicated that tumor might be regulated by similar pathways which modulate energy balance [37]. For example, Hou and associates noticed lower ALDH2 levels in tumor tissues, with a more pronounced decrease in tissues with increased invasive and migratory capacity. ALDH2 activation inhibited migration and invasion both in vivo and in vitro. Moreover, these investigators found that ALDH2 adjusted the redox status through increased phosphorylation of AMPK, which ultimately suppressed metastasis of

HCC cells [81]. In an analysis of ALDH2, polymorphisms found a significant positive direct effect and a protective indirect effect of the ALDH2 rs671 (A) allele against gastric cancer together with ALDH2-alcohol drinking interaction. In the following mediation analysis, they found ALDH2*2 mutation independently increased the risk for gastric cancer [82]. In addition, Jin and colleagues reported that ALDH2*2 knock-in mice presented decreased ALDH2 activity and increased DNA damage response in hepatocytes, pronounced liver injury, and accelerated hepatocellular carcinoma (HCC). The ALDH2*2 mutation rate in human 43 HCC samples was close to that of a normal liver. However, this sample size is relatively small to draw a solid conclusion [83].

5.3 *Metabolic Disease*

The hormonal secretory patterns and the sensitivity to regulatory signals altered with aging, featured by the disequilibrium of glucose, calcium, and sex hormonal homeostasis, which leads to a cluster of age-related metabolic disorders like obesity, hypertension, type 2 diabetes mellitus, and osteoporosis. [84] Both physiological and pathological process of metabolism are under the control of insulin-like growth factor 1 (IGF1) signaling pathway and the AMPK-SIRT1 pathway [30]. Traditionally, the alternations in hormone activity of the elderly are considered detrimental due to age-related decline in organ functions and may increase the chronic disease prevalence. However, some of these changes may be a beneficial adaptation to aging [84].

Glucose homeostasis is maintained through a set of procedures including glucose ingestion, absorption, utilization, and production under subtle control of insulin. Alternations due to aging perturb this balance, resulting in increased fasting plasma glucose and distorted insulin secretion. Type 2 diabetes mellitus without proper treatment triggers various forms of diabetic complications, including cardiovascular diseases, retinopathy, kidney failure, amputations, etc. Wang and colleagues showed that ALDH2 deficiency deteriorated diastolic function in the early stage of diabetic cardiomyopathy through activation of the AMPK-LKB1 pathway in response to increased AMP/ATP and ADP/ATP and decreased PCr/ATP ratio. They noticed perturbations of glucose homeostasis and an accumulation of lipid peroxidative product (4-HNE) in ALDH2 knockout mice [85]. In a multistage genome-wide association study (GWAS) involving 12,720 Han Chinese participants from China, interaction of ALDH2 rs671 mutation and alcohol consumption was found to be associated with metabolic syndrome. However, the association between ALDH2*2 and obesity was partly independent of alcohol intake [25].

A loss of bone mass owing to an imbalance between formation and resorption leads to osteoporosis [86]. By 2020, approximately 12.3 million individuals in the USA at the age of 50 or older are expected to suffer from osteoporosis [87]. In a study from Japan, Hoshi and coworkers found that ALDH2*2 (rs671) promoted osteoporosis due to impaired osteoblastogenesis. The ALDH2*2 transgenic mice exhibited an increased blood acetaldehyde and accumulation of 4-HNE and upregulated the expression of peroxisome proliferator-activated receptor gamma (PPARγ), a transcription factor that promotes adipogenesis and inhibits osteoblastogenesis. Antioxidant treatment could significantly reverse this process [7]. A cohort study in Japanese showed that individuals with homozygous ALDH2*2 had a higher morbidity rate of osteoporosis compared with ALDH2*1 individuals. Interestingly, females presented a more conspicuous result (OR: 4.31) [88].

5.4 *Conclusion*

Given the inevitable nature of aging, individual bears a witness to progressively declined cellular and organ functions with time. Emerging insights into the cellular and molecular mechanisms underlying aging- and age-related diseases are anticipated to expand health span. This review summarizes the effects of different ALDH2 enzyme activity in aging-related pathologies and genetic polymorphism distribution of ALDH2 in these diseases. Interestingly, the seemingly controversial role of ALDH2 in aging-related cardiac dysfunction virtually revealed several interacted pathways which regulate the process of aging. Predispositions and inducements such as increased oxidative stress trigger damage on cellular component during aging. A series of adaptive or survival program are designed to battle against aging, including autophagy as depicted in the schematic diagram (Fig. 2). Numerous studies have demonstrated a pivotal role of ALDH2 in this network, while the significant protective effects of ALDH2 still require some essential molecules; some of them has been consolidated such as SIRT1 (with NAD^+ possibly contributing to such beneficial process). Furthermore, other measures to regulate ALDH2 activity and autophagy potentially provide an innovated method for early intervention and treatment to aging-related conditions.

Acknowledgments This work was supported in part by NSFC91749128. We wish to express our apology for those authors whose important work was unable to be included due to space limitation.

References

1. Ren J, Sowers JR, Zhang Y (2018) Metabolic stress, autophagy, and cardiovascular aging: from pathophysiology to therapeutics. Trends Endocrinol Metab 29(10):699–711
2. Lutz W, Sanderson W, Scherbov S (2008) The coming acceleration of global population ageing. Nature 451(7179):716–719
3. Lakatta EG (1993) Cardiovascular regulatory mechanisms in advanced age. Physiol Rev 73(2):413–467
4. Boengler K, Schulz R, Heusch G (2009) Loss of cardioprotection with ageing. Cardiovasc Res 83(2):247–261
5. Rich MW, Quality of Care C, Heart Failure Society of A (2011) The year in quality of care in heart failure. J Card Fail 17(6):443–450

6. Sen P, Shah PP, Nativio R, Berger SL (2016) Epigenetic mechanisms of longevity and aging. Cell 166(4):822–839
7. Hoshi H, Hao W, Fujita Y, Funayama A, Miyauchi Y, Hashimoto K, Miyamoto K, Iwasaki R, Sato Y, Kobayashi T, Miyamoto H, Yoshida S, Mori T, Kanagawa H, Katsuyama E, Fujie A, Kitagawa K, Nakayama KI, Kawamoto T, Sano M, Fukuda K, Ohsawa I, Ohta S, Morioka H, Matsumoto M, Chiba K, Toyama Y, Miyamoto T (2012) Aldehyde-stress resulting from Aldh2 mutation promotes osteoporosis due to impaired osteoblastogenesis. J Bone Miner Res Off J Am Soc Bone Miner Res 27(9):2015–2023
8. Tosto G, Reitz C (2016) Genomics of Alzheimer's disease: value of high-throughput genomic technologies to dissect its etiology. Mol Cell Probes 30(6):397–403
9. Barnes PJ (2015) Mechanisms of development of multimorbidity in the elderly. Eur Respir J 45(3):790–806
10. Hartl FU (2016) Cellular homeostasis and aging. Annu Rev Biochem 85:1–4
11. Chen CH, Sun L, Mochly-Rosen D (2010) Mitochondrial aldehyde dehydrogenase and cardiac diseases. Cardiovasc Res 88(1):51–57
12. Wang S, Wang C, Turdi S, Richmond KL, Zhang Y, Ren J (2018) ALDH2 protects against high fat diet-induced obesity cardiomyopathy and defective autophagy: role of CaM kinase II, histone H3K9 methyltransferase SUV39H, Sirt1, and PGC-1alpha deacetylation. Int J Obes (2005) 42(5):1073–1087
13. Wu B, Yu L, Wang Y, Wang H, Li C, Yin Y, Yang J, Wang Z, Zheng Q, Ma H (2016) Aldehyde dehydrogenase 2 activation in aged heart improves the autophagy by reducing the carbonyl modification on SIRT1. Oncotarget 7(3):2175–2188
14. Zhang Y, Ren J (2011) ALDH2 in alcoholic heart diseases: molecular mechanism and clinical implications. Pharmacol Ther 132(1):86–95
15. Zhang Y, Wang C, Zhou J, Sun A, Hueckstaedt LK, Ge J, Ren J (2017) Complex inhibition of autophagy by mitochondrial aldehyde dehydrogenase shortens lifespan and exacerbates cardiac aging. Biochim Biophys Acta Mol basis Dis 1863(8):1919–1932
16. Gong D, Zhang H, Hu S (2013) Mitochondrial aldehyde dehydrogenase 2 activation and cardioprotection. J Mol Cell Cardiol 55:58–63
17. Chen CH, Ferreira JC, Gross ER, Mochly-Rosen D (2014) Targeting aldehyde dehydrogenase 2: new therapeutic opportunities. Physiol Rev 94(1):1–34
18. Ma C, Yu B, Zhang W, Wang W, Zhang L, Zeng Q (2017) Associations between aldehyde dehydrogenase 2 (ALDH2) rs671 genetic polymorphisms, lifestyles and hypertension risk in Chinese Han people. Sci Rep 7(1):11136
19. Chen Z, Foster MW, Zhang J, Mao L, Rockman HA, Kawamoto T, Kitagawa K, Nakayama KI, Hess DT, Stamler JS (2005) An essential role for mitochondrial aldehyde dehydrogenase in nitroglycerin bioactivation. Proc Natl Acad Sci U S A 102(34):12159–12164
20. Ebert AD, Kodo K, Liang P, Wu H, Huber BC, Riegler J, Churko J, Lee J, de Almeida P, Lan F, Diecke S, Burridge PW, Gold JD, Mochly-Rosen D, Wu JC (2014) Characterization of the molecular mechanisms underlying increased ischemic damage in the aldehyde dehydrogenase 2 genetic polymorphism using a human induced pluripotent stem cell model system. Sci Transl Med 6(255):255ra130
21. Liu X, Sun X, Liao H, Dong Z, Zhao J, Zhu H, Wang P, Shen L, Xu L, Ma X, Shen C, Fan F, Wang C, Hu K, Zou Y, Ge J, Ren J, Sun A (2015) Mitochondrial aldehyde dehydrogenase 2 regulates revascularization in chronic ischemia: potential impact on the development of coronary collateral circulation. Arterioscler Thromb Vasc Biol 35(10):2196–2206
22. Sun A, Zou Y, Wang P, Xu D, Gong H, Wang S, Qin Y, Zhang P, Chen Y, Harada M, Isse T, Kawamoto T, Fan H, Yang P, Akazawa H, Nagai T, Takano H, Ping P, Komuro I, Ge J (2014) Mitochondrial aldehyde dehydrogenase 2 plays protective roles in heart failure after myocardial infarction via suppression of the cytosolic JNK/p53 pathway in mice. J Am Heart Assoc 3(5):e000779
23. Zhong W, Zhang W, Li Q, Xie G, Sun Q, Sun X, Tan X, Sun X, Jia W, Zhou Z (2015) Pharmacological activation of aldehyde dehydrogenase 2 by Alda-1 reverses alcohol-induced hepatic steatosis and cell death in mice. J Hepatol 62(6):1375–1381

24. Sawada G, Niida A, Uchi R, Hirata H, Shimamura T, Suzuki Y, Shiraishi Y, Chiba K, Imoto S, Takahashi Y, Iwaya T, Sudo T, Hayashi T, Takai H, Kawasaki Y, Matsukawa T, Eguchi H, Sugimachi K, Tanaka F, Suzuki H, Yamamoto K, Ishii H, Shimizu M, Yamazaki H, Yamazaki M, Tachimori Y, Kajiyama Y, Natsugoe S, Fujita H, Mafune K, Tanaka Y, Kelsell DP, Scott CA, Tsuji S, Yachida S, Shibata T, Sugano S, Doki Y, Akiyama T, Aburatani H, Ogawa S, Miyano S, Mori M, Mimori K (2016) Genomic landscape of esophageal squamous cell carcinoma in a Japanese population. Gastroenterology 150(5):1171–1182
25. Zhu Y, Zhang D, Zhou D, Li Z, Li Z, Fang L, Yang M, Shan Z, Li H, Chen J, Zhou X, Ye W, Yu S, Li H, Cai L, Liu C, Zhang J, Wang L, Lai Y, Ruan L, Sun Z, Zhang S, Wang H, Liu Y, Xu Y, Ling J, Xu C, Zhang Y, Lv D, Yuan Z, Zhang J, Zhang Y, Shi Y, Lai M (2017) Susceptibility loci for metabolic syndrome and metabolic components identified in Han Chinese: a multi-stage genome-wide association study. J Cell Mol Med 21(6):1106–1116
26. Zhang X, Ye YL, Wang YN, Liu FF, Liu XX, Hu BL, Zou M, Zhu JH (2015) Aldehyde dehydrogenase 2 genetic variations may increase susceptibility to Parkinson's disease in Han Chinese population. Neurobiol Aging 36(9):2660.e2669–2660.e2613
27. Zhang Y, Mi SL, Hu N, Doser TA, Sun A, Ge J, Ren J (2014) Mitochondrial aldehyde dehydrogenase 2 accentuates aging-induced cardiac remodeling and contractile dysfunction: role of AMPK, Sirt1, and mitochondrial function. Free Radic Biol Med 71:208–220
28. Booth LN, Brunet A (2016) The aging epigenome. Mol Cell 62(5):728–744
29. Fontana L, Kennedy BK, Longo VD, Seals D, Melov S (2014) Medical research: treat ageing. Nature 511(7510):405–407
30. Kubben N, Misteli T (2017) Shared molecular and cellular mechanisms of premature ageing and ageing-associated diseases. Nat Rev Mol Cell Biol 18(10):595–609
31. Benayoun BA, Pollina EA, Brunet A (2015) Epigenetic regulation of ageing: linking environmental inputs to genomic stability. Nat Rev Mol Cell Biol 16(10):593–610
32. Bilkei-Gorzo A (2014) Genetic mouse models of brain ageing and Alzheimer's disease. Pharmacol Ther 142(2):244–257
33. Quiros PM, Langer T, Lopez-Otin C (2015) New roles for mitochondrial proteases in health, ageing and disease. Nat Rev Mol Cell Biol 16(6):345–359
34. He S, Sharpless NE (2017) Senescence in health and disease. Cell 169(6):1000–1011
35. Verdin E (2015) NAD(+) in aging, metabolism, and neurodegeneration. Science (New York, NY) 350(6265):1208–1213
36. Shay JW (2016) Role of telomeres and telomerase in aging and cancer. Cancer Discov 6(6):584–593
37. Falandry C, Bonnefoy M, Freyer G, Gilson E (2014) Biology of cancer and aging: a complex association with cellular senescence. J Clin Oncol Off J Am Soc Clin Oncol 32(24):2604–2610
38. Sies H (2015) Oxidative stress: a concept in redox biology and medicine. Redox Biol 4:180–183
39. Yee C, Yang W, Hekimi S (2014) The intrinsic apoptosis pathway mediates the pro-longevity response to mitochondrial ROS in C. elegans. Cell 157(4):897–909
40. Rogers LK, Cismowski MJ (2018) Oxidative stress in the lung – the essential paradox. Curr Opin Toxicol 7:37–43
41. Schottker B, Brenner H, Jansen EH, Gardiner J, Peasey A, Kubinova R, Pajak A, Topor-Madry R, Tamosiunas A, Saum KU, Holleczek B, Pikhart H, Bobak M (2015) Evidence for the free radical/oxidative stress theory of ageing from the CHANCES consortium: a meta-analysis of individual participant data. BMC Med 13:300
42. Garcia JJ, Lopez-Pingarron L, Almeida-Souza P, Tres A, Escudero P, Garcia-Gil FA, Tan DX, Reiter RJ, Ramirez JM, Bernal-Perez M (2014) Protective effects of melatonin in reducing oxidative stress and in preserving the fluidity of biological membranes: a review. J Pineal Res 56(3):225–237
43. Van Raamsdonk JM, Hekimi S (2012) Superoxide dismutase is dispensable for normal animal lifespan. Proc Natl Acad Sci U S A 109(15):5785–5790
44. Larsson NG (2010) Somatic mitochondrial DNA mutations in mammalian aging. Annu Rev Biochem 79:683–706

45. Finkel T (2015) The metabolic regulation of aging. Nat Med 21(12):1416–1423
46. Kim KH, Lee MS (2014) Autophagy – a key player in cellular and body metabolism. Nat Rev Endocrinol 10(6):322–337
47. Nunnari J, Suomalainen A (2012) Mitochondria: in sickness and in health. Cell 148(6):1145–1159
48. Bratic A, Larsson NG (2013) The role of mitochondria in aging. J Clin Invest 123(3):951–957
49. Haigis MC, Sinclair DA (2010) Mammalian sirtuins: biological insights and disease relevance. Annu Rev Pathol 5:253–295
50. Rodgers JT, Lerin C, Haas W, Gygi SP, Spiegelman BM, Puigserver P (2005) Nutrient control of glucose homeostasis through a complex of PGC-1alpha and SIRT1. Nature 434(7029):113–118
51. Riera CE, Dillin A (2015) Tipping the metabolic scales towards increased longevity in mammals. Nat Cell Biol 17(3):196–203
52. Gomes AP, Price NL, Ling AJ, Moslehi JJ, Montgomery MK, Rajman L, White JP, Teodoro JS, Wrann CD, Hubbard BP, Mercken EM, Palmeira CM, de Cabo R, Rolo AP, Turner N, Bell EL, Sinclair DA (2013) Declining NAD(+) induces a pseudohypoxic state disrupting nuclear-mitochondrial communication during aging. Cell 155(7):1624–1638
53. Stenesen D, Suh JM, Seo J, Yu K, Lee KS, Kim JS, Min KJ, Graff JM (2013) Adenosine nucleotide biosynthesis and AMPK regulate adult life span and mediate the longevity benefit of caloric restriction in flies. Cell Metab 17(1):101–112
54. Li Q, Ceylan-Isik AF, Li J, Ren J (2008) Deficiency of insulin-like growth factor 1 reduces sensitivity to aging-associated cardiomyocyte dysfunction. Rejuvenation Res 11(4):725–733
55. Lu JY, Lin YY, Sheu JC, Wu JT, Lee FJ, Chen Y, Lin MI, Chiang FT, Tai TY, Berger SL, Zhao Y, Tsai KS, Zhu H, Chuang LM, Boeke JD (2011) Acetylation of yeast AMPK controls intrinsic aging independently of caloric restriction. Cell 146(6):969–979
56. Orogo AM, Gustafsson AB (2015) Therapeutic targeting of autophagy: potential and concerns in treating cardiovascular disease. Circ Res 116(3):489–503
57. Frankel LB, Lubas M, Lund AH (2017) Emerging connections between RNA and autophagy. Autophagy 13(1):3–23
58. Morel E, Mehrpour M, Botti J, Dupont N, Hamai A, Nascimbeni AC, Codogno P (2017) Autophagy: a druggable process. Annu Rev Pharmacol Toxicol 57:375–398
59. Delbridge LMD, Mellor KM, Taylor DJ, Gottlieb RA (2017) Myocardial stress and autophagy: mechanisms and potential therapies. Nat Rev Cardiol 14(7):412–425
60. Amaravadi R, Kimmelman AC, White E (2016) Recent insights into the function of autophagy in cancer. Genes Dev 30(17):1913–1930
61. Rubinsztein DC, Bento CF, Deretic V (2015) Therapeutic targeting of autophagy in neurodegenerative and infectious diseases. J Exp Med 212(7):979–990
62. Lapierre LR, Gelino S, Melendez A, Hansen M (2011) Autophagy and lipid metabolism coordinately modulate life span in germline-less C. elegans. Curr Biol 21(18):1507–1514
63. Vidal RL, Matus S, Bargsted L, Hetz C (2014) Targeting autophagy in neurodegenerative diseases. Trends Pharmacol Sci 35(11):583–591
64. Ktistakis NT, Tooze SA (2016) Digesting the expanding mechanisms of autophagy. Trends Cell Biol 26(8):624–635
65. Chandra V, Bhagyaraj E, Parkesh R, Gupta P (2016) Transcription factors and cognate signalling cascades in the regulation of autophagy. Biol Rev Camb Philos Soc 91(2):429–451
66. Benjamin EJ, Virani SS, Callaway CW, Chamberlain AM, Chang AR, Cheng S, Chiuve SE, Cushman M, Delling FN, Deo R, de Ferranti SD, Ferguson JF, Fornage M, Gillespie C, Isasi CR, Jimenez MC, Jordan LC, Judd SE, Lackland D, Lichtman JH, Lisabeth L, Liu S, Longenecker CT, Lutsey PL, Mackey JS, Matchar DB, Matsushita K, Mussolino ME, Nasir K, O'Flaherty M, Palaniappan LP, Pandey A, Pandey DK, Reeves MJ, Ritchey MD, Rodriguez CJ, Roth GA, Rosamond WD, Sampson UKA, Satou GM, Shah SH, Spartano NL, Tirschwell DL, Tsao CW, Voeks JH, Willey JZ, Wilkins JT, Wu JH, Alger HM, Wong SS, Muntner P (2018) Heart disease and stroke statistics-2018 update: a report from the American Heart Association. Circulation 137(12):e67–e492

67. Camici GG, Savarese G, Akhmedov A, Luscher TF (2015) Molecular mechanism of endothelial and vascular aging: implications for cardiovascular disease. Eur Heart J 36(48):3392–3403
68. Shirakabe A, Ikeda Y, Sciarretta S, Zablocki DK, Sadoshima J (2016) Aging and autophagy in the heart. Circ Res 118(10):1563–1576
69. Nakayama H, Nishida K, Otsu K (2016) Macromolecular degradation systems and cardiovascular aging. Circ Res 118(10):1577–1592
70. Wenzel P, Schuhmacher S, Kienhofer J, Muller J, Hortmann M, Oelze M, Schulz E, Treiber N, Kawamoto T, Scharffetter-Kochanek K, Munzel T, Burkle A, Bachschmid MM, Daiber A (2008) Manganese superoxide dismutase and aldehyde dehydrogenase deficiency increase mitochondrial oxidative stress and aggravate age-dependent vascular dysfunction. Cardiovasc Res 80(2):280–289
71. Hua Y, Zhang Y, Ceylan-Isik AF, Wold LE, Nunn JM, Ren J (2011) Chronic Akt activation accentuates aging-induced cardiac hypertrophy and myocardial contractile dysfunction: role of autophagy. Basic Res Cardiol 106(6):1173–1191
72. Park JW, Ji YI, Choi YH, Kang MY, Jung E, Cho SY, Cho HY, Kang BK, Joung YS, Kim DH, Park SC, Park J (2009) Candidate gene polymorphisms for diabetes mellitus, cardiovascular disease and cancer are associated with longevity in Koreans. Exp Mol Med 41(11):772–781
73. Mizuno Y, Hokimoto S, Harada E, Kinoshita K, Nakagawa K, Yoshimura M, Ogawa H, Yasue H (2016) Variant aldehyde dehydrogenase 2 (ALDH2*2) is a risk factor for coronary spasm and ST-segment elevation myocardial infarction. J Am Heart Assoc 5(5):e003247
74. Mizuno Y, Harada E, Morita S, Kinoshita K, Hayashida M, Shono M, Morikawa Y, Murohara T, Nakayama M, Yoshimura M, Yasue H (2015) East Asian variant of aldehyde dehydrogenase 2 is associated with coronary spastic angina: possible roles of reactive aldehydes and implications of alcohol flushing syndrome. Circulation 131(19):1665–1673
75. Dassanayaka S, Zheng Y, Gibb AA, Cummins TD, McNally LA, Brittian KR, Jagatheesan G, Audam TN, Long BW, Brainard RE, Jones SP, Hill BG (2018) Cardiac-specific overexpression of aldehyde dehydrogenase 2 exacerbates cardiac remodeling in response to pressure overload. Redox Biol 17:440–449
76. Wyss-Coray T (2016) Ageing, neurodegeneration and brain rejuvenation. Nature 539(7628):180–186
77. Gan L, Cookson MR, Petrucelli L, La Spada AR (2018) Converging pathways in neurodegeneration, from genetics to mechanisms. Nat Neurosci 21(10):1300–1309
78. D'Souza Y, Elharram A, Soon-Shiong R, Andrew RD, Bennett BM (2015) Characterization of Aldh2 (−/−) mice as an age-related model of cognitive impairment and Alzheimer's disease. Mol Brain 8:27
79. Ohsawa I, Nishimaki K, Murakami Y, Suzuki Y, Ishikawa M, Ohta S (2008) Age-dependent neurodegeneration accompanying memory loss in transgenic mice defective in mitochondrial aldehyde dehydrogenase 2 activity. J Neurosci Off J Soc Neurosci 28(24):6239–6249
80. Kim JM, Stewart R, Shin IS, Jung JS, Yoon JS (2004) Assessment of association between mitochondrial aldehyde dehydrogenase polymorphism and Alzheimer's disease in an older Korean population. Neurobiol Aging 25(3):295–301
81. Hou G, Chen L, Liu G, Li L, Yang Y, Yan HX, Zhang HL, Tang J, Yang YC, Lin X, Chen X, Luo GJ, Zhu Y, Tang S, Zhang J, Liu H, Gu Q, Zhao LH, Li Y, Liu L, Zhou W, Wang H (2017) Aldehyde dehydrogenase-2 (ALDH2) opposes hepatocellular carcinoma progression by regulating AMP-activated protein kinase signaling in mice. Hepatology (Baltimore, Md) 65(5):1628–1644
82. Ishioka K, Masaoka H, Ito H, Oze I, Ito S, Tajika M, Shimizu Y, Niwa Y, Nakamura S, Matsuo K (2018) Association between ALDH2 and ADH1B polymorphisms, alcohol drinking and gastric cancer: a replication and mediation analysis. Gastric Cancer Off J Int Gastric Cancer Assoc Jpn Gastric Cancer Assoc 21:936–945
83. Jin S, Chen J, Chen L, Histen G, Lin Z, Gross S, Hixon J, Chen Y, Kung C, Chen Y, Fu Y, Lu Y, Lin H, Cai X, Yang H, Cairns RA, Dorsch M, Su SM, Biller S, Mak TW, Cang Y (2015) ALDH2(E487K) mutation increases protein turnover and promotes murine hepatocarcinogenesis. Proc Natl Acad Sci U S A 112(29):9088–9093

84. van den Beld AW, Kaufman JM, Zillikens MC, Lamberts SWJ, Egan JM, van der Lely AJ (2018) The physiology of endocrine systems with ageing. Lancet Diabetes Endocrinol 6(8):647–658
85. Wang C, Fan F, Cao Q, Shen C, Zhu H, Wang P, Zhao X, Sun X, Dong Z, Ma X, Liu X, Han S, Wu C, Zou Y, Hu K, Ge J, Sun A (2016) Mitochondrial aldehyde dehydrogenase 2 deficiency aggravates energy metabolism disturbance and diastolic dysfunction in diabetic mice. J Mol Med (Berlin, Germany) 94(11):1229–1240
86. Veldurthy V, Wei R, Oz L, Dhawan P, Jeon YH, Christakos S (2016) Vitamin D, calcium homeostasis and aging. Bone Res 4:16041
87. Curry SJ, Krist AH, Owens DK, Barry MJ, Caughey AB, Davidson KW, Doubeni CA, Epling JW Jr, Kemper AR, Kubik M, Landefeld CS, Mangione CM, Phipps MG, Pignone M, Silverstein M, Simon MA, Tseng CW, Wong JB (2018) Screening for osteoporosis to prevent fractures: US preventive services task force recommendation statement. JAMA 319(24):2521–2531
88. Yamaguchi J, Hasegawa Y, Kawasaki M, Masui T, Kanoh T, Ishiguro N, Hamajima N (2006) ALDH2 polymorphisms and bone mineral density in an elderly Japanese population. Osteoporos Int 17(6):908–913

The manufacturer's authorised representative in the EU is Springer Nature Customer Service Centre GmbH, Europaplatz 3, 69115 Heidelberg, Germany. If you have any concerns regarding our products, please contact ProductSafety@springernature.com

Printed and bound by CPI Group (UK) Ltd, Croydon, CR0 4YY

30/07/2026

02181310-0001